# Natural Computing Series

Series Editors: Thomas Bäck   Lila Kari

Natural Computing is one of the most exciting developments in computer science, and there is a growing consensus that it will become a major field in this century. This series includes monographs, textbooks, and state-of-the-art collections covering the whole spectrum of Natural Computing and ranging from theory to applications.

More information about this series at http://www.springer.com/series/4190

Hitoshi Iba • Nasimul Noman

Editors

# Deep Neural Evolution

## Deep Learning with Evolutionary Computation

 Springer

*Editors*
Hitoshi Iba
Graduate School of Information Science
and Technology
The University of Tokyo
Tokyo, Japan

Nasimul Noman
School of Electrical Engineering
and Computing
The University of Newcastle
Callaghan, NSW, Australia

ISSN 1619-7127
Natural Computing Series
ISBN 978-981-15-3687-8          ISBN 978-981-15-3685-4   (eBook)
https://doi.org/10.1007/978-981-15-3685-4

This Springer imprint is published by the registered company Springer Nature Singapore Pte Ltd.
The registered company address is: 152 Beach Road, #21-01/04 Gateway East, Singapore 189721,
Singapore

# Preface

The scientific interest in designing synergic systems using neural networks (NN) and evolutionary computation (EC) together, which began long ago, created the field of neuroevolution. The ultimate goal of this subfield of artificial intelligence is to create optimally designed neural networks, capable of exhibiting intelligent behavior, with no or minimum human intervention. Towards this aim, over the last decades, the concepts of neuroevolution have been effectively used with a wide range of neural network models and evolutionary algorithms resulting in many fruitful applications.

In the last decade, deep learning (DL) revolutionized the field of machine learning with its record-breaking performance in diverse application domains. The enormous research and investment in DL have resulted in many powerful models and algorithms progressing the field at an exponential rate. The extraordinary advancement in DL research has also uncovered some limitations and weaknesses that need immediate attention. The focus of this book is the collaborative efforts from the EC community to solve some of the challenges deep learning is facing today which gave birth to the field of deep neuroevolution.

The emerging field of deep neuroevolution is primarily concentrating on the design of optimal deep neural networks (DNN) automatically. The manual design process is laborious and intractable at that scale of network design. Therefore, the current efforts are largely applied in DNN hyper-parameter optimization, architectural design, and weight training. Lately, some exciting applications of EC have been seen with other aspects of DL research. The purpose of the book is to introduce the readers to the contemporary research topics of deep neuroevolution for the development of the field.

The concepts presented in this book are expected to be useful and valuable to different classes of readers. The contents are chosen to promote and facilitate the research in DL with EC in both theory and practice. The book also presents some interesting applications in real-world problems. The chapters cover cutting-edge research topics that encompass deep neural network and evolutionary computation. Therefore, a larger audience of researchers and practitioners from different fields of computer science (e.g., machine learning, artificial intelligence, evolutionary computation) will find this book interesting. Beginning with the essentials of EC and DL, the book covers state-of-the-art research methodology of the field as well as the growing research trends. Therefore, it is expected to be equally attractive to the novice, intermediate, and expert readers from related fields.

In order to cater to the needs of readers from diverse research backgrounds, we have organized the book into five parts. The outline of the book in terms of different parts is as follows.

The first part containing only two chapters gives an introductory background to the field. Taking into account the readers coming from different research backgrounds, a general introduction for beginners in EC is given in the first chapter. This chapter covers different classes of evolutionary algorithms and nature-inspired algorithms, many of which as well as their variants and hybrids are utilized in later chapters. Thereafter, Chap. 2 presents the relevant concepts and notions of deep learning necessary for readers who are novice to the area. This chapter narrates the origin of DNN from artificial neural networks (ANN) and makes readers acquainted with different DNN models, their architectural characteristics, and various algorithms used for learning those. Most of the DNN models discussed in this chapter appear in the subsequent chapters in the book.

The second part of the book presents the usage of EC and meta-heuristics based approaches for optimizing hyper-parameters in DNNs. These approaches usually work with a fixed DNN architecture and apply a meta-algorithm for finding the optimal setting of its hyper-parameters. EC approaches have proven to be effective for enhancing the performance of DNNs via hyper-parameter adjustments and three such works are presented in this section of the book. The third chapter presents a comparative study among various evolutionary and nature-inspired algorithms for tuning a deep belief network's hyper-parameters in the context of binary image reconstruction. Chapter 4 uses both single- and multi-objective evolutionary algorithms for the hyper-parameter optimization of DNN-based implementation of spoken language processing systems. The final chapter in this part shows the use of heuristic methods for structuring deep belief networks for time series forecasting.

The focus of the third part of the book is the architectural design of DNN using EC. Performance of a DNN is heavily dependent on the number and types of network layers used alongside their organization; however, identifying the optimal architecture of the DNN for a given task is perplexing. Chapter 6 describes an encoding strategy for effectively encoding the convolutional neural network (CNN) architectures and utilizes the particle swarm optimization (PSO) algorithm for learning variable-length CNN architecture. The method is used for image classification in both single- and multi-objective settings where the objectives are classification accuracy and computational cost. In Chap. 7, Cartesian genetic programming (CGP) is used for designing high-performing CNN architectures consisting of predefined modules. The success of the CGP-based CNN architecture design technique is verified by applying it in two types of computer vision tasks: image classification and image restoration. In Chap. 8, a genetic algorithm (GA) is used for optimizing the number of blocks and layers as well as some other hyper-parameters in the vanilla CNN structure for image classification.

The fourth part of the book compiled works on the application of EC in automatic construction and optimization of DNN architecture as well as weight training which is commonly known as deep neuroevolution. In deep neuroevolution, EC can be used as the sole method for weight training or can be used along with traditional backpropagation algorithm. In Chapter 9, the structure of long short-term memory (LSTM) nodes is optimized using a genetic programming-based approach. It is shown that evolutionary optimization of LSTM can result in design which is more complex and powerful compared to human design. Chapter 10 investigates the use of EC for evolving deep recurrent connections in recurrent neural networks (RNNs) consisting of a suite of memory cells and simple neurons. Chapter 11 explores the applicability of EC in handling the challenges in structural design and training of generative adversarial networks (GANs).

Finally, the last part of this book focuses on real-world and interesting applications of DNN with EC and other usages of EC in contemporary DNN research. These applications harness the representation power of DNN and the ability of evolutionary computation in creating complex systems. These usages have shown promising signs for a new research philosophy and methodology worth further investigation and exploration. Chapter 12 presents an evolutionary framework for automatic construction of network structure for detecting dangerous objects in an X-ray image. Chapter 13 utilizes an evolutionary search to determine the architecture and tune hyper-parameters of DNN for malware classification. In Chap. 14, the training of GANs with less data is investigated on a spatially distributed evolutionary GAN training framework. Chapter 15 focuses on the application of EC in designing adversarial examples as well as improving the robustness of DNNs against adversarial attacks.

All the chapters are authored by well-known researchers and experienced practitioners in the relevant topic. Therefore, their knowledge, experience, and guidance will reveal the current state and the future promises and obstacles of the field as well as the necessary solutions and workarounds to take the field to the next level.

We hope readers will expand the topics explained in this book and make an academic venture in EC and DL.

Tokyo, Japan  
Newcastle, Australia  
February 2020

Hitoshi Iba  
Nasimul Noman

# Contents

# Abbreviations

| | |
|---|---|
| ABC | Artificial bee colony optimization |
| ACO | Ant colony optimization |
| AI | Artificial intelligence |
| AMT | Automatic music transcription |
| ANN | Artificial neural network |
| APK | Android application package |
| BCE | Binary cross entropy |
| BN | Batch normalization |
| BPTT | Backpropagation through time |
| CD | Contrastive divergence |
| CGP | Cartesian genetic programming |
| CMA-ES | Covariance matrix adaptation evolution strategy |
| CNN | Convolutional neural network |
| COEGAN | Coevolutionary generative adversarial networks |
| CS | Cuckoo search |
| DE | Differential evolution |
| DNN | Deep neural network |
| DSF | Differential state framework |
| EC | Evolutionary computation |
| EMO | Embedded memory order |
| ES | Evolution strategy |
| EXAMM | Evolutionary eXploration of Augmenting Memory Models |
| FA | Firefly algorithm |
| FFNN | Feed-forward neural network |
| FID | Fréchet inception distance |
| FN | False negative |
| FP | False positive |
| FSM | Finite-state machine |
| GA | Genetic algorithms |
| GAN | Generative adversarial network |
| GP | Genetic programming |

| | |
|---|---|
| GRU | Gated recurrent unit |
| HS | Harmony search |
| IGD | Inverted generational distance |
| IoU | Intersection over union |
| IS | Inception score |
| JSD | Jensen–Shannon divergence |
| LSTM | Long short-time memory |
| mAP | mean Average Precision |
| MGU | Minimal gated unit |
| MLM | Music language model |
| MLP | Multilayer perceptron |
| Mustangs | Mutation spatial GANs |
| NAS | Neural architecture search |
| NEAT | Neuroevolution of augmented topologies |
| NES | Natural evolution strategies |
| NGAFID | National General Aviation Flight Information Database |
| PCD | Persistent contrastive divergence |
| PR curve | Precision-recall curve |
| PSO | Particle swarm optimization |
| PTB | Penn Tree Bank |
| RBM | Restricted Boltzmann machine |
| ReLU | Rectified linear unit |
| RL | Reinforcement learning |
| RNN | Recurrent neural network |
| RTRL | Real-time recurrent learning |
| SGD | Stochastic gradient descent |
| SMRNN | Segmented-memory recurrent neural network |
| SOC | Self-organizing classifier |
| SOM | Self-organizing map |
| SUNA | Spectrum-diverse unified neuroevolution architecture |
| TBPTT | Truncated BPTT |
| TP | True positive |
| TSP | Traveling salesman problem |
| UGRNN | Update gate RNN |

# Part I
# Preliminaries

# Chapter 1
# Evolutionary Computation and Meta-heuristics

Hitoshi Iba

**Abstract** This chapter presents several methods of evolutionary computation and meta-heuristics. Evolutionary computation is a computation technique that mimics the evolutionary mechanism of life to select, deform, and convolute data structures. Because of its high versatility, its applications are found in various fields. Meta-heuristics described in this chapter are considered as representatives of swarm intelligence, such as particle swarm optimization (PSO), artificial bee colony optimization (ABC), ant colony optimization (ACO), firefly algorithms, cuckoo search, etc. A benefit of these methods is global searching as well as local searching. Existence of local minima or saddle points could lead to a locally optimum solution when using gradient methods such as the steepest descent search. By contrast, the methods described in this chapter can escape from such local solutions by means of various kinds of operations. Methods of evolutionary computation and meta-heuristics are used in combination with deep learning to establish a framework of deep neural evolution, which will be described in later chapters.

## 1.1 Introduction

Let us consider the following questions:

- "Why are the peacock's feathers so incredibly beautiful?"
- "Why did the giraffe's neck become so long?"
- "If a worker bee cannot have any offspring of its own, why does it work so hard to serve the queen bee?"

If we make a serious effort to answer these mysteries, we realize that we are solving one of the problems of optimization for each species, i.e., the process of evolution of species. It is the objective of the bio-inspired method to exploit this concept

H. Iba (✉)
Graduate School of Information Science and Technology, University of Tokyo, Tokyo, Japan
e-mail: iba@iba.t.u-tokyo.ac.jp

© Springer Nature Singapore Pte Ltd. 2020
H. Iba, N. Noman (eds.), *Deep Neural Evolution*, Natural Computing Series,
https://doi.org/10.1007/978-981-15-3685-4_1

to establish an effective computing system. Evolutionary computation (EC) and meta-heuristics attempt to "borrow" Nature's methods of problem solving and have been widely applied to find solutions to optimization problems, to automatically synthesize programs, and to accomplish other AI (artificial intelligence) tasks for the sake of the effective learning and the formation of hypotheses. These methods imitate the evolutionary or biological mechanisms of living organisms to create, to combine, and to select data structures. They are widely applied in deep neural evolution fields.

In the following sections, we will explain these methodologies in details.

## 1.2 Evolutionary Algorithms: From Bullet Trains to Finance and Robots

Evolutionary computation is an engineering method that imitates the mechanism of evolution in organisms and applies this to the deforming, synthesis, and selection of data structures. Using this method, we aim to solve the problem of optimization and generate a beneficial structure. Common examples of this are the computational algorithms known as genetic algorithms (GA) and genetic programming (GP).

The basic data structures in evolutionary computation are based on knowledge of genetics. Hereafter, we shall provide an explanation of these.

The information used in evolutionary computation is formed from the two-layer structures of PTYPE and GTYPE. GTYPE (genotype, also called genetic codes, and equating to the chromosomes within the cells) are, in a genetic type analogy, a low-level, locally-regulating set. This is the evolutionary computation to be operated on, as described later. The PTYPE is a phenotype, and expresses the emergence of behavior and structures over a wide area, accompanied by development within a GTYPE environment. Fitness is determined by the PTYPE adapting to its environment, and selection relies on the fitness of the PTYPE (Fig. 1.1). For a time, the higher the fitness score taken, the better. Therefore, for individuals with a fitness of 1.0 and 0.3, the former can adapt better to their environment, and it is easier for them to survive (however, in other areas of this book, there are cases when it is better to have a smaller score).

We shall explain the basic framework of the evolution computation, based on the above description (Fig. 1.2). Here, we configure a set containing several dogs. We shall call this generation $t$. This dog has a genetic code for each GTYPE, and its fitness is determined according to the generated PTYPE. In the diagram, the fitness of each dog is shown as the value near the dog (remember that the larger the better). These dogs reproduce and create the descendants in the next generation $t + 1$. In terms of reproduction, the better (higher) the fitness, the more descendants they are able to create, and the worse (lower) the fitness, the easier it is for them to become extinct (in biological terminology, this refers to choice and selection). In the diagram, the elements undergoing slight change in the phenotypes

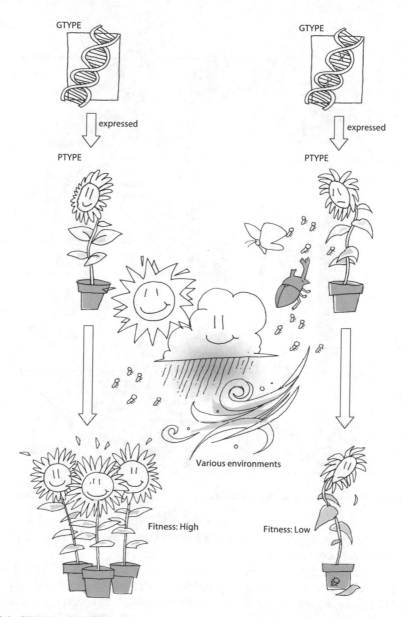

**Fig. 1.1** GTYPE and PTYPE

due to reproduction are drawn schematically. As a result of this, the fitness of each individual in the following generation $t + 1$ is expected to be better than that of the previous generation. Furthermore, the fitness as seen in the set as a whole also increases. In the same way, the dogs in the generation $t + 1$ become parents and produce the descendants in the generation $t + 2$. As this is repeated and the

**Fig. 1.2** Image of evolutionary computation

generations progress, the set as a whole improves and this is the basic mechanism of evolutionary computation.

In the case of reproduction, the operator shown in Fig. 1.3 is applied to the GTYPE, and produces the next generation of GTYPEs. To simplify things, here, the GTYPE is expressed as a 1-dimensional matrix. Each operator is an analogy for the genetic recombination and mutation, etc., in the organism. The application frequency and the application area of these operators are randomly determined in general.

Normally, the following kinds of methods are used for selection.

- **Roulette selection**: This is a method of selecting individuals in a ratio proportionate to their fitness. A roulette is created with an area proportionate to fitness. This roulette is spun, and individuals in the location where it lands are selected.

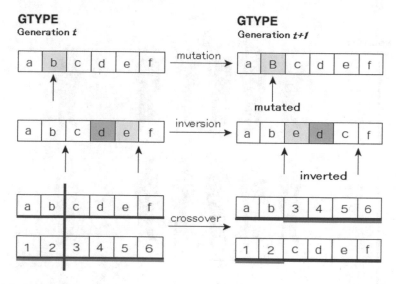

**Fig. 1.3** Genetic operators

- **Tournament selection**: Only the number of individuals from within the set (tournament size) are chosen at random and individuals from these with the highest fitness are selected. This process is repeated for the number of sets.
- **Elite strategy**: Several individuals with the highest fitness are left as is to the next generation. This can prevent individuals with the highest fitness not being selected coincidentally and left to perish. This strategy is used in combination with the above two methods.

With the elite strategy, the results will not get worse in the next generation as long as the environment does not change. For this reason, it is frequently applied in engineering applications. However, note that the flip side of this is that diversity is lost.

To summarize, in the evolutionary computation, generational change is as shown in Fig. 1.4. In the figure, $G$ is the elite rate (rate of upper level individuals that were copied and left results). We can refer to the reproduction rate as $1 - G$.

There are many riddles remaining biologically in terms of the evolution of mimicry. Research into solving these mysteries is flourishing with the use of computer simulation using evolutional computation.

Evolutional computation is used in a variety of areas of our everyday lives. For example, the front carriage model of the Japanese N700 series bullet train plays a major role in creating original forms (Fig. 1.5a). The N700 has the performance to take curves at 270 km, speeds 20 km faster than the previous model. However, in the traditional form of the front carriage, speeding up meant that the microbarometric waves in the tunnel increased, which are a cause of noise. To solve this difficulty, the original form known as "Aero double wing" has been derived from approximately 5000 simulations using evolutionary computation. Furthermore, in the wing design

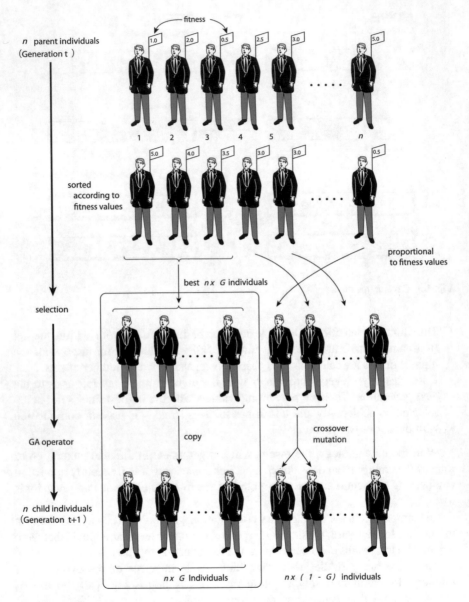

**Fig. 1.4** Selection and reproduction of GA

of the MRJ (Mitsubishi regional jet, which is the first domestic jet in Japan), a method known as multi-objective evolutionary computation was used (Fig. 1.5b). Using this method, the two objectives of improving the fuel efficiency of passenger jet devices and reduction in noise external to the engine were optimized simul-

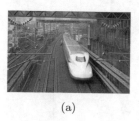  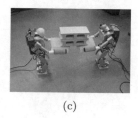

(a)                              (b)                              (c)

**Fig. 1.5** EC applications. (**a**) N700 series bullet train. (**b**) MRJ (Mitsubishi regional jet). (**c**) Collaborative transportation by humanoid robots

taneously, and they succeeded in improving performance compared to competing models.

In fields other than engineering, such as the financial field, the use of evolutionary computation methods is spreading. Investment funds are using this as a practical technology for portfolio construction and market prediction (see [10] for details). Furthermore, it has practical application in such fields as scheduling design to optimize the work shifts of nurses and allocating crews for aircraft.

Another field that is using evolutionary computation is the field of evolutionary robotics. For example, Fig. 1.5c is an example of cooperative work (collaborative transportation) of evolutionary humanoid robots. Here, a learning model is used that applies co-evolution to evolutionary computation. Furthermore, module robots, which modify themselves in accordance with geographical features, environment, and work content, by combining blocks, are gaining attention. This technology is even being used by NASA (National Aeronautics and Space Administration) for researching the form of robots optimized for surveying amidst the limited environment of Mars. The form of organisms we know about may be only those species that are remaining on earth. These may be types that match the earth environment, and it is not known if these are optimal. Through evolutionary computation, if we can reproduce the process of evolution on a computer, new forms may emerge that we do not yet know about. The result of this may be the evolution of robots compatible with Mars and unknown planets (see [17]).

## 1.3  Multi-Objective Optimization

An evolutionary algorithm can take competing goals into consideration. It complies to any policies from its users regarding limits and preferences on these goals. Evolutionary algorithms that deal with multiple objectives are usually called MOEAs (Multi-Objective Evolutionary Algorithms).

Assume you are engaged in transport planning for a town [6]. The means of reducing traffic accidents range from installing traffic lights, placing more traffic signs, and regulating traffic, to setting up checkpoints (Fig. 1.6). Each involves a different cost, and the number of traffic accidents will vary with the chosen

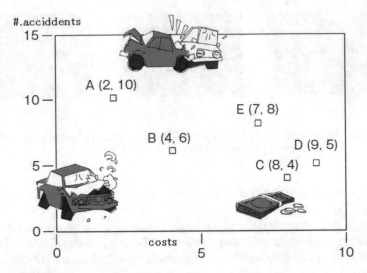

**Fig. 1.6** Cost vs. expected numbers of accidents

approach. Let us assume that five means (A, B, C, D, and E) are available, and that the cost and the predicted accident numbers are

$$A = (2, 10)$$

$$B = (4, 6)$$

$$C = (8, 4)$$

$$D = (9, 5)$$

$$E = (7, 8),$$

where the first element is the cost and the second is the predicted accident number, as plotted in Fig. 1.6. The natural impulse is to desire attainment of both goals in full: the lowest cost and the lowest predicted accident number. Unfortunately, it is not necessarily possible to attain both objectives by the same means and thus not possible to optimize both at the same time.

In such situations, the concept of "Pareto optimality" is useful. For a given developmental event to represent a Pareto optimal solution, it must be the case that no other developmental events exist which are of equal or greater desirability, for all evaluation functions, that is, fitness functions.

Let us look again at Fig. 1.6. Note that the points in the graph increase in desirability as we move toward the lower left. A, B, and C in particular appear to be good candidates. None of these three candidates is the best in both dimensions, that is, in both "evaluations," but for each there is no other candidate that is better in both evaluations. Such points are called "non-dominated" points. Points D and E, in contrast, are both "dominated" by other points and therefore less desirable. E is dominated by B, as B is better than E in both evaluations:

$$\text{Cost of B}(4) < \text{Cost of E}(7)$$

$$\text{Predicted accidents for B}(6) < \text{Predicted accidents for E}(8).$$

D is similarly dominated by C. In this example, therefore, the Pareto optimums are A, B, and C. As this suggests, the concept of the Pareto optimum cannot be used to select just one candidate from a group of candidates, and thus it cannot be concluded which of A, B, and C is the best.

Pareto optimality may be defined more formally as follows. Let two points $x = (x_1, \ldots, x_n)$ and $y = (y_1, \ldots, y_n)$ exist in an $n$-dimensional search space, with each dimension representing an objective (an evaluation) function, and with the objective being a minimization of each to the degree possible. The domination of $y$ by $x$ (written as $x <_p y$) may be, therefore, defined as

$$x <_p y \Longleftrightarrow (\forall i)(x_i \leq y_i) \wedge (\exists i)(x_i < y_i). \tag{1.1}$$

In the following, we will refer to $n$ (the number of different evaluation functions) as the "dimension number." Any point that is not inferior to any other point will be called "non-dominated" or "non-inferior," and the curve (or curved surface) formed by the set of Pareto optimal solutions will be called the "Pareto front."

The main problem that MOEAs face is how to combine the multiple objectives into a metric that can be used to perform selection. In other words, how to take into account all objectives when selecting individuals from one generation for crossover. The readers should refer to [2, 5, 13] for the studies on multi-objective optimization methods.

## 1.4 Genetic Programming and Its Genome Representation

### 1.4.1 Tree-based Representation of Genetic Programming

The aim of genetic programming (GP) is to extend genetic forms from genetic algorithm (GA) to the expression of trees and graphs and to apply them to the synthesis of programs and the formation of hypotheses or concepts. Researchers are using GP to attempt to improve their software for the design of control systems and structures for robots.

The procedures of GA are extended in GP in order to handle graph structures (in particular, tree structures). Tree structures are generally well described by S-expressions in LISP. Thus, it is quite common to handle LISP programs as "genes" in GP. As long as the user understands that the program is expressed in a tree format, then he or she should have little trouble reading a LISP program (the user should recall the principles of flow charts). The explanations below have been presented so as to be quickly understood by a reader who does not know LISP.

A tree is a graph with a structure as follows, incorporating no cycles:

$$
\begin{array}{c}
A \\
\diagup \mid \\
B \quad C \\
\mid \\
D
\end{array}
$$

More precisely, a tree is an acyclical connected graph, with one node defined as the root of the tree. A tree structure can be expressed as an expression with parentheses. The above tree would be written as follows:

```
(A (B)
    (C (D))).
```

In addition, the above can be simplified to the following expression:

```
(A B
    (C D)).
```

This notation is called an "S-expression" in LISP. Hereinafter, a tree structure will be identified with its corresponding S-expression. The following terms will be used for the tree structure:

- Node: Symbolized with A, B, C, D, etc.
- Root: A
- Terminal node: B, D (also called a "terminal symbol" or "leaf node")
- Non-terminal node: A, C (also called a "non-terminal symbol" and an "argument of the S-expression")
- Child: From the viewpoint of A, nodes B and C are children (also, "arguments of function A")
- Parent: The parent of C is A

Other common phrases will also be used as convenient, including "number of children," "number of arguments," "grandchild," "descendant," and "ancestor." These are not explained here, as their meanings should be clear from the context.

The following genetic operators acting on the tree structure will be incorporated:

1. **Gmutation** Alteration of the node label
2. **Ginversion** Reordering of siblings
3. **Gcrossover** Exchange of a subtree

These are natural extensions of existing genetic operators and act on sequences of bits. These operators are shown below in examples where they have been applied in LISP expression trees (S-expressions) (Fig. 1.7). The underlined portion of the statement is the expression that is acted upon:

**Fig. 1.7** Genetic operators in GP

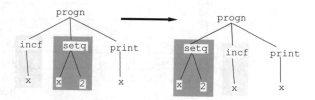

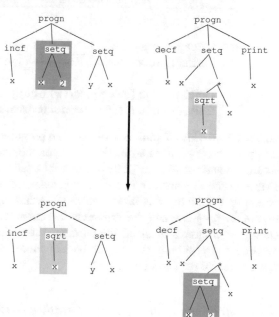

**Gmutation**   Parent:$(+ \ x \ \underline{y})$
$\Downarrow$
Child:$(+ \ x \ \underline{z})$

**Ginversion**   Parent:$(\text{progn} \ \underline{(\text{incf} \ x)} \ (\text{setq} \ x \ 2) \ (\text{print} \ x))$
$\Downarrow$
Child:$(\text{progn} \ \underline{(\text{setq} \ x \ 2) \ (\text{incf} \ x)} \ (\text{print} \ x))$

**Table 1.1** Program changes due to GP operators

| Operator | Program before operation | Program after operation |
|---|---|---|
| Mutation | Add $x$ and $y$ | Add $x$ and $z$ |
| Inversion | 1. Add 1 to $x$ | 1. Set $x = 2$ |
| | 2. Set $x = 2$ | 2. Add 1 to $x$ |
| | 3. Print $x(= 2)$ and return 2 | 3. Print $x(= 3)$ and return 3 |
| Crossover | Parent$_1$: | Child$_1$: |
| | 1. Add 1 to $x$ | 1. Add 1 to $x$ |
| | 2. Set $x = 2$ | 2. Take square root of $x$ |
| | 3. Set $y = x(= 2)$ and return 2 | 3. Set $y = x$ and return the value |
| | Parent$_2$: | Child$_2$: |
| | 1. Subtract 1 from $x$ | 1. Subtract 1 from $x$ |
| | 2. Set $x = \sqrt{x} \times x$ | 2. Set $x = 2$ and its value $(= 2)$ is multiplied by $x(= 2)$. The result value $(= 4)$ is set to $x$ again |
| | 3. Print $x$ and return the value. | 3. Print $x(= 4)$ and return 4. |

**Gcrossover**   Parent$_1$:(progn (incf $x$) $\overline{\text{(setq } x \text{ 2)}}$ (setq $y$ $x$))
Parent$_2$:(progn (decf $x$) $\overline{\text{(setq } x \text{ (* (sqrt } x) \ x))}$ (print $x$))
$\Downarrow$
Child$_1$:(progn (incf $x$) $\overline{\text{(sqrt } x)}$ (setq $y$ $x$))
Child$_2$:(progn (decf $x$) $\overline{\text{(setq } x \text{ (* (setq } x \text{ 2) } x))}$ (print $x$)).

Table 1.1 provides a summary of how the program was changed as a result of these operators. "progn" is a function acting on the arguments in the order of their presentation and returns the value of the final argument. The function "setq" sets the value of the first argument to the evaluated value of the second argument. It is apparent on examining this table that mutation has caused a slight change to the action of the program, and that crossover has caused replacement of the actions in parts of the programs of all of the parents. The actions of the genetic operators have produced programs that are individual children but that have inherited the characteristics of the parent programs.

## 1.4.2 Cartesian Genetic Programming (CGP)

CGP [18] is a genetic programming (GP) technique proposed by Miller et al. CGP represents a tree structure with a feed-forward-type network. It is a method by which all nodes are described genotypically beforehand to optimize connection relations. This is supposed to enable handling the problem of bloat, in which the tree structure becomes too large as a consequence of the number of GP genetic operations. Furthermore, by reusing a partial tree, the tree structure can be represented compactly.

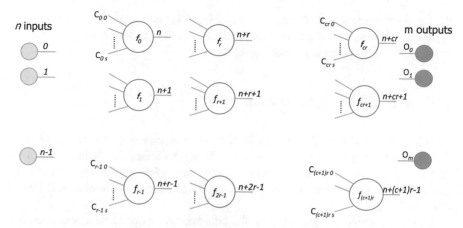

**Fig. 1.8** Example of a genotype and a phenotype

$$\underline{0}\,0\,1\ \ \underline{1}\,0\,0\ \ \underline{1}\,3\,1\ \ \underline{2}\,0\,1\ \ \underline{0}\,4\,4\ \ \underline{2}\,5\,4\ \ 2\,5\,7\,3$$

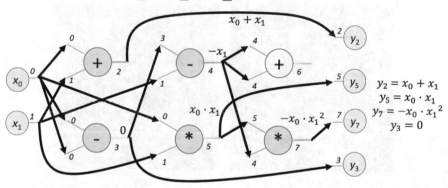

**Fig. 1.9** Example of a genotype and a phenotype

The CGP network comprises three varieties of node: input nodes, intermediate nodes, and output nodes. Figure 1.8 shows a CGP configuration with $n$ inputs, $m$ outputs, and $r \times c$ intermediate layers. Here, connecting nodes in the same column is not permitted. CGP networks are also restricted to being feed-forward networks.

CGP uses a one-dimensional numeric string for the genotypes. These describe the functional type and connection method of the intermediate nodes, and the connection method of the output nodes. Normally, all functions have the largest argument as the input and ignore unused connections. For example, consider the following genotypes for the CGP configuration shown in Fig. 1.9.

$$\underline{0}\,0\,1\qquad \underline{1}\,0\,0\qquad \underline{1}\,3\,1\qquad \underline{2}\,0\,1\qquad \underline{0}\,4\,4\qquad \underline{2}\,5\,4\qquad 2\,5\,7\,3$$

The function symbol numbers 0, 1, 2, and 3 (underlined above) correspond to addition, subtraction, multiplication, and division, respectively. The network corresponding to the genotype at this time is as shown in Fig. 1.9. For example, the inputs of the first node 0 are input 0 and input 1, and the addition computation is functional. Note that the output of the fifth node is not used anywhere (0 4 4), making it an intron (i.e., a non-coding region).

## 1.5 Ant Colony Optimization (ACO)

Ants march in a long line. There is food at one end, a nest at the other. This is a familiar scene in gardens and on roads, but the sophisticated distributed control by these small insects was recognized by humans only a few decades ago. Marching is a cooperative ant behavior that can be explained by the pheromone trail model (Fig. 1.10, [12]).

Optimization algorithms based on the collective behavior of ants are called ant colony optimization (ACO) [3]. ACO using a pheromone trail model for the TSP

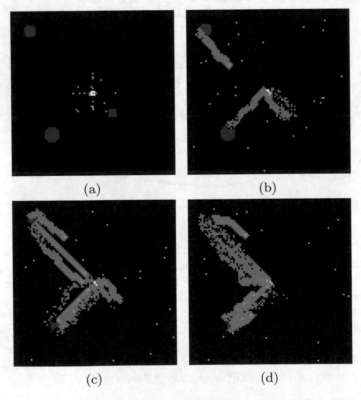

(a)          (b)

(c)          (d)

**Fig. 1.10** Pheromone trails of ants (**a**) The first random search phase. (**b**) The closer lower right and lower left food is found, and the pheromone trail is formed. The upper left is in the middle of the formation. (**c**) Pheromone trails are formed for all three sources, which makes the transport more efficient. The lower right source is almost exhaustively picked. (**d**) The lower right food source finishes, and the pheromone trail has already dissipated. As a result, a vigorous transportation for the two sources on the left is being done

uses the following algorithm to optimize the travel path:

| | |
|---|---|
| **Step 1** | Ants are placed randomly in each city. |
| **Step 2** | Ants move to the next city. The destination is probabilistically determined based on the information on pheromones and given conditions. |
| **Step 3** | Repeat until all cities are visited. |
| **Step 4** | Ants that make one full cycle secrete pheromones on the route according to the length of the route. |
| **Step 5** | Return to **Step** 1 if a satisfactory solution has not been obtained. |

The ant colony optimization (ACO) algorithm can be outlined as follows. Take $\eta_{ij}$ as the distance between cities $i$ and $j$ (Fig. 1.11). The probability $p_{ij}^k(t)$ that an ant $k$ in city $i$ will move to city $j$ is determined by the reciprocal of the distance $1/\eta_{ij}$ and the amount of pheromone $\tau_{ij}(t)$ as follows:

$$p_{ij}^k(t) = \frac{\tau_{ij}(t) \times \eta_{ij}^\alpha}{\sum_{h \in J_i^k} \tau_{ih}(t) \times \eta_{ih}^\alpha}. \tag{1.2}$$

Here, $J_i^k$ is the set of all cities that the ant $k$ in city $i$ can move to (has not visited). The condition that ants are more likely to select a route with more pheromone reflects the positive feedback from past searches as well as a heuristic for searching for a shorter path. The ACO can thereby include an appropriate amount of knowledge unique to the problem.

The pheromone table is updated by the following equations:

$$Q(k) = \text{the reciprocal of the path that the ant } k \text{ found} \tag{1.3}$$

$$\Delta\tau_{ij}(t) = \sum_{k \in A_{ij}} Q(k) \tag{1.4}$$

$$\tau_{ij}(t+1) = (1-\rho) \cdot \tau_{ij}(t) + \Delta\tau_{ij}(t). \tag{1.5}$$

**Fig. 1.11** Path selection rules of ants

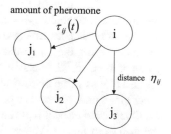

The amount of pheromone added to each path after one iteration is inversely proportional to the length of the paths that the ants found (Eq. (1.3)). The results for all ants that moved through a path are reflected in the path (Eq. (1.4)). Here, $A_{ij}$ is the set of all ants that moved on a path from city $i$ to city $j$. Negative feedback to avoid local solutions is given as an evaporation coefficient (Eq. (1.5)), where the amount of pheromone in the paths, or information from the past, is reduced by a fixed factor ($\rho$).

The ACO is an effective method to solve the traveling salesman problem (TSP) compared to other search strategies. The characteristic that specialized methods perform better in static problems is shared by many meta-heuristics (high-level strategies which guide an underlying heuristic to increase their performance). Complicated problems, such as TSPs where the distances between cities are asymmetric or where the cities change dynamically, do not have established programs and the ACO is considered to be one of the most promising methods.

## 1.6 Particle Swarm Optimization (PSO)

Kennedy et al. designed an effective optimization algorithm using the mechanism behind swarming boids [16]. This is called particle swarm optimization (PSO), and numerous applications are reported.

The classic PSO was intended to be applied to optimization problems. It simulates the motion of a large number of individuals (or "particles") moving in a multi-dimensional space [16]. Each individual stores its own location vector ($\mathbf{x_i}$), velocity vector ($\mathbf{v_i}$), and the position at which the individual obtained the highest fitness value ($\mathbf{p_i}$). All individuals also share information regarding the position with the highest fitness value for the group ($\mathbf{p_g}$).

As generations progress, the velocity of each individual is updated using the best overall location obtained up to the current time for the entire group and the best locations obtained up to the current time for that individual. This update is performed using the following formula:

$$\mathbf{v_i} = \chi \left( \omega \mathbf{v_i} + \phi_1 \cdot (\mathbf{p_i} - \mathbf{x_i}) + \phi_2 \cdot (\mathbf{p_g} - \mathbf{x_i}) \right). \tag{1.6}$$

The overall flow of the PSO is as shown in Fig. 1.12. Let us now consider the specific movements of each individual (see Fig. 1.13). A flock consisting of a number of birds is assumed to be in flight. We focus on one of the individuals (Step 1). In the figure, the ◯ symbols and linking line segments indicate the positions and paths of the bird. The nearby ◎ symbol (on its path) indicates the position with the highest fitness value on the individual's path (Step 2). The distant ◎ symbol (on the other bird's path) marks the position with the highest fitness value for the flock (Step 2). One would expect that the next state will be reached in the direction shown by the arrows in Step 3. Vector ① shows the direction followed in the previous steps; vector ② is directed towards the position with the highest fitness for the flock; and

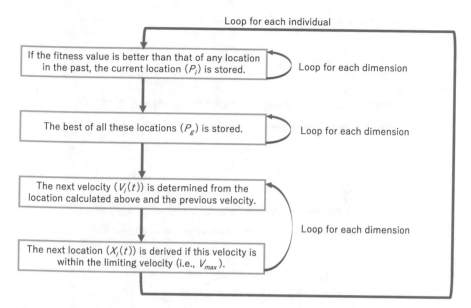

**Fig. 1.12** Flow chart of the PSO algorithm

vector ③ points to the location where the individual obtained its highest fitness value so far. Thus, all these vectors, ①, ②, and ③, in Step 3 are summed to obtain the actual direction of movement in the subsequent step (see Step 4).

The efficiency of this type of PSO search is certainly high because focused searching is available near optimal solutions in a relatively simple search space. However, the canonical PSO algorithm often gets trapped in local optimum in multimodal problems. Because of that, some sort of adaptation is necessary in order to apply PSO to problems with multiple sharp peaks.

To overcome the above limitation, a GA-like mutation can be integrated with PSO [8]. This hybrid PSO does not follow the process by which every individual of the simple PSO moves to another position inside the search area with a predetermined probability without being affected by other individuals, but leaves a certain ambiguity in the transition to the next generation due to Gaussian mutation. This technique employs the following equation:

$$mut(x) = x \times (1 + gaussian(\sigma)), \tag{1.7}$$

where $\sigma$ is set to be 0.1 times the length of the search space in one dimension. The individuals are selected at a predetermined probability and their positions are determined at the probability under the Gaussian distribution. Wide-ranging searches are possible at the initial search stage and search efficiency is improved at the middle and final stages by gradually reducing the appearance ratio of Gaussian mutation at the initial stage. Figure 1.14 shows the PSO search process with

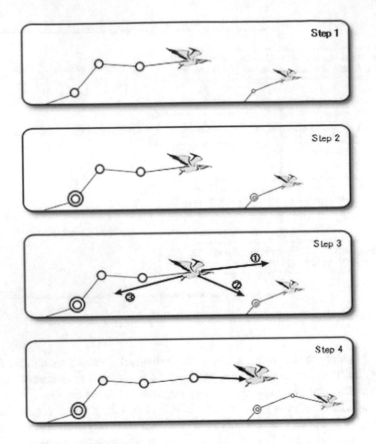

**Fig. 1.13** In which way do birds fly?

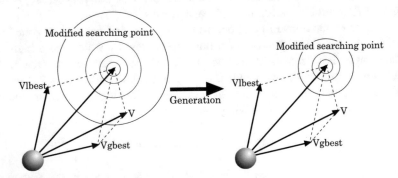

**Fig. 1.14** Concept of searching process by PSO with Gaussian mutation

Gaussian mutation. In the figure, $V_{lbest}$ represents the velocity based on the local best, i.e., $\mathbf{p_i} - \mathbf{x_i}$ in Eq. (1.6), whereas $V_{gbest}$ represents the velocity based on the global best, i.e., $\mathbf{p_g} - \mathbf{x_i}$.

## 1.7 Artificial Bee Colony Optimization (ABC)

Bees, along with ants, are well-known examples of social insects. Bees are classified into three types:

- employed bees
- onlooker bees
- and scout bees

Employed bees fly in the vicinity of feeding sites they have identified, sending information about food to onlooker bees. Onlooker bees use the information from employed bees to perform selective searches for the best food sources from the feeding site. When information about a feeding site is not updated for a given period of time, its employed bees abandon it and become scout bees that search for a new feeding site. The objective of a bee colony is to find the highest-rated feeding sites. The population is approximately half employed bees and scout bees (about 10–15% of the total), the rest are onlooker bees.

The waggle dance (a series of movements) performed by employed bees to transmit information to onlooker bees is well known (Fig. 1.15). The dance involves shaking the hindquarters and indicating the angle with which the sun will be positioned when flying straight to the food source, with the sun represented as straight up. For example, a waggle dance performed horizontally and to the right with respect to the nest combs means "fly with the sun at 90° to the left." The

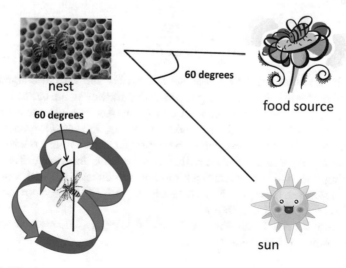

nest

60 degrees

60 degrees

food source

sun

**Fig. 1.15** Waggle dance

speed of shaking the rear indicates the distance to the food; when the rear is shaken quickly, the food source is very near, and when shaken slowly it is far away. Communication via similar dances is also performed with regard to pollen and water collection, as well as the selection of locations for new hives.

The artificial bee colony (ABC) algorithm [14, 15] initially proposed by Karaboga et al. is a swarm optimization algorithm that mimics the foraging behavior of honey bees. Since ABC was designed, it has been proved that ABC, with fewer control parameters, is very effective and competitive with other search techniques such as genetic algorithm (GA), particle swarm optimization (PSO), and differential evolution (DE [9]).

In ABC algorithms, an artificial swarm is divided into employed bees, onlooker bees, and scouts. $N$ $d$-dimensional solution candidates to the problem are randomly initialized in the domain and referred to as food sources. Each employed bee is assigned to a specific food source $\mathbf{x}_i$ and searches for a new food source $\mathbf{v}_i$ by using the following operator:

$$\mathbf{v}_{ij} = \mathbf{x}_{ij} + \text{rand}(-1, 1) \times (\mathbf{x}_{ij} - \mathbf{x}_{kj}), \tag{1.8}$$

where $k \in \{1, 2, \ldots, N\}$, $k \neq i$, and $j \in \{1, 2, \ldots, d\}$ are randomly chosen indices. $\mathbf{v}_{ij}$ is the $j$th element of the vector $\mathbf{v}_i$. If the trail to a food source is outside of the domain, it is reset to an acceptable value. The obtained $\mathbf{v}_i$ is then evaluated and put into competition with $\mathbf{x}_i$ for survival. The bee prefers the better food source. Unlike employed bees, each onlooker bee chooses a preferable source according to the food source's fitness to do further searches in the food space using Eq. (1.8). This preference scheme is based on the fitness feedback information from employed bees. In classic ABC [14], the probability of the food source $\mathbf{x}_i$ that can be exploited is expressed as

$$p_i = \frac{fit_i}{\sum_{j=1}^{N} fit_j}, \tag{1.9}$$

where $fit_i$ is the fitness of the $i$th food source, $\mathbf{x}_i$. For the sake of simplicity, we assume that the fitness value is non-negative and that the larger, the better. If the trail $\mathbf{v}_i$ is superior to $\mathbf{x}_i$ in terms of profitability, this onlooker bee informs the relevant employed bee associated with the $i$th food source, $\mathbf{x}_i$, to renew its memory and forget the old one. If a food source cannot be improved upon within a predetermined number of iterations, defined as Limit, this food source is abandoned. The bee that was exploiting this food site becomes a scout and associates itself with a new food site that is chosen via some principle. In canonical ABC [14], the scout looks for a new food site by random initialization.

The details of the ABC algorithm are described as below. The pseudocode of the algorithm is shown in Algorithm 1.

---

**Algorithm 1** The ABC algorithm

---

**Require:** $T_{max}$, #. of employed bees (= #. of onlooker bees), `Limit`
   Initialize food sources
   Evaluate food sources
   $i = 1$
   **while** $i < T_{max}$ **do**
      Use employed bees to produce new solutions
      Evaluate the new solutions and apply greedy selection process
      Calculate the probability values using fitness values
      Use onlooker bees to produce new solutions
      Evaluate new solutions and apply greedy selection process
      Determine abandoned solutions and use scouts to generate new ones randomly
      Remember the best solution found so far
      $i = i + 1$
   **end while**
   Return best solution

---

**Step 0: Preparation**   The total number of search points ($N$), total number of trips ($T_{max}$), and a scout control parameter (`Limit`) are initialized. The numbers of employed bees and onlooker bees are set to be the same as the total number of search points ($N$). The value of the objective function $f$ is taken to be non-negative, with larger values being better.

**Step 1: Initialization 1**   The trip counter $k$ is set to 1, and the number of search point updates $s_i$ is set to 0. The initial position vector for each search point $\mathbf{x}_i = (x_{i1}, x_{i2}, x_{i3}, \ldots, x_{id})^T$ is assigned random values. Here, the subscript $i$ ($i = 1, \ldots, N$) is the index of the search point, and $d$ is the number of dimensions in the search space.

**Step 2: Initialization 2**   Determine the initial best solution **best**.

$$i_g = \underset{i}{\mathrm{argmax}}\, f(\mathbf{x}_i) \tag{1.10}$$

$$\mathbf{best} = x_{i_g}. \tag{1.11}$$

**Step 3: Employed bee search**   The following equation is used to calculate a new position vector $\mathbf{v}_{ij}$ from the current position vector $\mathbf{x}_{ij}$:

$$\mathbf{v}_{ij} = \mathbf{x}_{ij} + \phi \cdot (\mathbf{x}_{ij} - \mathbf{x}_{kj}). \tag{1.12}$$

Here, $j$ is a randomly chosen dimensional number, $k$ is the index for some randomly chosen search point other than $i$, and $\phi$ is a uniform random number in the range $[-1, 1]$. The position vector $\mathbf{x}_i$ and the number of search point updates $s_i$ are determined according to the following equation:

$$I = \{i \mid f(\mathbf{x}_i) < f(\mathbf{v}_i)\} \tag{1.13}$$

$$\mathbf{x}_i = \begin{cases} \mathbf{v}_i & i \in I \\ \mathbf{x}_i & i \notin I \end{cases} \tag{1.14}$$

$$s_i = \begin{cases} 0 & i \in I \\ s_i + 1 & i \notin I. \end{cases} \tag{1.15}$$

**Step 4: Onlooker bee search**    The following two steps are performed.

1. Relative ranking of search points
   The relative probability $P_i$ is calculated from the fitness $fit_i$, which is based on the evaluation score of each search point. Note that $fit_i = f(\mathbf{x}_i)$. The onlooker bee search counter $l$ is set to 1.

$$P_i = \frac{fit_i}{\sum_{j=1}^{N} fit_j}. \tag{1.16}$$

2. Roulette selection and search point updating
   Search points are selected for updating based on the probability $P_i$, calculated above. After search points have been selected, perform a procedure as in **Step 3** to update the search point position vectors. Then, let $l = l + 1$ and repeat until $l = N$.

**Step 5: Scout bee search**    Given a search point for which $s_i \geq$ Limit, random numbers are used to exchange generated search points.

**Step 6: Update best solution**    Update the best solution **best**.

$$i_g = \underset{i}{\operatorname{argmax}}\, f(\mathbf{x}_i) \tag{1.17}$$

$$\mathbf{best} = \mathbf{x}_{i_g} \quad \text{when} \quad f(x_{i_g}) > f(\mathbf{best}). \tag{1.18}$$

**Step 7: End determination**    End if $k = T_{\max}$. Otherwise, let $k = k + 1$ and return to **Step 3**.

ABC has recently been improved in many aspects. For instance, we analyzed the mechanism of ABC to show a possible drawback of using parameter perturbation. To overcome this deficiency, we have proposed a new non-separable operator and embedded it in the main framework of the cooperation mechanism of bee foraging (see [7] for details).

## 1.8 Firefly Algorithms

Fireflies glow owing to a luminous organ and fly around. This glow is meant to attract females. The light generated by each firefly differs depending on the

---

**Algorithm 2** Firefly algorithm

---

Initialize a population of fireflies $\mathbf{x}_i$ ($i = 1, 2, \cdots, n$)      ▷ Minimizing objective function
$f(\mathbf{x})$, $\mathbf{x} = (x_1, \cdots, x_d)^T$.
Define light absorption coefficient $\gamma$
$t = 1$      ▷ Generation count.
**while** $t < MaxGeneration$ and the stop criterion is not satisfied **do**
     **for** $i = 1$ to $n$ **do**      ▷ for all $n$ fireflies
         **for** $j = 1$ to $n$ **do**      ▷ for all $n$ fireflies
             Light intensity $I_i$, $I_j$ at $\mathbf{x}_i$, $\mathbf{x}_j$ is determined by $f$
             **if** $I_i > I_j$ **then**
                 Move firefly $i$ towards $j$ in all $d$ dimensions
             **end if**
             Attractiveness varies with distance $r$ via $e^{-\gamma r}$
             Evaluate new solutions and update light intensity
         **end for**
     **end for**
     Rank the fireflies and find the current best
     $t = t + 1$
**end while**
Postprocess results and visualization

---

individual insect, and it is considered that they attract others following the rules described below:

- The extent of attractiveness is in proportion to the luminosity.
- Female fireflies are more strongly attracted by males that produce a strong glow.
- Luminosity decreases as a function of distance.

The firefly algorithm (FA) is a search method based on blinking fireflies [22]. This algorithm does not discriminate gender. That is, all fireflies are attracted to each other. In this case, the luminosity is determined by an objective function. To solve the minimization problem, fireflies at a lower functional value (with a better adaptability) glow much more strongly. The most glowing firefly moves around at random.

Algorithm 2 describes the outline of the FA. The moving formula for a firefly $i$ attracted by firefly $j$ is as follows:

$$\mathbf{x}_i^{new} = \mathbf{x}_i^{old} + \beta_{i,j} \left( \mathbf{x_j} - \mathbf{x}_i^{old} \right) + \alpha \left( rand(0, 1) - \frac{1}{2} \right), \tag{1.19}$$

where $rand(0, 1)$ is a uniform random numbers between 0 and 1. $\alpha$ is a parameter to determine the magnitude of the random numbers, and $\beta_{i,j}$ represents how attractive firefly $j$ is to firefly $i$, i.e.,

$$\beta_{i,j} = \beta_0 e^{-\gamma r_{i,j}^2}. \tag{1.20}$$

The variable $\beta_0$ represents how attractive fireflies are when $r_{i,j} = 0$, which indicates that the two are in the same position. Since $r_{i,j}$ represents the Euclid distance between firefly $i$ and $j$, their attractiveness varies depending on the distance between them.

The most glowing firefly moves around at random, according to the following formula:

$$\mathbf{x_k}(t + 1) = \mathbf{x_k}(t) + \alpha \left( rand(0, 1) - \frac{1}{2} \right). \tag{1.21}$$

The reason for this is that the entire population converges to the locally best solution in an initial allocation.

As the distance becomes greater, the attractiveness becomes weaker. Therefore, under the firefly algorithms, fireflies form groups with each other at a distance instead of gathering at one spot.

The firefly algorithms are suitable for optimization problems on multimodality and are considered to yield better results compared to those obtained using PSO. It has another extension that separates fireflies into two groups and limits the effect on those in the same group. This enables global solutions and local solutions to be searched simultaneously.

## 1.9 Cuckoo Search

The cuckoo search (CS) [23] is meta-heuristics based on brood parasitic behavior. Brood parasitism is an animal behavior in which an animal depends on a member of another species (or induces this behavior) to sit on its eggs. Some species of cuckoos are generally known to exhibit this behavior. They leave their eggs in the nests of other species of birds such as the great reed warblers, Siberian meadow buntings, bullheaded shrikes, azure-winged magpies, etc.[1] Before leaving, they demonstrate an interesting behavior referred to as egg mimicry: they take out one egg of a host bird (foster parent) already in the nest and lay an egg that mimics the other eggs in the nest, thus keeping the numbers balanced.[2] This is because a host bird discards an egg when it determines that the laid egg is not its own.

---

[1] Parasitized species are almost always fixed for each female cuckoo.

[2] Furthermore, a cuckoo chick having just been hatched expels all the eggs of its host. For this reason, a cuckoo chick has a pit in its back to place its host's egg, clambers up inside the nest and throw the egg out of the nest. This behavior was discovered by Edward Jenner, famous for smallpox vaccination.

A cuckoo chick has a remarkably large and bright bill; therefore, it is excessively fed by its foster parent. This is referred to as "supernormal stimulus." Furthermore, there is an exposed skin region at the back of the wings with the same color as its bill. When the foster parent carries foods, the chick spreads its wings to make the parent aware of the region. The foster parent mistakes it for its own chicks. Thus, the parent believes that it has more chicks to feed than it actually has and carries more food to the nest. It is considered to be an evolutional strategy for cuckoos to be fed corresponding to their size because a grown cuckoo is many times larger than the host.

The CS models the cuckoos' brood parasitic behavior based on three rules as described below:

---

**Algorithm 3** Cuckoo search

---

Initialize a population of $n$ host nests $\qquad$ ▷ Minimizing objective function
$f(\mathbf{x})$, $\mathbf{x} = (x_1, \cdots, x_d)^T$.
Produce one egg $x_i^0$ in each host $i = 1, \cdots, n$
$t = 1$ $\qquad$ ▷ Generation count.
**while** $t < MaxGeneration$ and the stop criterion is not satisfied **do**
$\quad$ Choose a nest $i$ randomly
$\quad$ Produce a new egg $\mathbf{x}_i^t$ by performing Lèvy flights $\qquad$ ▷ Brood parasite of the cuckoo.
$\quad$ Choose a nest $j$ randomly and let its egg be $\mathbf{x}_j^{t-1}$
$\quad$ **if** $f(\mathbf{x}_j^{t-1}) > f(\mathbf{x}_i^t)$ **then** $\qquad$ ▷ The new egg is better.
$\quad\quad$ Replace $j$'s egg by the new egg, i.e., $\mathbf{x}_i^t$
$\quad$ **end if**
$\quad$ Sort the nests according to their eggs' performance
$\quad$ A fraction $(p_a)$ of the worse nests are abandoned and new ones are built by performing Lèvy flights
$\quad$ $t = t + 1$
**end while**
Postprocess results and visualization

---

- A cuckoo lays one egg at a time and leaves it in a randomly selected nest.
- The highest quality egg (difficult to be noticed by the host bird) is carried over to the next generation.
- The number of nests is fixed, and a parasitized egg is noticed by a host bird with a certain probability. In this case, the host bird either discards the egg or rebuilds the nest.

Algorithm 3 shows the CS algorithm. Based on this algorithm, a cuckoo lays a new egg in a randomly selected nest, according to Lévy flight. This flight presents mostly a short distance random walk with no regularity. However, it sometimes exhibits a long-distance movement. This movement has been identified in several

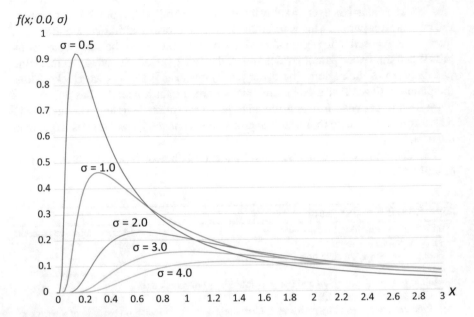

**Fig. 1.16** Lévy distribution

animals and insects. It is considered to be able to represent stochastic fluctuations observed in various natural and physical phenomena such as flight patterns, feeding behaviors, etc.

Specifically, Lévy distribution is represented by the following probability density function referred to in Fig. 1.16:

$$f(x; \mu, \sigma) = \begin{cases} \sqrt{\frac{\sigma}{2\pi}} \exp\left[-\frac{\sigma}{2(x-\mu)}\right](x - \mu)^{-3/2} & (\mu < x), \\ 0 & \text{(otherwise)}, \end{cases} \tag{1.22}$$

where $\mu$ represents a positional parameter, and $\sigma$ represents a scale parameter. Based on this distribution, Lévy flight mostly presents a short distance movement, while it also presents a random walk for a long-distance movement with a certain probability. For optimization, it facilitates an effective search compared to using random walk (Gaussian flight) according to a regular distribution [23].

Let us consider an objective function represented as $f(\mathbf{x})$, $\mathbf{x} = (x_1, \ldots, x_d)^T$. A cuckoo then creates a new solution candidate for the nest $i$ given by the following equation:

$$\mathbf{x}_i^{t+1} = \mathbf{x}_i^t + \alpha \otimes \text{Lévy}(\lambda), \tag{1.23}$$

where $\alpha(> 0)$ is related to the scale of the problem. In most cases, $\alpha = 1$. The operation $\otimes$ represents multiplication of each element by $\alpha$. Lévy($\lambda$) represents a random number vector whereby each element follows a Lévy distribution, and this is accomplished as follows:

$$\text{Lévy}(\lambda) \sim rand(0, 1) = t^{-\lambda} \quad 1 < \lambda \leq 3, \tag{1.24}$$

where $rand(0, 1)$ is a uniform random number between 0 and 1. This formula is essentially a random walk with the distribution achieved by powered steps with a heavy tail. Therefore, it includes infinite averages and infinite standard deviation. An exponential distribution with exponents from $-1$ to $-3$ is normally used for a long-distance movement of Lévy flight.

$p_a$ represents a parameter referred to as the switching probability, and its fraction of the worse nests are abandoned from the nest by a host bird and new ones are built by performing Lèvy flights. This probability strikes a balance between exploration and exploitation.

CS is considered to be robust, compared with PSO and ACO [1].

## 1.10 Harmony Search (HS)

Harmony search (HS) [4] is a meta-heuristic based on jazz session (generation process of human improvisation). Musicians are considered to perform improvisation mainly using any one of the methods as outlined below:

- Use already-known scales (stored in their memory).
- Partially change or modify the already-known scales. Play scales next to the one stored in their memory.
- Create new scales. Play random scales within their playable area.

A process whereby musicians combine various scales in their memory for the purpose of composition is regarded as a sort of optimization. While many meta-heuristics are based on swarm intelligence of life such as fish, insects, etc., HS significantly differs from them in terms of exploiting ideas from musical processes to search for harmony, according to an aesthetic standard.

Harmony search algorithms (referred to as "HS," hereinafter) search for the optimum solution by imitating a musician's processes according to the following three rules:

- Select an arbitrary value from HS memory.
- Select a value next to the arbitrary one from HS memory.
- Select a random value within a selectable range.

With HS, a solution candidate vector is referred to as a harmony, and a set of solution candidates is referred to as a harmony memory (HM). Solution candidates are replaced within HM by a specific order. This process is repeated a certain number of

---

**Algorithm 4** Harmony search

---

**for** $i = 1$ to $HMS$ **do**                                                        ▷ HM initialization.
  **for** $j = 1$ to $n$ **do**                                                   ▷ $n$: harmony length.
    Randomly initialize $x_j^i$ in HM                         ▷ $x_j^i$: $j$-th position of $i$-th harmony.
  **end for**
**end for**
**while** the stop criterion is not satisfied **do**                      ▷ Generate a new solution candidate **x**.
  **for** $j = 1$ to $n$ **do**
    **if** rand(0,1)$< HMCR$ **then**
      Let $x_j$ in **x** be the $j$-th dimension of a randomly selected HM member
      **if** rand(0,1)$< PAR$ **then**
        Apply pitch adjustment distance $bw$ to mutate $x_j$
        $x_j = x_j \pm \text{rand}(0, 1) \times bw$
      **end if**
    **else**
      Let $x_j$ in **x** be a random value
    **end if**
  **end for**
  Evaluate the fitness of **x** by $f(\mathbf{x})$
  **if** $f(\mathbf{x})$ is better than the fitness of the worst HM member **then**
    Replace the worst HM member with **x**                    ▷ HM update
  **else**
    Disregard **x**
  **end if**
**end while**
Postprocess results and visualization

---

times (or until the conditions for termination are met), and finally, the best harmony is selected among those that survive in HM as a final solution.

Algorithm 4 shows a harmony search algorithm. $HMCR$ (Harmony Memory Considering Rate) represents the probability of selecting a harmony from HM, while $PAR$ (Pitch Adjust Rate) represents the probability of amending a harmony selected from HM. $HMS$ is the number of harmonies (sets), which is normally set to between 50 and 100.

A new solution candidate (harmony) is generated from HM based on $HMCR$. $HMCR$ is the probability of selecting component elements[3] among the present HM. Thus, new elements are randomly generated by the probability of $1 - HMCR$. Subsequently, mutation occurs according to the probability of $PAR$. The $bw$ parameter (Bandwidth) represents the largest size of the mutation. In case a newly generated solution candidate (harmony) is better than the poorest solution of HM, they are replaced.

This method is also similar to a genetic algorithm (GA); however, it differs in that all the members of HM become a parent candidate in HS while only one or two existing range(s) of chromosomes (parent individual) is/are used to generate a child chromosome in GA.

---

[3]It corresponds to allele in genotype under GA.

The coefficients employed here are the convergence coefficient $\chi$ (a random value between 0.9 and 1.0) and the attenuation coefficient $\omega$, while $\phi_1$ and $\phi_2$ are random values unique to each individual and the dimension, with a maximum value of 2. When the calculated velocity exceeds some limit, it is replaced by a maximum velocity $V_{max}$. This procedure allows us to hold the individuals within the search region during the search.

The locations of each of the individuals are updated at each generation by the following formula:

$$\mathbf{x_i} = \mathbf{x_i} + \mathbf{v_i}. \tag{1.25}$$

## 1.11 Conclusion

This chapter introduced several methods of evolutionary computation and meta-heuristics used in deep neural evolution.

In concluding, we will describe some critical opinions against meta-heuristics and further discussions.

Meta-heuristics frequently uses unusual names, terms associating with nature, and metaphors. For example, in the harmony search, the following terms are used:

- harmony,
- pitch, note,
- sounds better.

However, these are just saying the following words listed below in another way:

- solution,
- decision variable,
- has a better objective function value.

Although critics insist that these replaced words cause confusion [19, 20], it is considered that use of metaphorical expressions itself does not influence the ease of understanding. For example, David Hilbert[4] is quoted as saying geometry does work even if a point, line, and face are expressed as a table, chair, and beer mug in discussing mathematical forms. That is to say, it does not matter at all in precise discussions when it is axiomatically defined. Nevertheless, we should be careful about insisting on the novelty of meta-heuristics. It is important to recognize the distinct difference with existing methods for further discussions. Wayland has

---

[4]David Hilbert (1862–1943): German mathematician. At the second International Congress of Mathematicians (ICM) in Paris in 1900, he made a speech on "problems in mathematics," where he stressed the importance of 23 unsolved problems and presented a prospect for future creative research through these problems. Some of them continue to be themes for research on mathematics and computer science.

criticized the harmony search as simply a special example of evolution strategy $(\mu + 1)ES$ [21].[5]

It does matter that researchers who propose meta-heuristics do not recognize other similar methods [19]. When developing a new method, we never fail to have discussions on the basis of past investigations (see [11] for further discussion).

## References

1. Civicioglu, P., Besdok, E.: A conceptual comparison of the Cuckoo-search, particle swarm optimization, differential evolution and artificial bee colony algorithms. Artif. Intell. Rev. **39**(4), 315–346 (2013)
2. Deb, K.D., Pratap, A., Agarwal, S., Meyarivan, T.: A fast and elitist multiobjective genetic algorithm: NSGA-II. IEEE Trans. Evol. Comput. **6**(2), 182–197 (2002)
3. Dorigo, M., Gambardella, L.M.: Ant colonies for the traveling salesman problem. Technical Report IRIDIA/97-12, Universite Libre de Bruxelles, Belgium (1997)
4. Geem, Z.W., Kim, J.H., Loganathan, G.V.: A new heuristic optimization algorithm: harmony search. Simulation **76**(2), 60–68 (2001). Physical Review E **79** (2009)
5. Ghosh, A., Dehuri, S., Ghosh, S. (eds.): Objective Evolutionary Algorithms for Knowledge Discovery from Databases. Springer, Berlin (2008)
6. Goldberg, D.E.: Genetic Algorithms in Search, Optimization and Machine Learning. Addison Wesley, Reading (1989)
7. He, C., Noman, N., Iba, H.: An improved artificial bee colony algorithm with non-separable operator. In: *Proceeding of International Conference on Convergence and Hybrid Information Technology*, pp. 203–210. Springer, Berlin (2012)
8. Higashi, N., Iba, H.: Particle swarm optimization with Gaussian mutation. In: *Proceedings of IEEE Swarm Intelligence Symposium (SIS03)*, pp.72–79. IEEE Press, New York (2003)
9. Iba. H., Noman, N.: New frontiers in evolutionary algorithms: theory and applications. World Scientific, Singapore (2011). ISBN-10:1848166818
10. Iba, H., Aranha, C.C.: Practical Applications of Evolutionary Computation to Financial Engineering: Robust Techniques for Forecasting, Trading and Hedging. Springer, Berlin (2012)
11. Iba, H.: Evolutionary Approach to Machine Learning and Deep Neural Networks—Neuro-Evolution and Gene Regulatory Networks. Springer, Berlin (2018). ISBN 978-981-13-0199-5
12. Iba, H.: AI and SWARM: Evolutionary Approach to Emergent Intelligence. CRC Press, West Palm Beach (2019). ISBN-13: 978-0367136314
13. Ishibuchi, H., Tsukamoto, N., Nojima, Y.: Evolutionary many-objective optimization: a short review. In: Proceeding of IEEE Congress on Evolutionary Computation, pp. 2419–2426 (2008)
14. Karaboga, D., Basturk, B.: A powerful and efficient algorithm for numerical function optimization: artificial bee colony (ABC) algorithm. J. Glob. Optim. **39**, 459–471 (2007)
15. Karaboga, D., Gorkemli, B., Ozturk, C., Karaboga, N.: A comprehensive survey: artificial bee colony (ABC) algorithm and applications. Artif. Intell. Rev. (2012). https://doi.org/10.1007/s10462-012-9328-0
16. Kennedy, J., Eberhart, R.C.: Particle swarm optimization. In: Proceedings of IEEE the International Conference on Neural Networks, pp.1942–1948 (1995)

---

[5]Further discussions have been presented, and there is an insistence on the part of some individuals that it is a different method. Refer to https://en.wikipedia.org/wiki/Harmony_search and [12] for details.

17. Lipson, H., Pollack, J.B.: Automatic design and manufacture of robotic lifeforms. Nature **406**, 974–978 (2000)
18. Miller, J.F. (ed.): Cartesian Genetic Programming. Springer, Berlin (2011)
19. Sörensen, K.: Metaheuristics–the metaphor exposed. Int. Trans. Oper. Res. **22**(1), 3–18 (2015)
20. Sörensen, K., Sevaux, M., Glover, F.: A history of metaheuristics. arXiv:1704.00853v1 [cs.AI] 4 Apr 2017, to appear in Mart, R., Pardalos, P., Resende, M., Handbook of Heuristics. Springer, Berlin.
21. Weyland, D.: A rigorous analysis of the harmony search algorithm—how the research community can be misled by a "novel" methodology. Int. J. Appl. Metaheuristic Comput. **1**(2), 50–60 (2010)
22. Yang, X.: Nature-Inspired Metaheuristic Algorithms, 2nd edn. Luniver Press, Frome (2010)
23. Yang, X.-S., Deb, S.: 2009 Cuckoo search via Levy flights. In: Proceeding of World Congress on Nature and Biologically Inspired Computing (NaBIC 2009), pp. 210–214. IEEE, New York (2009)

# Chapter 2
# A Shallow Introduction to Deep Neural Networks

Nasimul Noman

**Abstract** Deep learning is one of the two branches of artificial intelligence that merged to give rise to the field of deep neural evolution. The other one is evolutionary computation introduced in the previous chapter. Deep learning, the most active research area in machine learning, is a powerful family of computational models that learns and processes data using multiple levels of abstractions. Over the last years, deep learning methods have shown amazing performances in a diverse field of applications. This chapter familiarizes the readers with the major classes of deep neural networks that are frequently used, namely CNN (Convolutional Neural Network), RNN (Recurrent Neural Network), DBN (Deep Belief Network), Deep autoencoder, GAN (Generative Adversarial Network) and Deep Recursive Network. For each class of networks, we introduced the architecture, type of layers, processing units, learning algorithms and other relevant information. This chapter aims to provide the readers with necessary background information in deep learning for understanding the contemporary research in deep neural evolution presented in the subsequent chapters of the book.

## 2.1  Introduction

Deep learning (DL) is one of the top buzzwords of the moment. In 2013, it made its place in MIT's 10 breakthrough technologies for the first time, and since then almost every year deep learning and its related research have hit this top list, directly or indirectly. There is no doubt that the stunning success of DL at superhuman level in diverse fields that ranges from speech recognition to automatic game playing has brought it to the centre of attention.

N. Noman (✉)
The University of Newcastle, Callaghan, NSW, Australia
e-mail: nasimul.noman@newcastle.edu.au

© Springer Nature Singapore Pte Ltd. 2020
H. Iba, N. Noman (eds.), *Deep Neural Evolution*, Natural Computing Series,
https://doi.org/10.1007/978-981-15-3685-4_2

Deep learning is a branch of machine learning which falls under the domain of artificial intelligence. There are some problems which are effortlessly and robustly solved by people but it is difficult to explain how they solve those problems, like face and object recognition, interpretation of spoken language, etc. Traditional machine learning and pattern recognition approaches achieved limited success in these tasks, primarily, because of their dependence on data representation or features. Since the basic idea of machine learning is to "learn from data", the representation of the data is particularly important for the performance of a learning algorithm. Crafting the most effective features is not straightforward, moreover, it requires human expertise and prior knowledge related to the task which make the learning process human-dependent.

Representation learning which has emerged as an alternative to feature engineering is capable of automatically extracting useful representation [1]. Using representation learning the raw data can be mapped to useful representation that makes the subsequent learning task easier for the classifier or predictor. Deep learning is a formalism that learns a complex representation by aggregating simple representations which in turn depend on other simpler representations and so on. By utilizing this hierarchy of complex mapping the deep neural network can learn very complex concepts and exhibit human-like intelligence [2].

The advent of deep learning has triggered the resurgence of AI (Artificial Intelligence) hype. The volume of research being carried out on the topic and the number of research papers being published is enormous—contributing to the depth and breadth of the field. It is impossible to cover everything in a single book let alone a chapter. The purpose of the chapter, as the title suggests, is to give a very brief overview of the dominant methodologies in the field. The following chapters present a range of nature inspired algorithms for various kinds of optimization in deep neural networks. The readers who are new to deep learning will find the deep neural network concepts introduced in this chapter useful in understanding the subsequent chapters. There are many comprehensive and systematic reviews on the topic [3–6] as well as application specific reviews [7–14] which the reader can find very useful.

Keeping the above mentioned purpose of the chapter in mind, we have organized the contents of the chapter as follows. First in Sect. 2.2, we present the basic concepts of neural networks and the algorithm to train them. Section 2.3 explains the idea of generic deep neural network. Section 2.4 presents the prominent architectures in deep neural network followed by their application in Sect. 2.5. Finally we conclude the chapter in Sect. 2.6.

## 2.2 (Shallow) Neural Networks

The history of Neural Networks (NNs) can be traced back to 1940s when McCulloch and Pitts created the first mathematical model of the neural network to mimic the working of a human brain [15]. Starting from that elementary model, the capacity, capability and applications of NNs have seen tremendous progress until the end of the past century with two significant pauses in their development known as AI winters. Today NNs are the largest and the most developed branch of AI that has harvested most success in our quest for creating human-like intelligence in a machine. This section presents a very brief overview of shallow NNs before introducing their deep counterpart in the next section.

NNs a.k.a. artificial neural networks (ANNs) are computational models consisting of many simple information processing units called neurons. A neuron, $N_j$, like biological neuron, receives input signals ($x_i$; $i = 1, \ldots, n$) via connecting links. Each connection has a weight $w_{ij}$ to modulate the corresponding input $x_i$. The weighted sum of the inputs added with a bias term $b_j$ is mapped via an activation function $\varphi(\cdot)$ to generate the output $y_j$ of the neuron as shown below:

$$y_j = \varphi(z_j) = \varphi\left(\sum_{i=1}^{n} x_i w_{ij} + b_j\right). \tag{2.1}$$

The bias term $b_j$ is used to control the net input to the activation function. By renaming $b_j$ as $w_{0j}$ and setting $x_0 = 1$, Eq. (2.1) can be presented in a more compact form

$$y_j = \varphi(z_j) = \varphi\left(\sum_{i=0}^{n} x_i w_{ij}\right). \tag{2.2}$$

The activation function $\varphi(\cdot)$, also called transfer function, limits the output range of a neuron and a non-linear activation function is required for exhibiting complex behaviour. Traditionally, the sigmoid function (Eq. (2.3)), also called logistic function, has been used as the activation function. Other commonly used activation functions are hyperbolic tangent and rectified linear unit (ReLU). Figure 2.1a shows the model of the neuron.

$$\varphi(z_j) = \frac{1}{1 + exp(-z_j)}. \tag{2.3}$$

A feedforward neural network (FFNN) also loosely referred as multi-layer perceptrons (MLPs) consists of neurons organized into layers (Fig. 2.1b). The leftmost and the rightmost layer of the network are called the input layer and output layer, respectively, and the intermediate layers are called the hidden layers. A neuron in a particular layer is connected to all or a subset of neurons in the subsequent layer. Figure 2.1b shows a fully connected FFNN with a 3-neuron input layer, 4-neuron hidden layer and 2-neuron output layer (bias are not shown for simplicity). The neurons in the input layer (shown in dotted circles) do not perform any processing on the received data but just pass that to the next layer. When input data is placed to the input layer of a FFNN it is passed to the connected neurons which process the inputs and generate the outputs. The generated outputs work as the inputs to

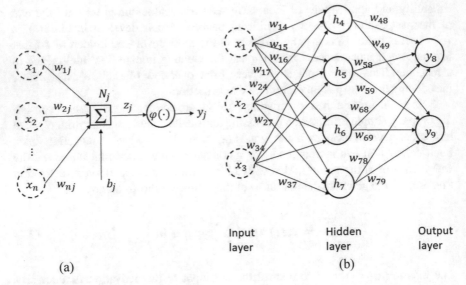

Input layer      Hidden layer      Output layer

(a)            (b)

**Fig. 2.1** Artificial neural network. (**a**) Model of a neuron (**b**) Structure of a fully connected feedforward neural network with one hidden layer

the neurons in the following layer. Thus passing through the intermediate (hidden) layers the network output is generated. A neural network actually implements a function $f(\cdot)$, characterized by the parameters $\mathbf{W}$, that maps the input to output.

## 2.2.1 Backpropagation Algorithm for Training NNs

In a supervised learning problem we are given a set of input–output paired samples $\{\mathbf{x}^{(k)}, \mathbf{y}^{(k)}\}; k = 1, \ldots, K$ and we are required to learn the underlying function in terms of NN parameters. An elegant algorithm called backpropagation lies in the core of learning parameters in different types of NNs. The backpropagation algorithm is a gradient based method that searches for network parameters to minimize the discrepancy between the NN predicted output and the target output in the given data. A commonly used error function to measure how far we are from the goal is

$$E^{(k)} = \frac{1}{2} \sum_{j \in L} \left( e_j^{(k)} \right)^2 = \frac{1}{2} \sum_{j \in L} \left( y_j^{(k)} - \hat{y}_j^{(k)} \right)^2 \tag{2.4}$$

where $y_j^{(k)}$ and $\hat{y}_j^{(k)}$ are the target output and the NN generated output at the $j$-th output unit in response to input $\mathbf{x}^{(k)}$; $e_j^{(k)}$ denotes the corresponding error and $L$ represents the set of neurons in the output layer. In backpropagation algorithm, this error signal is propagated through the network layer by layer in the backward direction making successive adjustment to the connection weights.

Suppose we have an $L+1$ layered FFNN, where $l = 0$ represents the input layer, $L$ represents the output layers and $l = 1, \ldots, L - 1$ represent the hidden layers. For the given training dataset $\{\mathbf{x}^{(k)}, \mathbf{y}^{(k)}\}$; $k = 1, \ldots, K$, the backpropagation algorithm can be described as follows:

**Step 1:**   *Initialization*: Without any prior information, the weights and biases of the networks are initialized to small random numbers sampled from a uniform distribution.

**Step 2:**   *Iterate through the batch of training examples*: Until termination criteria are not satisfied, for each example $\{\mathbf{x}^{(k)}, \mathbf{y}^{(k)}\}$ in the given $K$ training samples, chosen in some specific order, perform the forward pass and the backward pass described in Step 2.1 and 2.2, respectively.

***Step 2.1:***   *Forward Pass*: In this step, the input $\mathbf{x}$ is presented to the network and the outputs of the different neurons are calculated (layer by layer) going from the input layer towards the output layer.

$$y_j = \begin{cases} x_j & : \text{if } y_j \text{ is a neuron in input layer} \\ \varphi(z_j) & : \text{if } y_j \text{ is a neuron in hidden or output layer,} \end{cases} \qquad (2.5)$$

where $z_j$ represents the weighted sum of neuron inputs and the bias as shown in Eq. (2.2). Next calculate the error signal from the network outputs

$$e_j = y_j - \hat{y}_j. \qquad (2.6)$$

***Step 2.2:***   *Backward Pass*: In this phase, the error is propagated backward by adjusting the network weights and biases (layer by layer) going from the output layer towards the input layer. The weight, $w_{ij}$, of the network connection to the node $N_j$ in layer $l$ is adjusted by

$$w_{ij} = w_{ij} + \eta \delta_j y_i, \qquad (2.7)$$

where $\eta$ is the learning rate parameter typically having a value between 0.0 and 1.0. $\delta_j$ is the local gradient of the network defined as

$$\delta_j = \begin{cases} \varphi'_j(z_j) e_j & : \text{if } N_j \text{ is a neuron in output layer} \\ \varphi'_j(z_j) \sum_i \delta_i w_{ji} & : \text{if } N_j \text{ is a neuron in hidden layer,} \end{cases} \qquad (2.8)$$

where $\varphi'_j(\cdot)$ represents differentiation with respect to the argument. For the sigmoid function Eq. (2.8) takes the following form:

$$\delta_j = \begin{cases} y_j(1 - y_j)e_j & : \text{if } N_j \text{ is a neuron in output layer} \\ y_j(1 - y_j)\sum_i \delta_i w_{ji} & : \text{if } N_j \text{ is a neuron in hidden layer,} \end{cases}$$

(2.9)

where $y_j(1 - y_j)$ is the derivative of the sigmoid function.

The weight adjustment method in the above algorithm, where the network weights are updated for each training sample, is known as *online updating*. Alternatively, the weights can be update after presenting a batch of samples to the network. This setup is known as *batch updating*. After all the $K$ samples of training dataset have been presented to the network it is called one *epoch* of training. Networks are trained epoch after epoch until the termination criteria is satisfied.

In this section, we only introduced FFNN—the most widely known type of NN. There are several other types of NNs such as Recurrent Neural Networks (RNN), Autoencoder (AE), Restricted Boltzmann Machine (RBM), Time Delay Neural Network (TDNN), etc. which have been studied by the researchers for many years. Many of these models have been extended to their deep architectures. The relevant models of those NNs will be introduced along with their deep architectures in Sect. 2.4.

## 2.3   Deep Neural Networks: What, Why and How?

Based on the brief introduction of NN in the previous section, a deep learning architecture can be defined as a representation learning technique that draws attention to successively learning more meaningful representations in a sequence of layers of processing units (neurons). Therefore, the central idea of deep learning lies in exploiting many non-linear layers of information processing for useful feature extraction and transformation for supervised or unsupervised learning. So it is a logical extension of the classical MLPs. Today, a deep neural network can utilize hundreds even thousands of layers those collectively learn a hierarchy of representations. There is no strict division between shallow and deep NN based on the number of layers; however, any architecture more than two or three layers can be considered as deep.

Ideally, we would like the NNs to learn a large volume of information at different abstract levels and utilize those to exhibit human-like intelligence in solving different complex problems. Let us consider the well-known cats and dogs photos classification problem from Kaggle[1] competition. We intend to capture different

---

[1]www.kaggle.com.

common features, e.g. eyes, ears, and discriminating features, e.g. shape of the face, relative position of eyes and nose etc., in our network and utilize those in solving the task. NNs, whether shallow or deep, are universal, i.e. they can approximate arbitrarily well any continuous function on a compact domain of $R^n$. Now what are the merits/demerits of using a deep architecture over a shallow architecture?

Using representation learning, we want to extract useful information from the raw data that can be used for our purpose. One immediate benefit of creating the abstract representation in an hierarchical manner is the sharing and reusability of the lower level representations [3]. By using many layers of incremental abstractions, a deep architecture can learn very complex and effective representation for the task. The experimental results in speech-to-text transcription problem show that for a fixed number of parameters, a deep model is clearly better than a shallow one [16]. Besides, the ability of learning these incremental representations at different levels simultaneously makes deep learning an efficient learner. The depth of the NN architecture is also related to the "compactness" of the expression of the learned function and the necessary data to learn it. According to theoretical results, a function that can be compactly represented in a deep architecture may need exponential number of elements to represent in a shallow architecture. Furthermore, an increasing number of elements in the representation may need a lot of training examples to tune them [17]. It has been also suggested that deep architectures are more suitable for representing highly varying functions (functions that requires a large number of elements for its piece-wise approximation) [3].

Although the theory of deep learning was familiar to the AI community in the 1980s, it essentially took off only recently. The beginning of the era of the modern deep learning is marked by the proposal of Hinton et al. in which they showed how to learn a deep belief network layer by layer [18]. But the practical applications and successes of deep learning become noticeable only after 2012. Progress in three areas has worked as the driving force behind this accomplishment: data availability, computing power and algorithmic advancement. Like in any machine learning method, data is the key ingredient for converting an abstract model into intelligence machine. Over the past 20 years, an exponential increase in capacity as well as decrease in price of storage hardware and a tremendous growth in internet and communication technology have made the collection, storage and distribution of very large dataset possible. The speed of CPU has increased by more than thousands times from 1990s. However, the major leap forward came from the usage of GPUs (graphical processing units) which are capable of doing massive parallel operations necessary for deep learning. Today the training of a deep learning network with millions of parameters is possible in less than an hour using a standard GPU. The availability of data and hardware made the experimentation with deep NN feasible and the researchers achieved remarkable progress in developing improved activation functions, optimization schemes and weight-initialization schemes. These algorithmic improvements helped to overcome a key hurdle in training deep neural networks—gradient propagation through the stacks of layers in deep neural networks.

## 2.4  Architectures of Deep Networks

In the previous section, deep neural network (DNN) was introduced as a logical extension of FFNN; however, deep networks are frequently built by using specialized types of layers or by assembling larger networks from smaller networks using them as building blocks. Based on their architectural characteristics, DNNs can be broadly categorized into different types. In this section, we will introduce the dominant variants of DNNs based on their architectural design.

### 2.4.1  Convolutional Neural Network

Convolutional Neural Network (CNN) is one of the most well-known and widely used DNN architectures which has shown tremendous success in various computer vision applications. CNNs are inspired by the animal visual system, in particular, by the model, proposed by Hube and Wiesel, consisting of simple and complex cells [19, 20]. Neocognitron, a model of ANN that works by hierarchical organized transformation of image, is considered as the predecessor of CNN [21].

CNNs are feedforward neural networks in which inputs are transformed through many layers in sequence for generating the outputs, yet there are some noteworthy architectural differences between them. A CNN consists of multiple layer types in contrast to one layer type in FFNN. Another key difference between CNN and FFNN is that the layers in a CNN consist of neurons organized in three-dimensional volumes. Among the other distinguishing features of CNNs are local connectivity of neurons and weight sharing among connections which will be explained in the following subsections.

A CNN consists of three main types of layer (1) convolutional layers, (2) pooling layers and (3) fully connected layers. Figure 2.2 shows pipeline of the vanilla CNN architecture for the image classification task. As shown in Fig. 2.2, the input itself is

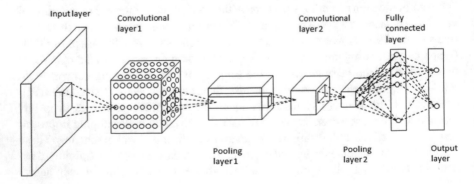

**Fig. 2.2**  Architecture of a basic convolutional neural network

considered as a three-dimensional entity, e.g. in case of an image the dimensions are width, height and colour channels. Once the raw input date (image) is placed in the network input layer, it passes through several stages of convolutional and pooling layers then the generated representation is processed by one or more fully connected layers to generate the final output of the network for classification of the image.

### 2.4.1.1 Convolutional Layers

Convolutional layers, which are considered to be the core building blocks of a CNN, are essentially feature-extractors. Each neuron in the convolutional layer is connected to at a small region in the previous layer called *receptive field* via a set of trainable weights. The set of trainable weights for a neuron in the convolutional layer is called *filter* or *kernel* which connects the neuron through the full depth of previous layer. By replicating the same filter across the entire visual field (i.e. across the width and height of the previous layer), it creates a *feature map*. By convolving the input with the learned filter the feature map is generated which actually represents the response of a given convolutional layer to the same feature. By using multiple filters, various features can be extracted from the same receptive field and multiple feature maps can be created. The number of learned filters determines the extent of the depth dimension of the convolutional layer.

Designing a convolutional layer requires choosing the values of many hyperparameters. The depth of the convolutional layer depends on the number of filters we want to learn from the previous layer. The width and height of the output volume are controlled by *filter size*, *stride size* and *zero-padding*. The spatial dimension of the filter is called filter size or kernel size. Stride size determines the stride with which we slide the filter to create the feature map and padding the input with zeros on the border of the input, we can control the spatial size of the output volume. Although the sigmoid and hyperbolic tangent functions are used for ANN traditionally, ReLU and its successors such as Leaky ReLU (LReLU), Parametric ReLU (PReLU), Exponential Linear Unit (ELU) can contribute to improve CNN performance. Therefore, the choice of the best activation function has become another design parameter for CNN.

### 2.4.1.2 Pooling Layers

Pooling layers are used to achieve spatial invariance to translations. Generally, pooling layers are inserted between successive convolution layers to reduce the spatial size of the feature maps. Pooling layers reduce the number of parameters and the amount of computation progressively and also help to control overfitting. Pooling function operates along the spatial dimension of the input volume and downsample it but keeps the depth unchanged.

Max-pooling and average-pooling are the most common types of pooling used in CNN. Like convolution, pooling operations are performed with a filter size and a stride size. A max-pooling operation with filter size $2 \times 2$ and stride size two will reduce a feature map from $8 \times 8$ to $4 \times 4$. This operation will move the filter spatially performing the max operation on four inputs (in the $2 \times 2$ region of the filter). The average-pooling, under the same setting, will only replace the max operation with the average. Other well-known approaches related to pooling layers are stochastic pooling, spatial pyramid pooling and def-pooling [8].

### 2.4.1.3  Fully Connected Layers

After multiple runs of successive convolution and pooling layers, CNNs extract very high-level features from the input data. In order to perform high-level reasoning based on the extracted representation a few fully connected layers are used at the end. Similar to FFNN, fully connected layers are one dimensional and all neurons of fully connected layer are connected to every neuron in the previous layer. Fully connected layers contain most of the parameters of a CNN and impose a computational burden for training [8].

Besides these three layers, many types of normalization layers have been used in CNNs among which the batch normalization is most common. Some studies have shown that batch normalization can help to speed up the training and reduce the sensitivity of training toward weight initialization [22].

### 2.4.1.4  Training Strategies

CNNs can be trained using backpropagation algorithm introduced before. However, overfitting is a major challenge faced in training deep neural network because of their large number of parameters. In order to deal with the issue of overfitting, various regularization techniques have been proposed. Dropout is one of the most effective regularization methods in which the hidden units are randomly omitted with some probability, called *dropout rate*, during the training period. This prohibits complex co-adaptation of features on training data and thereby improves generalizations. During testing, all hidden units are multiplied by the dropout rate which generates a strong regularization effect reducing the overfitting. Data augmentation is another popular technique for creating additional data without extra cost of labelling. By applying translation, reflection and changing image channel intensities supplementary samples are created which can improve the classification performance. Besides the regularization techniques, pre-training and fine-tuning for network parameters, utilizing weight-decay and weight-tying can improve the generalization of the network [8, 9].

### 2.4.1.5   Popular CNN Models

Over the last few years various CNN architectures came out with some innovative ideas. Utilization of novel architectural design, regularization tricks and reorganization of processing units have presented some powerful CNN models which have become widely known standards in the field. Some of the most prominent CNN architectures are briefly introduced in this section.

AlexNet, proposed by Krizhevsky et al. [23], won the ILSVRC (ImageNet Large Scale Visual Recognition Challenge) 2012 and popularized CNN in computer vision. AlexNet has relatively simple architecture consisting of five convolutional layers, max-pooling layers, ReLU and three fully connected layers and dropout. Szegedy et al. introduced the concept of inception module in their developed architecture called GoogLeNet which won ILSVRC 2014 [24]. Their proposal showed that CNN layers can be arranged in ways other than traditional sequential manner. VGGNet is a CNN model proposed by the Visual Geometry Group (VGG) from the University of Oxford [25]. Among several versions of the model VGG-16 and VGG-19 consisting of 16 and 19 weight layers, respectively, became very popular. ResNet, developed by Microsoft, won ILSVRC 2016 [26]. ResNet network is well known for its very deep architecture (152 layers) and the introduction of residual blocks. In their Xception model, Google has taken the inception hypothesis to eXtreme from which the name was derived [27]. In this architecture, the inception modules are replaced with modified depthwise separable convolutions which is pointwise convolution followed by depthwise convolution. Khan et al. introduced the idea of channel boosting to improve the representative capacity of DNN by learning from multiple input channels and transfer learning. Their designed Channel Boosted CNN (CB-CNN) architecture, evaluated on complex classification problem, exhibited superior performance compared to many existing methods [28].

## 2.4.2   Recurrent Neural Network

Recurrent Neural Networks (RNNs) differ from the other FFNNs in their capability to selectively pass information over time. Cyclic connection, a distinguishing characteristic feature of RRN, enables them to update the current state based on the previous state and the current input. Therefore, they can be used to model a sequence of elements which are not independent. Naturally, RRNs have been widely used in applications related to sequence processing such as text, audio and video.

### 2.4.2.1   RNN Architecture

An RNN is a feedforward neural networks in which the hidden layers are replaced by recurrent layers consisting of *recurrent nodes* a.k.a. *recurrent cells*. The recurrent cells have connections that span adjacent time steps incorporating the concept of

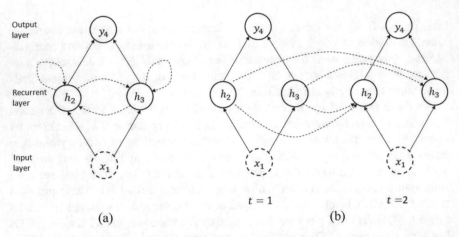

Output layer

Recurrent layer

Input layer

$t = 1$            $t = 2$

(a)                 (b)

**Fig. 2.3** Recurrent neural network: firm arrows represent feedfoward connections and dashed arrows represent recurrent connections. (**a**) RNN with one recurrent layer with two recurrent cells (**b**) RNN unrolled along time axis

time to the model. Design of a cell determines its capacity in remembering previous information. Additionally, the recurrent layers can be organized various ways giving different RNN architectures. Therefore, it is possible to design different RNNs by blending different types of cells and architectures which obviously leads to RNNs with different characteristics [29].

A simple recurrent cell is a standard neuron augmented with recurrent connections which causes its state to be affected by the networks past state (via recurrent connection) as well as by the current inputs it receives (via feedforward connection). In other words, a recurrent connection connects the output of a recurrent neuron as an input to the recurrent neurons in the same layer (Fig. 2.3a). The mathematical expression for a simple recurrent cell is given by

$$h(t) = \varphi \left( W^{xh} x(t) + W^{hh} h(t-1) + b^h \right) \tag{2.10}$$

$$y(t) = h(t) \tag{2.11}$$

where $x(t)$, $h(t)$ and $y(t)$ denote the input, recurrent information and the cell output at time $t$, respectively. $W^{xh}$ is the weight matrix between the input and the hidden layer and $W^{hh}$ is the recurrent weight matrix between the hidden layer and itself. $b^h$ represents the bias vector. Figure 2.3b illustrates the dynamics of the network across time steps to visually understand it by unrolling the network of Fig. 2.3a. The unrolled network can be considered as a deep network in which information flows through in a feedforward manner over time but the weights remain constant.

### 2.4.2.2  RNN Training

Considering the unfolded representation of RNN, it is understandable that the network can be trained across many time steps in backpropagation algorithm. This algorithm is called backpropagation through time (BPTT). BPTT is called 'through time' because it will have error signals flowing backward from future time steps as well as from layer above. Training RNN with BPTT is computationally expensive, moreover, two well-known phenomena in training deep neural network are *vanishing gradient* and *exploding gradient* problems (which occur when backpropagating error across many layers) are obvious. A variant of BPTT called Truncated BPTT (TBPTT) that places a limit on the number of time steps the error can be propagated, can better handle some of the above problems. However, by limiting the propagation of errors, TBPTT can end up reducing the length of the dependency learned [30].

### 2.4.2.3  Memory Cells

In general, it is difficult for the simple recurrent cells to capture the long-term dependency in the data—as the gap between the related inputs increases it becomes difficult to learn it. A class of improved cells have been designed to enhance the memorizing capacity of a simple recurrent cell by incorporating the gating mechanism in the cell. The first and the most well-known member of this family is Long Short Time Memory (LSTM) proposed by Hochreiter and Schmidhuber [31]. Figure 2.4 shows the most widely used variant of LSTM: LSTM with a forget gate [32]. Among the other well-known variants are LSTM without a forget gate and LSTM with peephole connection.

The LSTM cell, shown in Fig. 2.4, consists of three gates: forget gate, input gate and output gate. The forget gate decides what information to be discarded from its internal cell state, the input gate chooses what information to be stored in the cell state and the output gate decides what information to be exposed based on the cell state. The mathematical model for the LSTM cell with forget gate is

$$f(t) = \sigma \left( W^{fh} h(t-1) + W^{fx} x(t) + b^f \right) \tag{2.12}$$

$$i(t) = \sigma \left( W^{ih} h(t-1) + W^{ix} x(t) + b^i \right) \tag{2.13}$$

$$\tilde{c}(t) = \tanh \left( W^{\tilde{c}h} h(t-1) + W^{\tilde{c}x} x(t) + b^{\tilde{c}} \right) \tag{2.14}$$

$$c(t) = f(t) \cdot c(t-1) + i(t) \cdot \tilde{c}(t) \tag{2.15}$$

$$o(t) = \sigma \left( W^{oh} h(t-1) + W^{ox} x(t) + b^o \right) \tag{2.16}$$

$$h(t) = o(t) \cdot \tanh(c(t)). \tag{2.17}$$

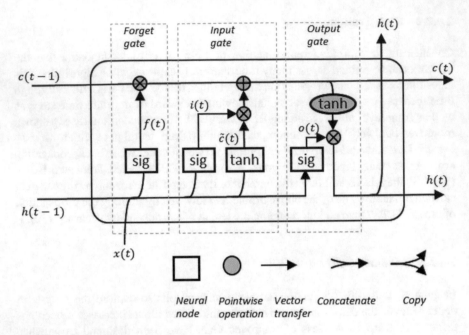

**Fig. 2.4** LSTM cell with a forget gate

Several study found that RNN constructed with LSTM cells is better in handling long-term dependency than simple cells. However, learning with LSTM is computationally expensive because of the additional parameters the cell has. Another memory cell, named Gated Recurrent Unit (GRU), proposed by Cho et al. [33] has become popular because of its simpler architecture but competitive performance in many applications. The mathematical equations for GRU is

$$r(t) = \sigma \left( W^{rh} h(t-1) + W^{rx} x(t) + b^r \right) \tag{2.18}$$

$$z(t) = \sigma \left( W^{zh} h(t-1) + W^{zx} x(t) + b^z \right) \tag{2.19}$$

$$\tilde{h}(t) = \tanh \left( W^{\tilde{h}h} (r(t) \cdot h(t-1)) + W^{\tilde{h}x} x(t) + b^{\tilde{h}} \right) \tag{2.20}$$

$$h(t) = (1 - z(t)) \cdot h(t-1) + z(t) \cdot \tilde{h}(t). \tag{2.21}$$

A GRU cell actually consists of two gates: reset gate and update gate, where the update gate is essentially combination of LSTM's input gate and forget gate. A couple of variants of GRU and some other memory cells have also came out in recent years.

When unrolled in time, RNNs can be considered as a deep neural network. These indefinite number of layers in RNN are intended for memory not for hierarchical processing which is the case in other DNNs [34]. The most obvious way to build deep RNN is via stacking up recurrent layers, however, there are other ways a RNN can be extended to a deep RNN. Pascanu et al. proposed new deep neural network architectures by extending the input-to-hidden function, hidden-to-hidden transition and hidden-to-output function of a RNN [35].

### 2.4.3   Deep Autoencoder

An autoencoder is a special type of ANN that is used to learn a compressed representation of a dataset. They are popular models for dimensionality reduction or learning efficient encoding for a given dataset. An autoencoder is an ANN that is used to reconstruct its input to its output in unsupervised learning. Learning to copy the input to output may seem to be trivial but by imposing some constraints on the network structure, autoencoders capture the most important aspects of the data filtering the noise.

Structurally, autoencoders are very similar to FFNN: in the simplest form they have one input layer, one hidden layer and one output layer. The key structural difference from the FFNN is that autoencoders have the same number of nodes in their input and output layers. Figure 2.5 shows an autoencoder with one hidden layer. As shown in Fig. 2.5, an autoencoder consists of two parts: an *encoder* and a *decoder*. The encoder maps the input data $x$ into a different representation $h$ using a neural network

$$h = \varphi \left( W^1 x + b^1 \right) \tag{2.22}$$

**Fig. 2.5** An autoencoder with one input, one output and one hidden layer

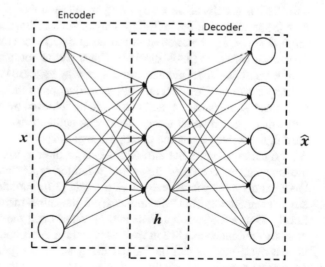

where $W^1$ and $b^1$ represent weights and biases of the encoder network, respectively. The job of the decoder is to map $h$ to $\hat{x}$ which is of the same dimension as $x$

$$\hat{x} = \varphi\left(W^2 h + b^2\right) \tag{2.23}$$

where $W^2$ and $b^2$ represent weights and biases of the decoder network, respectively. With the aim of reconstructing $x$, the autoencoder's weights are adjusted to minimize difference between $x$ and $\hat{x}$, also known as loss

$$J = \sum_k ||x - \hat{x}||^2, \tag{2.24}$$

where $k$ is the number of input samples. When the data is highly non-linear, we can design autoencoders using more than one hidden layers, and we call them deep autoencoder.

One major difference between autoencoder and FFNN lies in their learning process–autoencoders utilize unlabelled data instead of labelled data as in FFNN. Autoencoders are trained using unsupervised training with backpropagation algorithm. Training deep autoencoders with many hidden layers is problematic using backpropagation. The errors, as those are backpropagated towards the input layer, become very small and fail to perform effective adjustments to the network weights. This results in very slow and poor learning. Deep autoencoders can be learned using a layer-wise approach which is explained in the next section.

Different variants of autoencoder have been developed to improve its performance in extracting more effective and complex representation. Compression autoencoders (CAE) are used to learn a compact or compressed representation of the data. This is usually achieved by using smaller number of nodes in its hidden layer than the number of nodes in the input layer. Such bottleneck forces a CAE to extract input features, e.g. correlation in the input data before expanding back to output. Sparse autoencoders (SAE) create the information bottleneck without reducing the number of nodes in the hidden layer rather may increase it. However, they impose a sparsity constraint in which the average activation of hidden units is required to be very small. The sparsity of activation can be achieved by exploiting KL divergence and using regularization terms. Denoising autoencoders (DAE) are trained using partially corrupted input and attempt to reconstruct the original uncorrupted input. As the name suggests the aim of DAE is to filter noise from the input. Through the mapping of corrupted input into a hidden representation and retrieval of it, DAE extract more robust features of the data [36]. Variational autoencoders (VAE) are mathematically most dissimilar from the other members in the autoencoder family. Figure 2.6 shows a VAE and elucidates that the similarity between VAEs and other autoencoders lies in their architecture. A second similarity between them is that these models can be trained using backpropagation algorithm. VAEs are generative models that can be used to generate examples of input data by learning its distribution. To generate a sample from the model, VAE first draws a sample

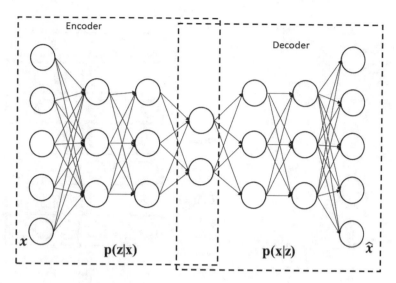

**Fig. 2.6** Architecture of variational autoencoder

$z^i \sim p(z)$ from a prior distribution and then the data instance is generated from a conditional distribution $x^l \sim p(x|z)$. In VAE the encoder is used to approximate the distribution $p(z|x)$ and the decoder is used to approximate $p(x|z)$. There are other types of autoencoders like contractive autoencoder, non-linear predictive autoencoder, etc.

### 2.4.4   Deep Belief Network (DBN)

Deep Belief Networks (DBN) are graphical models which can be used to learn a model of the input data using a greedy layer-wise learning algorithm. The layers of DBN are constructed using Restricted Boltzmann machines (RBMs) which are probabilistic graphical models. Boltzmann machines are a class of stochastic recurrent neural networks invented by Geoffrey Hinton and Terry Sejnowski. They are based on physical systems and consists of stochastic neurons which can have one of the two possible states 0 or 1. RBMs are simplified version of Boltzmann machines that imposes restriction on the connectivity.

An RBM is an undirected graphical model which consists of two types of neurons: visible units and hidden units. The two layers of neurons in RBM form a bipartite graph as shown in Fig. 2.7a. In other words RBM does not allow connections between the neurons in the same layer—this is the restriction it imposes on the network architecture. Let us consider the RBM shown in Fig. 2.7a with $m$ visible nodes $\mathbf{v} = \{v_1, v_2, \ldots, v_m\}$ and $n$ hidden nodes $\mathbf{h} = \{h_1, h_2, \ldots, h_n\}$ where $v_i, h_j \in \{0, 1\}$. Being an energy based model, RBMs associate a scalar energy to

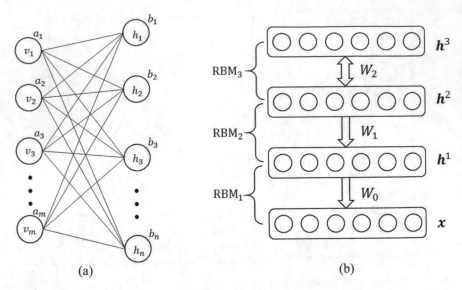

**Fig. 2.7** (a) An RBM with $m$ visible nodes and $n$ hidden nodes. (b) A DBN constructed from three RBMs

each configuration of variables to capture dependencies between variables [37]. The energy function is given by

$$E(\boldsymbol{v}, \boldsymbol{h}, \boldsymbol{\theta}) = -\sum_{i=1}^{m} a_i v_i - \sum_{j=1}^{n} b_j h_j - \sum_{i=1}^{m} \sum_{j=1}^{n} w_{ij} v_i h_j, \qquad (2.25)$$

where $\boldsymbol{\theta} = \{W, a, b\}$ are the model parameters: $W$, $a$ and $b$ represent the connection weights and biases on visible and hidden nodes, respectively. The probability that the model assigns to a particular visible vector $v$ is given by

$$p(\boldsymbol{v}|\boldsymbol{\theta}) = \frac{1}{Z(\boldsymbol{\theta})} \sum_{h} exp\left(-E(\boldsymbol{v}, \boldsymbol{h}, \boldsymbol{\theta})\right), \qquad (2.26)$$

where $Z(\boldsymbol{\theta})$ is a normalizing term given by

$$Z(\boldsymbol{\theta}) = \sum_{v} \sum_{h} exp\left(-E(\boldsymbol{v}, \boldsymbol{h}, \boldsymbol{\theta})\right). \qquad (2.27)$$

The conditional probability of activating a single visible neuron $(v_i)$ or hidden neuron $(h_j)$ is given by

$$p(v_i = 1|\boldsymbol{h}) = \varphi \left( \sum_{j=1}^{n} w_{ij} h_j + a_i \right) \tag{2.28}$$

$$p(h_j = 1|\boldsymbol{v}) = \varphi \left( \sum_{i=1}^{m} w_{ij} v_i + b_j \right), \tag{2.29}$$

where $\varphi$ represents the sigmoid function (Eq. (2.3)). RBMs have the property that given the visible units, all the hidden units become independent and vice versa. Therefore, the conditional probabilities of hidden and visible variables become

$$p(\boldsymbol{h}|\boldsymbol{v}) = \prod_{j=1}^{n} p(h_j|\boldsymbol{v}) \quad \text{and} \quad p(\boldsymbol{v}|\boldsymbol{h}) = \prod_{i=1}^{m} p(v_i|\boldsymbol{h}) \tag{2.30}$$

The objective of RBM training is to adjust the weights and biases in such way that maximizes the average log-likelihood of the given data. This is usually achieved by using gradient descent on the log-likelihood. The derivative of the log-likelihood with respect to a weight takes a simple form giving a very simple learning rule for parameter update [38]:

$$\Delta w_{ij} = \eta \left( \langle v_i h_j \rangle_{data} - \langle v_i h_j \rangle_{model} \right), \tag{2.31}$$

where $\eta$ denotes learning rate and $\langle \cdot \rangle$ represents expectations with which visible unit $v_i$ and hidden unit $h_j$ are on together under the distribution of data and model. For the bias term a similar but simple update rule can be derived. Calculation of $\langle v_i h_j \rangle_{data}$ is straightforward but $\langle v_i h_j \rangle_{model}$ can be estimated using Gibbs sampling. However, each parameter update using Gibbs sampling may take very long time and can be computationally expensive.

Contrastive Divergence (CD) [39] presents a much faster learning mechanism in which two tricks are used to speed up the sampling process (1) Gibbs chain is started by setting the visible units to a training sample and (2) the Gibbs chain is run for only $k$-steps (in practice $k = 1$ is usually used). Persistent contrastive divergence (PCD) [40] is a direct descendant of CD. PCD uses a persistent Markov chain (i.e. the chain is not reinitialized between parameter updates). For each parameter update, the chain is run for k-steps to obtain new samples; then the status of the chain is retained for the following update. PCD makes the training more efficient by allowing the Markov chain to approach thermal equilibrium faster.

DBNs consist of several middle layers of RBMs with which they model many layers of hidden causal variables. An RBM can be used for extracting features from the training data. The output of its hidden layer can be used as the input to the visible layer of another RBM. This process can be considered as a higher-level feature extraction from the previously extracted features from the data. By repeating this process DBN can extract hierarchical features from the training data. Figure 2.7b shows a DBN constructed by stacking three RBMs (DBNs are usually drawn top to bottom rather left to right). The top two layers have undirected symmetric connections and the lower layers have top to bottom directed connections.

Hinton et al. [18] have developed a greedy layer by layer unsupervised training procedure for DBN which works very efficiently. The method first trains the first layer of DBN as an RBM that models the input data $x = h^0$. After learning the parameters $W^0$ it freezes it and uses the output of this RBM as the input data of the next layer and train the second layer as an RBM. By repeating this process, we get a DBN whose parameters are trained for extracting higher-level complex features of the data. This algorithm can be followed by other learning procedure for fine-tuning the network weights for better generalization. The training of DBN is unsupervised; however, the learned representation can be used for supervised prediction. For example, by adding an additional network layer and training it using backpropagation the extracted features of DBN can be used for classification.

### 2.4.5 Generative Adversarial Network (GAN)

Generative Adversarial Networks (GANs), which are deep generative models based on game theory, have attracted much attention as an innovative machine learning approach. Like in any generative model, the goal of a GAN is to learn the underlying distribution of the given training dataset as accurately as possible. Generative models try to learn the probability distribution either *explicitly* (i.e. the method operates by using an explicit probability density function, therefore, evaluation of its likelihood is straightforward) or *implicitly* (i.e. the method does not represent the likelihood explicitly but able to generate data using some sampling mechanism) [41]. GANs primarily fall under the second category and utilize two adversarial networks called a generator and a discriminator in the form of a minimax game. The goal of the generator is to fool the discriminator by creating realistic samples that are intended to come from the distribution of the training dataset. On the other hand the job of the discriminator is to correctly classify the presented samples as real or fake. By simultaneously training both the generator and the discriminator networks, the GAN framework drives them to improve in their respective objectives until the fake samples are indistinguishable from the real samples [42]. Figure 2.8 shows the GAN framework with interactions between its components.

#### 2.4.5.1  GAN Architecture

GANs are based on the principle of game theory in which the discriminator ($D$) competes with the generator ($G$) until they achieve the Nash Equilibrium, i.e. each

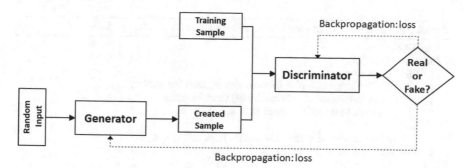

**Fig. 2.8** Illustration of the Generative Adversarial Network (GAN) framework

network attains its best performance with respect to the other. Given a distribution $z \sim p_z$, $G$ learns the probability distribution $p_g$ over data $x$. The discriminator $D$ and the generator $G$ are represented by two functions $D(x; \theta_d)$ and $G(z; \theta_g)$ with $\theta_d$ and $\theta_g$ as parameters, respectively. $G$ maps the noise vector $z$ into the fake sample, $G(z)$, which is a multi-dimensional vector, whereas $D$ maps the real samples and the fake samples from $G$ into a single scalar. Both $D$ and $G$ are represented by deep neural networks and can be trained by backpropagation algorithm. The loss function of the discriminator can be given by [43]

$$J^{(D)} = -\frac{1}{2}\mathbb{E}_{x \sim p_{\text{data}}}\log D(x) - \frac{1}{2}\mathbb{E}_{z \sim p_z}\log(1 - D(G(z))), \tag{2.32}$$

where $p_{\text{data}}$ represents the dataset used. In zero sum game, in which the sum of losses of both players is zero, the generator's loss function is given by

$$J^G = -J^{(D)}. \tag{2.33}$$

Consequently, the optimization of a GAN is transformed into the two-player minimax game with value function $V(D, G)$:

$$\min_G \max_D V(D, G) = \mathbb{E}_{x \sim p_{\text{data}}}\log D(x) + \mathbb{E}_{z \sim p_z}\log(1 - D(G(z))). \tag{2.34}$$

#### 2.4.5.2 GAN Training

For training a GAN, the minimax game is implemented using an iterative approach (Algorithm 1) [42]. As shown in Algorithm 1, the generator is kept constant during the training of the discriminator. In the same way, the discriminator does not change during the training of the generator. The training procedure alternates between the $k$ steps of optimizing the discriminator and one step of optimizing the generator. In the original proposal, Goodfellow et al. have used $k = 1$ to keep to computational

---

**Algorithm 1:** Training of Generative Adversarial Networks (GANs)

---

initialization;
**for** *i iterations* **do**
    **for** *k steps* **do**
        Sample a minibatch of *m* fake samples ($z$) from noise prior.
        Sample minibatch of *m* examples ($x$) from the dataset.
        Update $\theta_d$ to reduce discriminator's loss function.
    **end**
    Sample a minibatch of *m* fake samples ($z$) from noise prior.
    Update $\theta_g$ to reduce generator's loss function.
**end**

---

cost minimum [42]. The update of the generator and discriminator weights can be done using any stochastic gradient descent method.

In GAN training, if the generator's performance becomes better, then the discriminator's performance becomes poorer because the discriminator struggles to differentiate between the real and fake samples. If the generator achieves the perfection in generating fake samples, then the discriminator's accuracy drops to 50%. This is problematic for the convergence of GAN training because if the discriminator's accuracy drops, then the feedback the generator receives becomes poor and in turn its quality drops. Therefore, in practice, GANs often seem to oscillate without converging to an equilibrium.

### 2.4.5.3 Progresses in GAN Research

Over the last few years, a significant amount of research effort has been noticed in GAN which basically focuses on overcoming some well-known GAN problems. The most well-known problem in GAN is mode collapse in which the generator fails to learn the distribution of the complete dataset and concentrates on a few or even on one model. The result is lack of diversity in the generated samples. As mentioned earlier, if one of the networks, e.g. the discriminator, becomes very accurate, then the training of the generator may fail due to vanishing gradients. Maintaining the equilibrium between the discriminator and the generator is another challenge. The solutions to these problems usually came in two forms (1) designing new architectures and (2) proposing novel loss functions. We will briefly introduce some of the prominent proposals.

Goodfellow et al. [42] used fully connected NNs for both generator and discriminator in their original proposal. Following them, several architectural variants of GANs have been proposed for improving performance in different applications. Deep Convolutional GAN (DCGAN) was proposed by replacing the fully connected layers with a number of modifications, like use of strided convolution (discriminator) and fractional-strided convolutions (generator), batch normalization, use of ReLU activation (generator) and LeakyReLU activation (discriminator), etc., for generating high resolution images [44]. Boundary Equilibrium GAN (BEGAN) proposed to use an autoencoder as a discriminator and aims to match autoencoder's loss distribution instead of data distribution directly [45]. The key idea behind Progressive GAN is to expand the network architecture by adding new layers to both generator and discriminator progressively. The method allows to generate high quality images at higher speed and stable training [46]. Self-Attention GAN (SAGAN) incorporates self-attention mechanism into the GAN framework which was shown to be effective in modelling long-range dependencies [47]. Motivated by style transfer literature, the researchers from NVIDIA proposed an alternative generative architecture for GANs [48]. In contrast to traditional generator, the style-based generator first maps the input latent code into an intermediate latent space which then uses adaptive instance normalization for controlling the generator.

Wasserstein GAN (WGAN) is a variant of the vanilla GAN proposal which uses Wasserstein distance, instead of Jensen–Shannon divergence, to measure the difference between the model and target distributions [49]. To overcome the vanishing gradients problem, the author of Least Squares GAN (LSGAN) proposed to use the least squares loss function for the discriminator instead of sigmoid cross-entropy loss used in the original proposal [50]. Unrolled GAN (UGAN) uses the unrolling optimization of the discriminator objective during training for solving the problem of unstable optimization and mode collapse [51]. The basic idea behind UGAN is to update the generator by adding a gradient term that captures the reaction of the discriminator in response to the change in the generator. In Loss-Sensitive Generative Adversarial Network (LS-GAN) the authors introduce a loss function to measure the quality of the generated samples. Utilizing a constraint, it makes sure that the loss of a real sample should be smaller than that of a generated sample [52]. A comprehensive list of different GAN architectures and loss functions can be found in GAN zoo.[2]

## 2.4.6  Recursive Neural Networks

Recursive Neural Network is a family of neural network models that operates on structured inputs, particularly on directed acyclic graphs. Like Recurrent Neural Networks (RNN), recursive neural networks are used to process variable-length inputs, however, they can be seen as a generalization over RNN [53]. Recursive neural networks can identify the hierarchical structure in dataset, therefore, have been applied in natural language processing, image analysis, theorem proving, bioinformatics, etc. [54].

The simplest member in the class of recursive neural networks is the standard recursive neural network in which the same set of weights is recursively applied within a structural setting to generate a structured prediction. Given a positional

---

[2]https://github.com/hindupuravinash/the-gan-zoo.

**Fig. 2.9** Architecture of
standard recursive neural
network (reproduced from
[55])

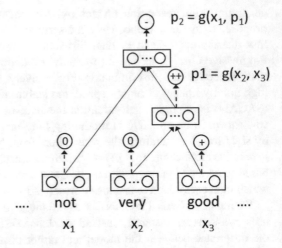

directed acyclic graph (e.g. binary tree), it will generate the parent representation by recursively applying transformations on the children representations in a bottom-up fashion. Figure 2.9 shows a model of recursive neural network for sentiment analysis. The recursive neural network model will compute the representation of an internal node if its children are already computed. Therefore, in the example of Fig. 2.9, it will compute parent representation $p_1$ and $p_2$ in order

$$p_1 = \varphi\left(W\begin{bmatrix} x_2 \\ x_3 \end{bmatrix}\right) \quad \text{and} \quad p_2 = \varphi\left(W\begin{bmatrix} x_1 \\ p_1 \end{bmatrix}\right), \tag{2.35}$$

where $W$ is the weight matrix (bias is not shown for simplicity) and $\varphi(\cdot)$ represents the activation function often hyperbolic tangent. The parent representations must have the same dimensionality so that those can be recursively used to generate higher-level representations. The model also applies a task-specific output from the representation given by

$$y_i = \varphi'\left(W' p_i\right), \tag{2.36}$$

where $W'$ represents the output weight matrix and $y_i$ represents the prediction of the class label in supervised learning. Like many other deep architectures, deep recursive neural networks are constructed by stacking recursive layers one on top of the other.

Recursive neural networks are trained using a variant of backpropagation called backpropagation through structure (BPTS) [56]. The forward pass starts from the leaves and proceeds bottom-up and the backward pass begins at the root and proceeds top-down. Similar to standard backpropagation, BPTS can operate in batch or online mode. Training of a deep recursive neural network can be considered as interleaved application of standard backpropagation that operates across multiple stacked layers and BPTS that operates through the structure of a layer [53]. There are a number of variants of recursive neural networks. The most prominent ones are recursive autoencoder and recursive neural tensor network.

## 2.5  Applications of Deep Learning

By virtue of its envious performance, deep learning has made its way in a wide variety of domains. In a very short span of time, the scope of deep learning has been expanded from the area of computer vision and natural language processing, to include diverse applications in the domains of science, engineering, business, medicine even art. Deep learning is being applied in new problems on a regular basis and exhibiting sensational performance outperforming other traditional machine learning and non-machine learning methods. In this section, we will enumerate only the major application areas of deep learning.

*Computer Vision:* Applications are image classification, image segmentation, object detection/localization/tracking, human face/action/pose recognition, style transfer, image reconstruction, image synthesis, image enhancement, image captioning. Because of their unique capability in feature learning, majority of the methods use CNN on different applications in computer vision. Among the other DL models used in computer vision are deep autoencoder, deep belief network, deep Boltzmann machines and GANs.

*Natural language processing:* Applications are Name Entity Recognition (NER), sentence classification, sentiment analysis, sarcasm detection, query-document matching, semantic matching, natural language generation, language modelling, etc. Because of their capability to process sequences, RNNs and their variants are widely used in various NLP applications. CNN and its variants have been used simple NLP tasks such NER and more complex tasks involving varying lengths of texts. Recursive neural networks have been used for parsing, semantic relationship classification, sentence relatedness.

*Speech and audio processing:* Applications are phone recognition, speech recognition, speech synthesis, music signal processing, music information retrieval, music analysis, music similarity estimation, acoustic surveillance, localization and tracking of sound sources, source separation, audio enhancement. Different types of deep learning models like DNN, CNN, RNN, RBM, DBN, GAN, VAE and their hybrids have been applied in these applications.

*Biology:* Deep learning found many applications in high-throughput biology and 'omics' studies, e.g. molecular target prediction, protein structures prediction, protein residue-residue contacts prediction, functional genomics elements (e.g. promoter) prediction, gene expression data analysis, etc. Among the most frequently used DL model in these applications are CNN, deep RNN and denoising autoencoder.

*Health and medicine:* Applications are medical, biological image classification, drug discovery, novel drug-target prediction, disease sub-typing, early diagnosis of disease, patient categorization, information retrieval, electronic health record analysis, etc. CNN, deep autoencoder, DBN and RNN are the most commonly used models in these applications.

*Business applications:* Among the most prominent applications of DNN in business are stock market analysis, financial fraud detection, mobile advertising, customer relationship management, social marketing, recommender systems, anomaly detection, risk prediction, cybersecurity. Most of these applications use CNN and RNN.

Besides, DNNs have found many applications in control engineering, robotics, smart manufacturing, automatic game playing, networking, cybersecurity, big data analysis, visual art processing, military and defence.

## 2.6 Conclusion

Deep learning is the fastest growing branch of machine learning that has shown remarkable achievements in different real-life applications in which machines had limited success so far. The surge of thriving applications of deep learning in diverse fields is sufficient to advocate its strength and versatility. However, the ultimate goal of this field is far-reaching and more general than its superhuman performance in specific tasks. Towards this goal we need integration and unification of our gathered knowledge from other branches of machine learning and artificial intelligence.

Although deep learning has yielded astonishing performance outperforming the state-of-the-art methods in numerous fields, it is not a panacea to all problems which we want artificial intelligence to solve for us. As researchers are working to understand and enhance the capabilities of deep learning, they are identifying the challenges and limitations the field needs to overcome for achieving human-like intelligence. Some of the well-known criticisms for deep learning are requirement of large volume of training data, uninterpretable knowledge, non-transferable extracted patterns/knowledge and lack of common sense.

In this chapter, we presented a very general and preliminary introduction to deep learning. The main objective of the chapter is to walk the readers through the principal categories of deep neural networks used in various applications. Starting with the vanilla DNN we introduced CNN, RNN, DBN, deep autoencoder, GAN and recursive neural network. The architectures of these deep neural networks were presented along with the different types of layers and processing units used in them. The major variants of different architectures and basic learning algorithms used for each architecture were also introduced. In the rest of the book, the readers will get to know about the state-of-the-art research in which different evolutionary and meta-heuristics algorithms, introduced in the previous chapter, are applied in designing these deep architectures and optimizing their various aspects.

# References

1. Bengio, Y., Courville, A., Vincent, P.: Representation learning: a review and new perspectives. IEEE Trans. Pattern Anal. Mach. Intell. **35**(8), 1798–1828 (2013)
2. LeCun, Y., Bengio, Y., Hinton, G.: Deep learning. Nature **521**(7553), 436 (2015)
3. Bengio, Y., et al.: Learning deep architectures for AI. Found. Trends® Mach. Learn. **2**(1), 1–127 (2009)
4. Deng, L.: A tutorial survey of architectures, algorithms, and applications for deep learning. APSIPA Trans. Signal Inf. Process. **3**, e2 (2014). https://doi.org/10.1017/atsip.2013.9
5. Schmidhuber, J.: Deep learning in neural networks: an overview. Neural Netw. **61**, 85–117 (2015)
6. Vargas, R., Mosavi, A., Ruiz, R.: Deep learning: a review. Adv. Intell. Syst. Comput. (2017). https://eprints.qut.edu.au/127354/
7. Ling, Z.-H., Kang, S.-Y., Zen, H., Senior, A., Schuster, M., Qian, X.-J., Meng, H.M., Deng, L.: Deep learning for acoustic modeling in parametric speech generation: a systematic review of existing techniques and future trends. IEEE Signal Process. Mag. **32**(3), 35–52 (2015)
8. Guo, Y., Liu, Y., Oerlemans, A., Lao, S., Wu, S., Lew, M.S.: Deep learning for visual understanding: a review. Neurocomputing **187**, 27–48 (2016)
9. Rawat, W., Wang, Z.: Deep convolutional neural networks for image classification: a comprehensive review. Neural Comput. **29**(9), 2352–2449 (2017)
10. Miotto, R., Wang, F., Wang, S., Jiang, X., Dudley, J.T.: Deep learning for healthcare: review, opportunities and challenges. Brief. Bioinform. **19**(6), 1236–1246 (2017)
11. Garcia-Garcia, A., Orts-Escolano, S., Oprea, S., Villena-Martinez, V., Garcia-Rodriguez, J.: A review on deep learning techniques applied to semantic segmentation (2017). arXiv preprint:1704.06857
12. Ain, Q.T., Ali, M., Riaz, A., Noureen, A., Kamran, M., Hayat, B., Rehman, A.: Sentiment analysis using deep learning techniques: a review. Int. J. Adv. Comput. Sci. Appl. **8**(6), 424 (2017)
13. Voulodimos, A., Doulamis, N., Doulamis, A., Protopapadakis, E.: Deep learning for computer vision: a brief review. Comput. Intell. Neurosci. **2018** (2018). https://doi.org/10.1155/2018/7068349
14. Fawaz, H.I., Forestier, G., Weber, J., Idoumghar, L., Muller, P.-A.: Deep learning for time series classification: a review. Data Min. Knowl. Disc. **33**(4), 917–963 (2019)
15. McCulloch, W.S., Pitts, W.: A logical calculus of the ideas immanent in nervous activity. Bull. Math. Biophys. **5**(4), 115–133 (1943)
16. Seide, F., Li, G., Yu, D.: Conversational speech transcription using context-dependent deep neural networks. In: Twelfth Annual Conference of the International Speech Communication Association (2011)
17. Mhaskar, H., Liao, Q., Poggio, T.: When and why are deep networks better than shallow ones?. In: Thirty-First AAAI Conference on Artificial Intelligence (2017)
18. Hinton, G.E., Osindero, S., Teh, Y.-W.: A fast learning algorithm for deep belief nets. Neural Comput. **18**(7), 1527–1554 (2006)
19. Hubel, D.H., Wiesel, T.N.: Receptive fields of single neurones in the cat's striate cortex. J. Physiol. **148**(3), 574–591 (1959)
20. Hubel, D.H., Wiesel, T.N.: Receptive fields, binocular interaction and functional architecture in the cat's visual cortex. J. Physiol. **160**(1), 106–154 (1962)
21. Fukushima, K.: Neocognitron: a self-organizing neural network model for a mechanism of pattern recognition unaffected by shift in position. Biol. Cybern. **36**(4), 193–202 (1980)
22. Ioffe, S., Szegedy, C.: Batch normalization: accelerating deep network training by reducing internal covariate shift (2015). arXiv preprint:1502.03167
23. Krizhevsky, A., Sutskever, I., Hinton, G.E.: ImageNet classification with deep convolutional neural networks. In: Advances in Neural Information Processing Systems, pp. 1097–1105 (2012)

24. Szegedy, C., Liu, W., Jia, Y., Sermanet, P., Reed, S., Anguelov, D., Erhan, D., Vanhoucke, V., Rabinovich, A.: Going deeper with convolutions. In: Proceedings of the IEEE Conference on Computer Vision and Pattern Recognition, pp. 1–9 (2015)
25. Simonyan, K., Zisserman, A.: Very deep convolutional networks for large-scale image recognition (2014). arXiv preprint:1409.1556
26. He, K., Zhang, X., Ren, S., Sun, J.: Deep residual learning for image recognition. In: Proceedings of the IEEE Conference on Computer Vision and Pattern Recognition, pp. 770–778 (2016)
27. Chollet, F.: Xception: deep learning with depthwise separable convolutions. In: Proceedings of the IEEE Conference on Computer Vision and Pattern Recognition, pp. 1251–1258 (2017)
28. Khan, A., Sohail, A., Ali, A.: A new channel boosted convolutional neural network using transfer learning (2018). arXiv preprint:1804.08528
29. Yu, Y., Si, X., Hu, C., Zhang, J.: A review of recurrent neural networks: LSTM cells and network architectures. Neural Comput. 31(7), 1235–1270 (2019)
30. Lipton, Z.C., Berkowitz, J., Elkan, C.: A critical review of recurrent neural networks for sequence learning (2015). arXiv preprint:1506.00019
31. Hochreiter, S., Schmidhuber, J.: Long short-term memory. Neural Comput. 9(8), 1735–1780 (1997)
32. Gers, F.A., Schmidhuber, J., Cummins, F.: Learning to forget: continual prediction with LSTM. Neural Comput. 12(10), 2451–2471 (2000)
33. Cho, K., Van Merriënboer, B., Gulcehre, C., Bahdanau, D., Bougares, F., Schwenk, H., Bengio, Y.: Learning phrase representations using RNN encoder-decoder for statistical machine translation (2014). arXiv preprint:1406.1078
34. Hermans, M., Schrauwen, B.: Training and analysing deep recurrent neural networks. In: Advances in Neural Information Processing Systems, pp. 190–198 (2013)
35. Pascanu, R., Gulcehre, C., Cho, K., Bengio, Y.: How to construct deep recurrent neural networks (2013). arXiv preprint:1312.6026
36. Marchi, E., Vesperini, F., Squartini, S., Schuller, B.: Deep recurrent neural network-based autoencoders for acoustic novelty detection. Comput. Intell. Neurosci. 2017 (2017). https://doi.org/10.1155/2017/4694860
37. LeCun, Y., Chopra, S., Hadsell, R., Ranzato, M., Huang, F.: A tutorial on energy-based learning. Predicting Struct. Data 1(0) (2006). http://yann.lecun.com/exdb/publis/orig/lecun-06.pdf
38. Hinton, G.E.: A practical guide to training restricted Boltzmann machines. In: Neural Networks: Tricks of the Trade, pp. 599–619. Springer, Berlin (2012)
39. Hinton, G.E.: Training products of experts by minimizing contrastive divergence. Neural Comput. 14(8), 1771–1800 (2002)
40. Tieleman, T.: Training restricted Boltzmann machines using approximations to the likelihood gradient. In: Proceedings of the 25th International Conference on Machine Learning, pp. 1064–1071. ACM, New York (2008)
41. Goodfellow, I., Bengio, Y., Courville, A.: Deep Learning. MIT Press, Cambridge (2016). http://www.deeplearningbook.org
42. Goodfellow, I., Pouget-Abadie, J., Mirza, M., Xu, B., Warde-Farley, D., Ozair, S., Courville, A., Bengio, Y.: Generative adversarial nets. In: Advances in Neural Information Processing Systems, pp. 2672–2680 (2014)
43. Goodfellow, I.: Nips 2016 tutorial: generative adversarial networks (2016). arXiv preprint:1701.00160
44. Radford, A., Metz, L., Chintala, S.: Unsupervised representation learning with deep convolutional generative adversarial networks (2015). arXiv preprint:1511.06434
45. Berthelot, D., Schumm, T., Metz, L.: BEGAN: boundary equilibrium generative adversarial networks (2017). arXiv preprint:1703.10717
46. Karras, T., Aila, T., Laine, S., Lehtinen, J.: Progressive growing of GANs for improved quality, stability, and variation (2017). arXiv preprint:1710.10196

47. Zhang, H., Goodfellow, I., Metaxas, D., Odena, A.: Self-attention generative adversarial networks (2018). arXiv preprint:1805.08318
48. Karras, T., Laine, S., Aila, T.: A style-based generator architecture for generative adversarial networks. In: Proceedings of the IEEE Conference on Computer Vision and Pattern Recognition, pp. 4401–4410 (2019)
49. Arjovsky, M., Chintala, S., Bottou, L.: Wasserstein GAN (2017). arXiv preprint:1701.07875
50. Mao, X., Li, Q., Xie, H., Lau, R.Y., Wang, Z., Paul Smolley, S.: Least squares generative adversarial networks. In: Proceedings of the IEEE International Conference on Computer Vision, pp. 2794–2802 (2017)
51. Metz, L., Poole, B., Pfau, D., Sohl-Dickstein, J.: Unrolled generative adversarial networks. CoRR, vol. abs/1611.02163 (2016). http://arxiv.org/abs/1611.02163
52. Qi, G.-J.: Loss-sensitive generative adversarial networks on Lipschitz densities (2017). arXiv preprint:1701.06264
53. Irsoy, O., Cardie, C.: Deep recursive neural networks for compositionality in language. In: Advances in Neural Information Processing Systems, pp. 2096–2104 (2014)
54. Bianchini, M., Maggini, M., Sarti, L., Hawkes, P.: Recursive neural networks and their applications to image processing. Adv. Imaging Electron Phys. **140**, 1 (2006)
55. Socher, R., Perelygin, A., Wu, J., Chuang, J., Manning, C.D., Ng, A., Potts, C.: Recursive deep models for semantic compositionality over a sentiment treebank. In: Proceedings of the 2013 Conference on Empirical Methods in Natural Language Processing, pp. 1631–1642 (2013)
56. Goller, C., Kuchler, A.: Learning task-dependent distributed representations by backpropagation through structure. In: Proceedings of International Conference on Neural Networks (ICNN'96), vol. 1, pp. 347–352. IEEE, Piscataway (1996)

# Part II
# Hyper-Parameter Optimization

# Chapter 3
# On the Assessment of Nature-Inspired Meta-Heuristic Optimization Techniques to Fine-Tune Deep Belief Networks

**Leandro Aparecido Passos, Gustavo Henrique de Rosa, Douglas Rodrigues, Mateus Roder, and João Paulo Papa**

**Abstract** Machine learning techniques are capable of talking, interpreting, creating, and even reasoning about virtually any subject. Also, their learning power has grown exponentially throughout the last years due to advances in hardware architecture. Nevertheless, most of these models still struggle regarding their practical usage since they require a proper selection of hyper-parameters, which are often empirically chosen. Such requirements are strengthened when concerning deep learning models, which commonly require a higher number of hyper-parameters. A collection of nature-inspired optimization techniques, known as meta-heuristics, arise as straightforward solutions to tackle such problems since they do not employ derivatives, thus alleviating their computational burden. Therefore, this work proposes a comparison among several meta-heuristic optimization techniques in the context of Deep Belief Networks hyper-parameter fine-tuning. An experimental setup was conducted over three public datasets in the task of binary image reconstruction and demonstrated consistent results, posing meta-heuristic techniques as a suitable alternative to the problem.

## 3.1 Introduction

In the past years, multimedia-based applications fostered the generation of a massive amount of data. These data provide a wide range of opportunities for machine learning applications in several areas of knowledge, such as medicine, financial market, intelligent manufacturing, and event classification. Among such machine

L. A. Passos (✉) · G. H. de Rosa · M. Roder · J. P. Papa
Department of Computing, São Paulo State University, Bauru, Brazil
e-mail: leandro.passos@unesp.br; gustavo.rosa@unesp.br; mateus.roder@unesp.br; joao.papa@unesp.br

D. Rodrigues
Department of Computing, São Carlos Federal University, São Carlos, Brazil

© Springer Nature Singapore Pte Ltd. 2020
H. Iba, N. Noman (eds.), *Deep Neural Evolution*, Natural Computing Series,
https://doi.org/10.1007/978-981-15-3685-4_3

learning approaches, deep learning methods have received significant attention due to their excellent results, often surpassing even humans.

Deep learning models try to simulate the human-brain behavior on how the information is processed. The basic idea is to use multiple layers to extract higher-level features progressively, where each layer learns to transform input data into a more abstract representation. Regarding applications in the image processing area, lower layers may identify edges, while higher layers may identify human-meaningful items such as human faces and objects. Among the most employed methods, one can include Convolutional Neural Networks (CNNs) [8], Deep Belief Networks (DBNs) [5], and Deep Boltzmann Machines (DBMs) [23], among others.

Since "deep" in deep learning refers to the architecture complexity, the more complex it becomes, the higher the number of hyper-parameters to fit. Yosinski and Lipson [36], for instance, highlighted some approaches for visualizing the behavior of a single Restricted Boltzmann Machine (RBM) [24], which is an energy-based model that can be used to build DBNs and DBMs, during its learning procedure, and provided an overview toward such complexities comprehension. Such a problem was usually tackled using auto-learning tools, which combine parameter fine-tuning with feature selection techniques [26]. Despite, it can also be posed as an optimization task in which one wants to choose suitable hyper-parameters.

Therefore, meta-heuristic algorithms have become a viable alternative to solve optimization problems due to their simple implementation. Kuremoto et al. [7], for instance, employed the Particle Swarm Optimization (PSO) [6] to the context of hyper-parameter fine-tuning concerning RBMs, while Liu et. al [10] and Levy et al. [9] applied Genetic Algorithms (GA) [29] for model selection and automatic painter classification using RBMs, respectively. Later, Rosa et al. [22] addressed the Firefly Algorithm to fine-tune DBN hyper-parameters. Finally, Passos et al. [15, 16] proposed a similar approach comparing several meta-heuristic techniques to fine-tune hyper-parameters in DBMs, infinity Restricted Boltzmann Machines [13, 18], and RBM-based models in general [14].

Following this idea, this chapter presents a comparison among ten different swarm- and differential evolution-based meta-heuristic algorithms in the context of fine-tuning DBN hyper-parameters. We present a discussion about the viability of such approaches in three public datasets, as well as the statistical evaluation through the Wilcoxon signed-rank test. The remainder of this chapter is organized as follows. Section 3.2 introduces the theoretical background concerning RBMs and DBNs. Sections 3.4 and 3.5 present the methodology and the experimental results, respectively. Finally, Sect. 3.6 states conclusions and future works.

## 3.2 Theoretical Background

In this section, we present a theoretical background concerning Restricted Boltzmann Machines and Deep Belief Networks.

### 3.2.1 Restricted Boltzmann Machines

Restricted Boltzmann Machines are well-known stochastic-nature neural networks inspired by physical laws of statistical mechanics and parameterized by concepts like energy and entropy. These networks are commonly employed in the field of unsupervised learning, having at least two layers of neurons, i.e., one visible and one hidden.

The Restricted Boltzmann Machine basic architecture is composed of a visible layer $\mathbf{v} = \{v_1, v_2, \ldots, v_m\}$ with $m$ units and a hidden layer $\mathbf{h} = \{h_1, h_2, \ldots, h_n\}$ with $n$ units. Furthermore, a real-valued matrix $\mathbf{W}_{m \times n}$ is responsible for modeling the restricted connections, i.e., the weights, between the visible and hidden neurons, where $w_{ij}$ represents the connection between the visible unit $v_i$ and the hidden unit $h_j$. Figure 3.1 describes the vanilla RBM architecture.

Regarding the learning process, a layer composed of visible units represents the input data to be processed, while the hidden layer is employed to extract deep-seated patterns and information from this data. Besides, both visible and hidden units assume only binary values, i.e., $\mathbf{v} \in \{0, 1\}^m$ and $\mathbf{h} \in \{0, 1\}^n$, once sampling process is derived from a Bernoulli distribution [4]. Finally, the training process is performed by minimizing the system's energy considering both the visible and hidden layers units, as well as the biases associated with each layer. The energy can be computed as follows:

$$E(\mathbf{v}, \mathbf{h}) = - \sum_{i=1}^{m} a_i v_i - \sum_{j=1}^{n} b_j h_j - \sum_{i=1}^{m} \sum_{j=1}^{n} v_i h_j w_{ij}, \tag{3.1}$$

where $\mathbf{a}$ and $\mathbf{b}$ represent the biases of visible and hidden units, respectively.

Computing the system's probability is an intractable task due to the computational cost. However, one can estimate the probability of activating a single visible neuron $i$ given the hidden units through Gibbs sampling over a Markov chain, as follows:

$$P(v_i = 1 | \mathbf{h}) = \phi \left( \sum_{j=1}^{n} w_{ij} h_j + a_i \right), \tag{3.2}$$

**Fig. 3.1** Vanilla RBM architecture

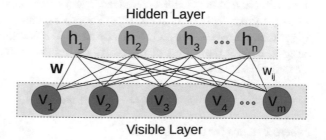

and, in a similar fashion, the probability of activating a single hidden neuron $j$ given the visible units is stated as follows:

$$P(h_j = 1|\mathbf{v}) = \phi\left(\sum_{i=1}^{m} w_{ij}v_i + b_j\right), \tag{3.3}$$

where $\phi(\cdot)$ stands for the logistic-sigmoid function.

The training process consists of maximizing the product of probabilities given a set of parameters $\theta = (W, a, b)$ and the data probability distribution over the training samples. Such a process can be easily computed using either the Contrastive Divergence (CD) [3] or the Persistent Contrastive Divergence (PCD) [27] algorithms.

### 3.2.2 Contrastive Divergence

Hinton [3] introduced a faster methodology to compute the energy of the system based on contrastive divergence. The idea is to initialize the visible units with a training sample, to compute the states of the hidden units using Eq. (3.3), and then to compute the states of the visible unit (reconstruction step) using Eq. (3.2). In short, this is equivalent to perform Gibbs sampling using $k = 1$ and to initialize the chain with the training samples.

Therefore, the equation below leads to a simple learning rule for updating the weights matrix $\mathbf{W}$, and biases $\mathbf{a}$ and $\mathbf{b}$ at iteration $t$:

$$\mathbf{W}^{t+1} = \mathbf{W}^t + \underbrace{\eta(P(\mathbf{h}|\mathbf{v})\mathbf{v}^T - P(\tilde{\mathbf{h}}|\tilde{\mathbf{v}})\tilde{\mathbf{v}}^T) + \Phi}_{=\Delta\mathbf{W}^t}, \tag{3.4}$$

$$\mathbf{a}^{t+1} = \mathbf{a}^t + \underbrace{\eta(\mathbf{v} - \tilde{\mathbf{v}}) + \varphi\Delta\mathbf{a}^{t-1}}_{=\Delta\mathbf{a}^t}, \tag{3.5}$$

$$\mathbf{b}^{t+1} = \mathbf{b}^t + \underbrace{\eta(P(\mathbf{h}|\mathbf{v}) - P(\tilde{\mathbf{h}}|\tilde{\mathbf{v}})) + \varphi\Delta\mathbf{b}^{t-1}}_{=\Delta\mathbf{b}^t}, \tag{3.6}$$

where $\eta$ stands for the learning rate, $\varphi$ denotes the momentum, $\tilde{\mathbf{v}}$ stands for the reconstruction of the visible layer given $\mathbf{h}$, and $\tilde{\mathbf{h}}$ denotes an estimation of the hidden vector $\mathbf{h}$ given $\tilde{\mathbf{v}}$. In a nutshell, Eqs. (3.4), (3.5), and (3.6) show the optimization algorithm, the well-known Gradient Descent. The additional term $\Phi$ in Eq. (3.4) is used to control the values of matrix $\mathbf{W}$ during the convergence process, and it is described as follows:

$$\Phi = -\lambda\mathbf{W}^t + \varphi\Delta\mathbf{W}^{t-1}, \tag{3.7}$$

where $\lambda$ stands for the weight decay.

### 3.2.3   Persistent Contrastive Divergence

Most of the issues related to the Contrastive Divergence approach concern the number of iterations employed to approximate the model to the real data. Although the approach proposed by Hinton [3] takes $k = 1$ and works well for real-world problems, one can settle different values for $k$ [1].[1]

Notwithstanding, Contrastive Divergence provides a good approximation to the likelihood gradient, i.e., it gives a reasonable estimation of the model to the data when $k \to \infty$. However, its convergence might become poor when the Markov chain has a "low mixing," as well as a good convergence only on the early iterations, getting slower as iterations go by, thus, demanding the use of parameters decay.

Therefore, Tieleman [27] proposed the Persistent Contrastive Divergence, an interesting alternative for contrastive divergence using higher values for $k$ while keeping the computational burden relatively low. The idea is quite simple: on CD-1, each training sample is employed to start an RBM and rebuild a model after a single Gibbs sampling iteration. Once every training sample is presented to the RBM, we have a so-called epoch. The process is repeated for each next epoch, i.e., the same training samples are used to feed the RBM, and the Markov chain is restarted at each epoch.

### 3.2.4   Deep Belief Networks

Deep Belief Networks [5] are graphical models composed of a visible and $L$ hidden layers, where each layer is connected to the latter through a weight matrix $\mathbf{W}_l$, $l \in \{1, 2, \ldots, L\}$, and there is no connection between units from the same layer. In a nutshell, one can consider each set of two subsequent layers as an RBM trained in a greedy fashion such that the trained hidden layer of the bottommost RBM feeds the next RBM's visible layer, and so on. Figure 3.2 depicts the model. Notice $\mathbf{v}$ and $\mathbf{h}_l$ stand for the visible and the $l$-th hidden layers.

Although this work focuses on image reconstruction, one can use DBNs for supervised classification tasks. Such an approach requires, after the greedy feed-forward pass mentioned above, fine-tuning the network weights using either backpropagation or gradient descent. Afterward, a softmax layer is added at the top of the model to attribute the predicted labels.

---

[1] Usually, contrastive divergence with a single iteration is called CD-1.

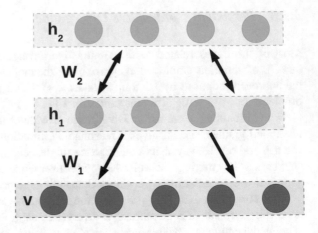

**Fig. 3.2** DBN architecture with two hidden layers

## 3.3 Meta-heuristic Optimization Algorithms

This section presents a brief description of the meta-heuristic optimization techniques employed in this work.

- Improved Harmony Search (IHS) [11]: an improved version of the Harmony Search optimization algorithm that employs dynamic values for both the Pitch Adjusting Rate (PAR), considering values in the range [$PAR_{min}$, $PAR_{max}$], and the Harmony Memory Considering Rate (HMCR), which assumes values in the range [$HMCR_{min}$, $HMCR_{max}$]. Additionally, the algorithm uses the bandwidth variable $\varrho$ in the range [$\varrho_{min}$, $\varrho_{max}$] to calculate PAR.
- Particle Swarm Optimization with Adaptive Inertia Weight (AIWPSO) [12]: an improved version of the Particle Swarm Optimization that employs self-adjusting inertia weights $w$ over each particle along with the search space aiming to balance the global exploration and local exploitation. Notice the method uses the variables $c_1$ and $c_2$ to control the particles' acceleration.
- Flower Pollination Algorithm (FPA) [21, 35]: a meta-heuristic optimization algorithm that tries to mimic the pollination process performed by flowers. The algorithm employs four basic rules: (1) the cross-pollination, which stands for the pollination performed by birds and insects, (2) the self-pollination, representing the pollination performed by the wind diffusion or similar approaches, (3) the constancy of birds/insects, representing the probability of reproduction, and (4) the interaction of local and global pollination, controlled by the probability parameter $p$. Additionally, the algorithm employs an additional parameter $\beta$ to control the amplitude of the distribution.
- Bat Algorithm (BA) [34]: based on the bats' echolocation system while searching for food and prey. The algorithm employs a swarm of virtual bats randomly flying in the search space at different velocities, even following a random walk approach for local search intensification. Additionally, it applies a dynamically updated wavelength frequency in the range {$f_{min}$, $f_{max}$} according to the distance from the objective, as well as loudness $A$ and the pulse rate $r$.

- Firefly Algorithm (FA) [31]: the algorithm is based on the fireflies' approach for attracting potential preys and mating partners. It employs the attractiveness $\beta$ parameter, which influences the brightness of each agent, depending on its position and light absorption coefficient $\gamma$. Moreover, the model employs a random perturbation $\alpha$ used to perform a random walk and avoid local optima.
- Cuckoo Search (CS) [20, 32, 33]: the model combines some cuckoo species parasitic behavior with a $\tau$-step random walk over a Markov chain. It employs three basic concepts: (1) each cuckoo lays a single egg for iteration at another bird's randomly chosen nest, (2) $p_a \in [0, 1]$ defines the probability of this bird discover and discard the cuckoo's egg or abandon it and create a new chest, i.e., a new solution, and (3) the nests with best eggs will carry over to the next generations.
- Differential Evolution (DE) [25]: evolution algorithm maintains a population of candidate solutions which are combined and improved in following generations aiming to find the characteristics that best fit the problem. The algorithm employs a mutation factor to control the mutation amplitude, as well as a parameter to control the crossover probability.
- Backtracking Search Optimization Algorithm (BSA) [2, 17]: an evolution algorithm that employs a random selection of a historical population for mutation and crossover operations to generate a new population of individuals based on past experiences. The algorithm controls the number of elements to be mutated using a mixing rate ($mix\_rate$) parameter, as well as the amplitude of the search-direction matrix with the parameter $F$.
- Differential Evolution Based on Covariance Matrix Learning and Bimodal Distribution Parameter Setting Algorithm (CoBiDE) [28]: a differential evolution model that represents the search space coordinate system using a covariance matrix according to the probability parameter $P_b$, and the proportion of individuals employed in the process using the $P_s$ variable. Moreover, it employs a binomial distribution to control the mutation and crossover rates, aiming a better trade-off between exploitation and exploration.
- Adaptive Differential Evolution with Optional External Archive (JADE) [19, 37]: JADE is a differential evolution-based algorithm that employs the "DE/current-to-$p$-best" strategy, i.e., only the $p - best$ agents are used in the mutation process. Further, the algorithm employs both a historical population and a control parameter, which is adaptively updated. Finally, it requires a proper selection of the rate of adaptation parameter $c$, as well as the mutation greediness parameter $g$.

## 3.4  Methodology

This section introduces the intended procedure for DBN hyper-parameter fine-tuning. Additionally, it describes the employed datasets and the experimental setup.

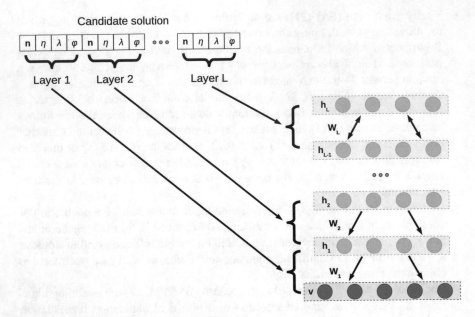

**Fig. 3.3** DBN hyper-parameter optimization approach

### 3.4.1 Modeling DBN Hyper-parameter Fine-tuning

The learning procedure of each RBM employs four hyper-parameters, as specified in Sect. 3.2.1: the learning rate $\eta$, weight decay $\lambda$, momentum $\varphi$, and the number of hidden units $n$. Since DBNs are built over RBM blocks, they employ a similar process to fine-tune each of their layers individually. In short, a four-dimensional search space composed of three real- and one integer-valued variables should be selected for each layer. Notice the variable values are intrinsically real numbers, thus requiring a type casting to obtain the nearest integer. Such an approach aims at electing the assortment of DBN hyper-parameters that minimizes the training images reconstruction error, denoted by the minimum squared error (MSE). Subsequently, the selected set of parameters is applied to reconstruct the unseen images of the test set. Figure 3.3 depicts the procedure.

### 3.4.2 Datasets

We employed three datasets, as described below:

- MNIST dataset[2]: a dataset composed of "0"–"9" handwritten digits images. Regarding the pre-processing, the images were converted from gray-scale to

---

[2]http://yann.lecun.com/exdb/mnist/.

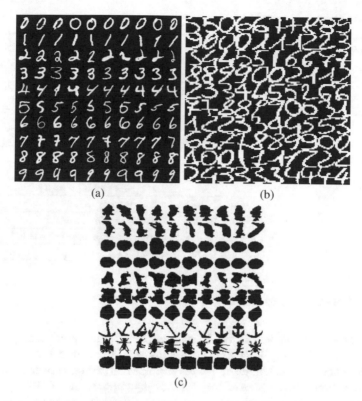

**Fig. 3.4** Some training examples from (**a**) MNIST, (**b**) Semeion, and (**c**) CalTech 101 Silhouettes datasets

binary, as well as resized to $14 \times 14$. Additionally, the training was performed over 2% of the training set, i.e., 1200 images, due to the demanded computational burden. Moreover, the complete set of 10,000 was employed for testing.

- Semeion Handwritten Digit Dataset[3]: similar to the MNIST, Semeion is also a dataset composed of "0"–"9" handwritten digits images formed by 1593 images. In this paper, we resized the samples to $16 \times 16$ and binarized each pixel.
- CalTech 101 Silhouettes Dataset[4]: a dataset composed of 101 classes of silhouettes with a resolution of $28 \times 28$. No pre-processing step was applied to the image samples.

Figure 3.4 displays some training examples from the above datasets.

---

[3]https://archive.ics.uci.edu/ml/datasets/Semeion+Handwritten+Digit.

[4]https://people.cs.umass.edu/~marlin/data.shtml.

**Table 3.1** Meta-heuristic
algorithms' parameter
configuration

| Algorithm | Parameters |
|-----------|------------|
| IHS | $HMCR = 0.7 \mid PAR_{MIN} = 0.1$ |
| | $PAR_{MAX} = 0.7 \mid \varrho_{MIN} = 1$ |
| | $\varrho_{MAX} = 10$ |
| AIWPSO | $c_1 = 1.7 \mid c_2 = 1.7$ |
| | $w = 0.7 \mid w_{MIN} = 0.5 \mid w_{MAX} = 1.5$ |
| FPA | $\beta = 1.5 \mid p = 0.8$ |
| BA | $f_{\min} = 0 \mid f_{\max} = 100 \mid A = 1.5 \mid r = 0.5$ |
| FA | $\alpha = 0.2 \mid \beta = 1.0 \mid \gamma = 1.0$ |
| CS | $\beta = 1.5 \mid p = 0.25 \mid \alpha = 0.8$ |
| BSA | $mix\_rate = 1.0 \mid F = 3$ |
| CoBiDE | $P_b = 0.4 \mid P_s = 0.5$ |
| DE | $mutation\_factor = 0.8$ |
| | $cross\_over\_probability = 0.7$ |
| JADE | $c = 0.1 \mid g = 0.05$ |

### 3.4.3 Experimental Setup

Experiments were conducted over 20 runs and a 2-fold cross-validation for statistical analysis using the Wilcoxon signed-rank test [30] with 5% of significance. Each meta-heuristic technique employed five agents (particles) over 50 iterations for convergence purposes over the three configurations, i.e., DBNs with 1, 2, and 3 layers. Additionally, the paper compares different techniques ranging from music composition process, swarm-based, and evolutionary-inspired methods, in the context of DBN hyper-parameter fine-tuning, as presented in Sect. 3.3:

Table 3.1 exhibits the parameter configuration for every meta-heuristic technique.[5]

Finally, each DBN layer is composed of an RBM whose hyper-parameters are randomly initialized according to the following ranges: $n \in [5, 100]$, $\eta \in [0.1, 0.9]$, $\lambda \in [0.1, 0.9]$, and $\varphi \in [10^{-5}, 10^{-1}]$. Additionally, the experiments were conducted over three different depth configurations, i.e., DBNs composed of 1, 2, and 3 RBM layers, which implies on fine-tuning a $4-$, $8-$, and $12-$dimensional set of hyper-parameters. We also have employed $T = 10$ as the number of epochs for DBN learning weights procedure with mini-batches of size 20. In order to present a more in-depth experimental validation, all DBNs were trained with the Contrastive Divergence (CD) [3] and Persistent Contrastive Divergence (PCD) [27]. Figure 3.5 depicts the pipeline proposed in this paper.

---

[5]Note that these values were empirically chosen according to their author's definition.

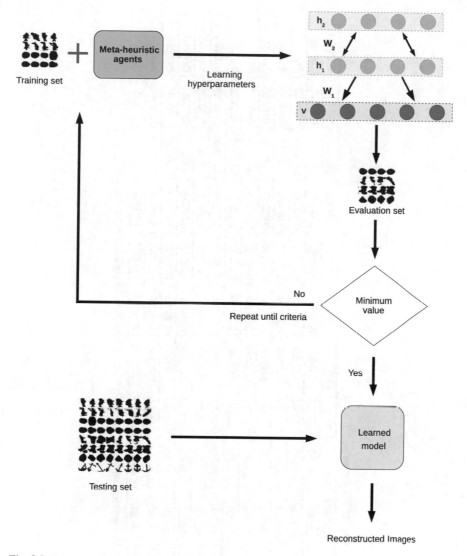

**Fig. 3.5** Proposed pipeline to the task of DBN hyper-parameter fine-tuning

## 3.5  Experimental Results

This section introduces the results obtained during the experiments. Further, a detailed discussion about them is provided. Tables 3.2, 3.3, and 3.4 present the average MSE, and their standard deviation regarding MNIST, Semeion Handwritten Digit, and CalTech 101 Silhouettes datasets, respectively. The best results accordingly to the Wilcoxon signed-rank test with 5% of significance level are presented in bold.

**Table 3.2** Average MSE values considering MNIST dataset

| Layer | Alg | Statistics | IHS | AIWPSO | BA | CS | FA | FPA | BSA | CoBiDE | DE | JADE | RANDOM |
|---|---|---|---|---|---|---|---|---|---|---|---|---|---|
| 1 | CD | Mean | **0.08759** | 0.08765 | 0.08762 | 0.08764 | 0.08764 | **0.08759** | 0.08763 | 0.08763 | 0.08762 | 0.08761 | 0.08763 |
| | | Std | 0.00008 | 0.00006 | 0.00007 | 0.00005 | 0.00007 | 0.00009 | 0.00005 | 0.00006 | 0.00005 | 0.00007 | 0.00006 |
| | PCD | Mean | 0.08762 | 0.08762 | 0.08763 | 0.08764 | 0.08765 | 0.08761 | 0.08762 | 0.08762 | 0.08763 | 0.08763 | 0.08764 |
| | | Std | 0.00008 | 0.00005 | 0.00006 | 0.00007 | 0.00006 | 0.00007 | 0.00007 | 0.00007 | 0.00006 | 0.00006 | 0.00005 |
| 2 | CD | Mean | 0.08762 | 0.08764 | 0.08763 | 0.08764 | 0.08764 | 0.08762 | 0.08763 | 0.08764 | 0.08763 | 0.08763 | 0.08763 |
| | | Std | 0.00005 | 0.00006 | 0.00007 | 0.00007 | 0.00006 | 0.00006 | 0.00007 | 0.00006 | 0.00005 | 0.00006 | 0.00004 |
| | PCD | Mean | 0.08763 | 0.08763 | **0.08763** | 0.08765 | 0.08763 | 0.08766 | 0.08764 | 0.08764 | 0.08762 | 0.08765 | 0.08764 |
| | | Std | 0.00006 | 0.00005 | 0.00007 | 0.00006 | 0.00006 | 0.00006 | 0.00006 | 0.00005 | 0.00005 | 0.00006 | 0.00005 |
| 3 | CD | Mean | 0.08763 | 0.08763 | 0.08765 | 0.08764 | 0.08764 | 0.08762 | 0.08763 | 0.08763 | 0.08763 | 0.08763 | 0.08764 |
| | | Std | 0.00007 | 0.00004 | 0.00005 | 0.00007 | 0.00006 | 0.00006 | 0.00006 | 0.00006 | 0.00006 | 0.00007 | 0.00007 |
| | PCD | Mean | 0.08763 | **0.08762** | 0.08765 | 0.08765 | 0.08763 | 0.08763 | 0.08763 | 0.08763 | 0.08763 | 0.08764 | 0.08764 |
| | | Std | 0.00006 | 0.00005 | 0.00007 | 0.00008 | 0.00006 | 0.00004 | 0.00007 | 0.00006 | 0.00007 | 0.00007 | 0.00006 |

Bold values denote the lowest average MSE or values whose Wilcoxon's p-value is above 0.05, i.e., values that are statistically similar

**Table 3.3** Average MSE values considering Semeion Handwritten Digit dataset

| Layer | Alg | Statistics | IHS | AIWPSO | BA | CS | FA | FPA | BSA | CoBiDE | DE | JADE | RANDOM |
|---|---|---|---|---|---|---|---|---|---|---|---|---|---|
| 1 | CD | Mean | 0.19359 | 0.20045 | 0.20743 | 0.20529 | 0.20638 | 0.19526 | 0.19571 | **0.19328** | 0.19470 | 0.19893 | 0.19711 |
| | | Std | 0.00137 | 0.00686 | 0.00449 | 0.00495 | 0.00492 | 0.00456 | 0.00365 | 0.00133 | 0.00513 | 0.00789 | 0.00313 |
| | PCD | Mean | 0.20009 | 0.20274 | 0.20763 | 0.20648 | 0.20895 | 0.19895 | 0.20003 | 0.19896 | 0.19950 | 0.20166 | 0.20362 |
| | | Std | 0.00197 | 0.00399 | 0.00274 | 0.00356 | 0.00209 | 0.00223 | 0.00254 | 0.00148 | 0.00377 | 0.00532 | 0.00184 |
| 2 | CD | Mean | 0.20961 | 0.20959 | 0.20963 | 0.20966 | 0.20966 | 0.20961 | 0.20962 | 0.20963 | 0.20962 | 0.20963 | 0.20962 |
| | | Std | 0.00037 | 0.00035 | 0.00039 | 0.00040 | 0.00041 | 0.00035 | 0.00036 | 0.00035 | 0.00037 | 0.00036 | 0.00036 |
| | PCD | Mean | 0.20962 | 0.20961 | 0.20965 | 0.20961 | 0.20966 | 0.20964 | 0.20959 | 0.20961 | 0.20960 | 0.20960 | 0.20959 |
| | | Std | 0.00036 | 0.00039 | 0.00038 | 0.00036 | 0.00039 | 0.00036 | 0.00037 | 0.00036 | 0.00037 | 0.00036 | 0.00035 |
| 3 | CD | Mean | 0.20961 | 0.20964 | 0.20964 | 0.20965 | 0.20965 | 0.20961 | 0.20962 | 0.20960 | 0.20961 | 0.20964 | 0.20961 |
| | | Std | 0.00037 | 0.00038 | 0.00038 | 0.00037 | 0.00035 | 0.00035 | 0.00037 | 0.00036 | 0.00038 | 0.00036 | 0.00037 |
| | PCD | Mean | 0.20963 | 0.20962 | 0.20966 | 0.20963 | 0.20966 | 0.20958 | 0.20963 | 0.20962 | 0.20959 | 0.20960 | 0.20961 |
| | | Std | 0.00038 | 0.00036 | 0.00038 | 0.00036 | 0.00041 | 0.00036 | 0.00038 | 0.00041 | 0.00036 | 0.00037 | 0.00036 |

Bold values denote the lowest average MSE or values whose Wilcoxon's p-value is above 0.05, i.e., values that are statistically similar

**Table 3.4** Average MSE values considering CalTech 101 Silhouettes dataset

| Layer | Alg | Statistics | IHS | AIWPSO | BA | CS | FA | FPA | BSA | CoBiDE | DE | JADE | RANDOM |
|---|---|---|---|---|---|---|---|---|---|---|---|---|---|
| 1 | CD | Mean | **0.15554** | 0.15641 | 0.15923 | 0.15923 | 0.16003 | 0.15607 | 0.15600 | 0.15638 | 0.15726 | 0.15608 | 0.15676 |
| | | Std | 0.00211 | 0.00241 | 0.00167 | 0.00171 | 0.00156 | 0.00230 | 0.00154 | 0.00191 | 0.00172 | 0.00184 | 0.00162 |
| | PCD | Mean | 0.15731 | 0.15825 | 0.16009 | 0.15993 | 0.15957 | 0.15810 | 0.15775 | 0.15800 | 0.15726 | 0.15790 | 0.15846 |
| | | Std | 0.00158 | 0.00231 | 0.00076 | 0.00103 | 0.00118 | 0.00192 | 0.00151 | 0.00121 | 0.00172 | 0.00135 | 0.00122 |
| 2 | CD | Mean | 0.16058 | 0.16056 | 0.16059 | 0.16057 | 0.16061 | 0.16062 | 0.16056 | 0.16059 | 0.16058 | 0.16058 | 0.16060 |
| | | Std | 0.00020 | 0.00020 | 0.00020 | 0.00019 | 0.00021 | 0.00024 | 0.00020 | 0.00023 | 0.00020 | 0.00020 | 0.00020 |
| | PCD | Mean | 0.16054 | 0.16060 | 0.16058 | 0.16062 | 0.16058 | 0.16065 | 0.16057 | 0.16058 | 0.16058 | 0.16058 | 0.16062 |
| | | Std | 0.00029 | 0.00022 | 0.00021 | 0.00023 | 0.00021 | 0.00038 | 0.00022 | 0.00022 | 0.00020 | 0.00020 | 0.00019 |
| 3 | CD | Mean | 0.16059 | 0.16058 | 0.16058 | 0.16060 | 0.16061 | 0.16058 | 0.16058 | 0.16060 | 0.16057 | 0.16059 | 0.16057 |
| | | Std | 0.00022 | 0.00022 | 0.00019 | 0.00021 | 0.00023 | 0.00022 | 0.00021 | 0.00021 | 0.00021 | 0.00019 | 0.00020 |
| | PCD | Mean | 0.16059 | 0.16058 | 0.16060 | 0.16061 | 0.16058 | 0.16060 | 0.16058 | 0.16059 | 0.16057 | 0.16058 | 0.16057 |
| | | Std | 0.00021 | 0.00021 | 0.00020 | 0.00020 | 0.00021 | 0.00024 | 0.00020 | 0.00020 | 0.00021 | 0.00021 | 0.00022 |

Bold values denote the lowest average MSE or values whose Wilcoxon's p-value is above 0.05, i.e., values that are statistically similar

Table 3.2 presents the results concerning the MNIST dataset. IHS obtained the lowest errors using the Contrastive Divergence algorithm over one single layer. BA and AIWPSO obtained statistically similar results using the PCD algorithm over two and three layers, respectively. One can notice that FPA using CD over a single layer also obtained the same average errors as the IHS, although the Wilcoxon signed-rank test does not consider both statistically similar. Moreover, the evolutionary algorithms also obtained good results, though not statistically similar as well.

Regarding Semeion Handwritten Digit dataset, Table 3.3 demonstrates the best results were obtained using CoBiDe technique over the CD algorithm with one layer. Worth pointing that none of the other methods achieved similar statistical results, which confirms the robustness of evolutionary-based meta-heuristic optimization algorithms.

Similar to MNIST dataset, the best results over CalTech 101 Silhouettes dataset was obtained using the IHS method with the CD algorithm over a single-layered DBN, as presented in Table 3.4. IHS was also the sole technique to achieve the lowest errors since none of the other methods obtained statistically similar results.

### 3.5.1 Training Evaluation

Figure 3.6 depicts the learning steps considering MNIST dataset. Except for the BA algorithm (and the random search), all techniques converged equally to the same point since the initial iterations. Notice FA outperformed such results, achieving the lowest error at iteration number 20. However, the training error regresses to the initial values, which suggests the problem presents a local optimum hard to be overpassed, given the set of optimized parameters.

An interesting behavior is depicted in Fig. 3.7. One can observe AIWPSO converges faster than the other techniques obtaining an average MSE of 0.2 after ten iterations. However, AIWPSO gets stuck at this time step and is outperformed by both JADE and DE after approximately 15 iterations. Moreover, DE still improves its performance until reaching its optimum at nearly 40 iterations. The behavior is not observed over the testing set, where although DE obtained good results, CoBiDE was the most accurate technique.

Regarding the Caltech 101 Silhouettes, the learning curve depicted in Fig. 3.8 showed that AIWPSO presented a similar behavior as presented over Semeion dataset, and a faster convergence in the 15 initial iterations, being outperformed by JADE afterward. Notice that IHS and FPA also demonstrated a good convergence, which is expected since IHS obtained the best results over the testing set and FPA achieved very close results. Additionally, CoBiDE and BSA are also among the best techniques together with JADE and DE, confirming the robustness of evolution techniques to the task of DBN meta-parameter fine-tuning.

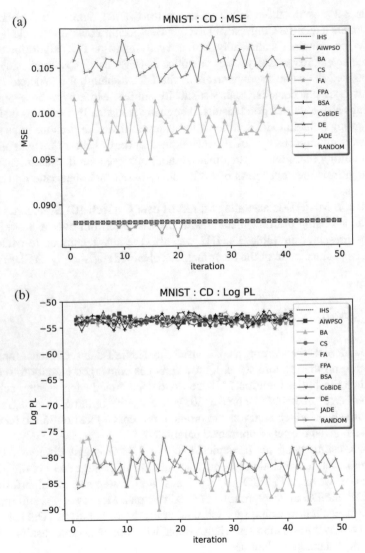

**Fig. 3.6** Training convergence (**a**) MSE and (**b**) log pseudo-likelihood using the CD algorithm and a single layer of hidden units over the MNIST dataset

### 3.5.2 Time Analysis

Tables 3.5, 3.6, and 3.7 present the computational burden, in hours, regarding MNIST, Semeion Handwritten Digit, and Caltech 101 Silhouettes datasets, respectively. One can observe that CS is the fastest technique, followed by IHS. Such a result is expected since IHS evaluates a single solution per iteration, and CS employs a probability of evaluating or not each solution. On the other hand, the

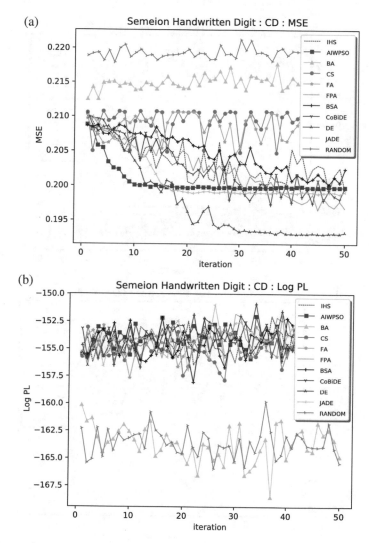

**Fig. 3.7** Training convergence (**a**) MSE and (**b**) log pseudo-likelihood using the CD algorithm and a single layer of hidden units over the Semeion Hand Written Digit dataset

remaining techniques evaluate every solution for each iteration, contributing to a higher computational burden.

Additionally, evolutionary algorithms, in general, present a higher computation burden than swarm-based approaches. AIWPSO stands for an exception, offering itself as the most costly technique among all the others, due to its updating mechanism.

In most cases, the best results were obtained using a single layer as well as the CD algorithm. Such behavior is probably related to the limited number of epochs

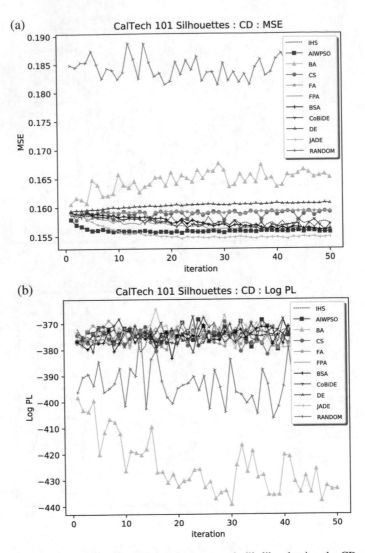

**Fig. 3.8** Training convergence (**a**) MSE and (**b**) log pseudo-likelihood using the CD algorithm and a single layer of hidden units over the CalTech 101 Silhouettes dataset

employed for training, i.e., more complex models composed of a more significant amount of layers would require a higher number of epochs for convergence than the 10 epochs employed in this work. However, running the experiments over such conditions is not plausible in this context due to the massive amount of executions performed for the comparisons presented in the chapter. The same is valid for the PCD algorithm.

**Table 3.5** Average computational burden (in hours) considering MNIST dataset

| Layer | Alg | IHS | AIWPSO | BA | CS | FA | FPA | BSA | CoBiDE | DE | JADE |
|---|---|---|---|---|---|---|---|---|---|---|---|
| 1 | CD | 0.18229 | 1.15104 | 0.53797 | **0.15625** | 0.39062 | 0.58773 | 0.58982 | 0.65554 | 0.92451 | 0.52083 |
|   | PCD | 0.13020 | 1.18750 | 0.54166 | 0.23437 | 0.77604 | 0.66907 | 0.59709 | 0.67061 | 0.95791 | 0.84895 |
| 2 | CD | 0.31250 | 1.76562 | 0.57388 | 0.25520 | 0.71354 | 0.78310 | 0.58333 | 0.80579 | 1.09327 | 1.00520 |
|   | PCD | 0.27083 | 1.39583 | 0.78111 | **0.23437** | 0.67708 | 0.80880 | 0.36979 | 0.83888 | 1.28069 | 0.91666 |
| 3 | CD | 0.28125 | 2.24479 | 1.30138 | 0.24479 | 1.16145 | 1.03157 | 0.77083 | 1.14248 | 1.40536 | 0.69791 |
|   | PCD | 0.29166 | 2.46354 | 1.53472 | **0.15104** | 1.15625 | 0.85460 | 0.74479 | 1.09691 | 1.80823 | 0.88020 |

Bold values denote the lowest average MSE or values whose Wilcoxon's p-value is above 0.05, i.e., values that are statistically similar

**Table 3.6** Average computational burden (in hours) considering Semeion Handwritten Digit dataset

| Layer | Alg | IHS | AIWPSO | BA | CS | FA | FPA | BSA | CoBiDE | DE | JADE |
|---|---|---|---|---|---|---|---|---|---|---|---|
| 1 | CD | **0.08421** | 0.59375 | 0.44277 | 0.13756 | 0.25925 | 0.40037 | 0.31366 | 0.41404 | 0.70628 | 0.28571 |
|   | PCD | 0.10052 | 0.52910 | 0.26055 | 0.09523 | 0.39153 | 0.39609 | 0.32504 | 0.38522 | 0.52661 | 0.11640 |
| 2 | CD | 0.12169 | 0.85185 | 0.58083 | 0.13756 | 0.32804 | 0.30717 | 0.28571 | 0.41388 | 0.53883 | 0.13227 |
|   | PCD | **0.10416** | 0.74603 | 0.65111 | 0.12698 | 0.51851 | 0.41691 | 0.23280 | 0.40367 | 0.45277 | 0.42328 |
| 3 | CD | 0.14814 | 1.13756 | 0.73333 | 0.14834 | 0.94382 | 0.44263 | 0.43915 | 0.46728 | 0.57517 | 0.19576 |
|   | PCD | 0.14583 | 0.95238 | 0.80250 | **0.12169** | 0.60317 | 0.37777 | 0.48677 | 0.47605 | 0.69325 | 0.68253 |

Bold values denote the lowest average MSE or values whose Wilcoxon's p-value is above 0.05, i.e., values that are statistically similar

**Table 3.7** Average computational burden (in hours) considering CalTech 101 Silhouettes dataset

| Layer | Alg | IHS | AIWPSO | BA | CS | FA | FPA | BSA | CoBiDE | DE | JADE |
|---|---|---|---|---|---|---|---|---|---|---|---|
| 1 | CD | 0.93181 | 5.03977 | 2.16003 | 0.88068 | 1.92045 | 3.47451 | 3.75112 | 3.34347 | 6.12850 | 4.68181 |
| | PCD | 0.83522 | 5.36363 | 2.37611 | **0.57386** | 2.99431 | 3.63909 | 3.13693 | 3.20528 | 6.12850 | 2.44886 |
| 2 | CD | 0.72727 | 5.34659 | 3.41027 | **0.52840** | 3.55113 | 2.68115 | 3.43181 | 3.12311 | 5.45633 | 4.38068 |
| | PCD | 0.77840 | 5.59832 | 2.12305 | 1.42613 | 1.85795 | 2.95302 | 3.18181 | 3.14890 | 5.45633 | 2.32954 |
| 3 | CD | 0.64204 | 6.34659 | 4.78611 | 0.82954 | 2.03409 | 2.73834 | 2.40340 | 3.19263 | 5.90374 | 2.59659 |
| | PCD | **0.60227** | 4.51704 | 3.98444 | 0.76704 | 4.59090 | 3.03666 | 2.57386 | 3.34851 | 5.90374 | 5.35795 |

Bold values denote the lowest average MSE or values whose Wilcoxon's p-value is above 0.05, i.e., values that are statistically similar

### 3.5.3  Hyper-Parameters Analysis

This section provides a complete list of the average values of hyper-parameters obtained during the execution of every possible experimental configuration. Notice that values in bold stand for the configuration that obtained the best results accordingly to the Wilcoxon signed-rank test.

Table 3.8 presents the average hyper-parameter values considering the MNIST dataset. The similarity between both IHS and FPA considering a single layer over the Contrastive Divergence algorithm is evident, which is expected since both obtained similar results. However, comparing these results with the ones obtained with a higher number of layers, i.e., BA with 2 layers and AIWPSO with 3 layers, over the PCD algorithm denotes a harder task, since the number of hyper-parameters is also higher, and each one exerts a degree of influence over the others.

Regarding the Semeion Handwritten Digit dataset, presented in Table 3.9, one can once again identify some relation between the set of hyper-parameters and the final results. Although IHS did not obtain the best results over the CD algorithm with a single layer, its results are pretty close to the best obtained using the CoBiDE algorithm. The resemblance is reflected in their hyper-parameter sets. Another example of this resemblance is observed in the 1-layered FPA and BSA over the CD algorithm: a close set of hyper-parameters leads to close results in the experiments.

An analogous behavior is observed in Table 3.10 regarding Caltech 101 Silhouettes dataset. Although FPA, BSA, and CoBiDE did not obtain statistically similar results to IHS according to the Wilcoxon signed-rank test, their results are very much alike, which is perceptible in their selected sets of hyper-parameter. Regarding more complex models, i.e., with two and three layers, one can still observe some likeness. Notice, for instance, the similarity between AIWPSO trained with both CD and PCD, and BSA trained with CD over three layers. However, since they require a larger number of hyper-parameters to be fine-tuned, the combination is exponentially larger, thus providing more diverse combination sets.

## 3.6  Conclusions and Future Works

This chapter dealt with the problem of Deep Belief Network's hyper-parameter parameter fine-tuning through meta-heuristic approaches. Experiments were conducted using three architectures, i.e., one (naïve RBM), two, and three layers, which were trained using both the Contrastive Divergence and the Persistent Contrastive Divergence algorithms. Further, the performance of ten techniques, as well as a random search, were compared over three public binary image datasets. Results demonstrated that Improved Harmony Search obtained the best results in two out of three datasets, while CoBiDE obtained the best values regarding Semeion Handwritten Digit dataset, denoting the efficiency of differential evolution-based techniques. Concerning the training steps, in general, AIWPSO converges faster

**Table 3.8** Average hyper-parameter values considering MNIST dataset

| Layer | Alg | L. | Statistics | IHS | AIWPSO | BA | CS | FA | FPA | BSA | CoBiDE | DE | JADE | RANDOM |
|---|---|---|---|---|---|---|---|---|---|---|---|---|---|---|
| 1 | CD | 1 | Hidden units | **24** | 47 | 39 | 53 | 32 | **22** | 29 | 25 | 27 | 30 | 40 |
| | | | Learning rate | **0.40108** | 0.58222 | 0.39295 | 0.49517 | 0.45688 | **0.45357** | 0.53070 | 0.55669 | 0.49457 | 0.52729 | 0.60503 |
| | | | Weight decay | **0.50896** | 0.63658 | 0.50958 | 0.56223 | 0.55476 | **0.46455** | 0.62393 | 0.66229 | 0.62942 | 0.59766 | 0.64491 |
| | | | Momentum | **0.00538** | 0.00468 | 0.00438 | 0.00575 | 0.00412 | **0.00496** | 0.00422 | 0.00458 | 0.00517 | 0.00514 | 0.00471 |
| | PCD | 1 | Hidden units | 31 | 49 | 58 | 54 | 45 | 26 | 30 | 33 | 28 | 35 | 36 |
| | | | Learning rate | 0.58896 | 0.49638 | 0.46768 | 0.56154 | 0.60276 | 0.54862 | 0.59345 | 0.47164 | 0.59364 | 0.63215 | 0.44548 |
| | | | Weight decay | 0.61429 | 0.70234 | 0.68814 | 0.42895 | 0.61434 | 0.57970 | 0.68172 | 0.54690 | 0.56728 | 0.64276 | 0.57213 |
| | | | Momentum | 0.00532 | 0.00497 | 0.00563 | 0.00355 | 0.00570 | 0.00558 | 0.00480 | 0.00469 | 0.00526 | 0.00586 | 0.00575 |
| 2 | CD | 1 | Hidden units | 33 | 38 | 35 | 57 | 39 | 39 | 29 | 23 | 23 | 34 | 47 |
| | | | Learning rate | 0.63873 | 0.48991 | 0.52931 | 0.53659 | 0.51596 | 0.46246 | 0.48493 | 0.40542 | 0.53679 | 0.52998 | 0.56503 |
| | | | Weight decay | 0.65397 | 0.67948 | 0.60555 | 0.61635 | 0.60650 | 0.69145 | 0.72092 | 0.64380 | 0.73774 | 0.73380 | 0.67747 |
| | | | Momentum | 0.00518 | 0.00552 | 0.00511 | 0.00568 | 0.00414 | 0.00475 | 0.00317 | 0.00529 | 0.00542 | 0.00452 | 0.00401 |
| | | 2 | Hidden units | 61 | 55 | 53 | 60 | 62 | 52 | 56 | 54 | 65 | 49 | 55 |
| | | | Learning rate | 0.55104 | 0.55017 | 0.50505 | 0.45789 | 0.59360 | 0.50327 | 0.43345 | 0.46479 | 0.44132 | 0.49408 | 0.42773 |
| | | | Weight decay | 0.52753 | 0.50562 | 0.45094 | 0.51444 | 0.51906 | 0.41652 | 0.51250 | 0.41738 | 0.54725 | 0.42865 | 0.48087 |
| | | | Momentum | 0.00503 | 0.00490 | 0.00464 | 0.00492 | 0.00414 | 0.00412 | 0.00457 | 0.00481 | 0.00597 | 0.00387 | 0.00526 |
| | PCD | 1 | Hidden units | 36 | 40 | **40** | 43 | 47 | 35 | 35 | 25 | 24 | 47 | 43 |
| | | | Learning rate | 0.56116 | 0.59384 | **0.51544** | 0.54896 | 0.60736 | 0.53856 | 0.49305 | 0.55222 | 0.69303 | 0.47322 | 0.48475 |
| | | | Weight decay | 0.74886 | 0.65189 | **0.64484** | 0.49600 | 0.61311 | 0.64874 | 0.63995 | 0.58538 | 0.72703 | 0.66851 | 0.71593 |
| | | | Momentum | 0.00406 | 0.00559 | **0.00516** | 0.00417 | 0.00500 | 0.00477 | 0.00459 | 0.00512 | 0.00406 | 0.00528 | 0.00430 |
| | | 2 | Hidden units | 51 | 58 | **44** | 47 | 59 | 45 | 56 | 53 | 56 | 52 | 44 |
| | | | Learning rate | 0.52316 | 0.55521 | **0.43453** | 0.55172 | 0.55607 | 0.44427 | 0.46964 | 0.49459 | 0.49107 | 0.46367 | 0.60980 |
| | | | Weight decay | 0.48901 | 0.54542 | **0.44236** | 0.44567 | 0.53144 | 0.51670 | 0.51330 | 0.51885 | 0.52582 | 0.48421 | 0.54849 |
| | | | Momentum | 0.00455 | 0.00410 | **0.00375** | 0.00575 | 0.00500 | 0.00490 | 0.00486 | 0.00424 | 0.00611 | 0.00428 | 0.00471 |

(continued)

**Table 3.8** (continued)

| Layer | Alg | L. | Statistics | IHS | AIWPSO | BA | CS | FA | FPA | BSA | CoBiDE | DE | JADE | RANDOM |
|---|---|---|---|---|---|---|---|---|---|---|---|---|---|---|
| 3 | CD | 1 | Hidden units | 29 | 40 | 44 | 51 | 43 | 30 | 30 | 30 | 29 | 38 | 43 |
| | | | Learning rate | 0.33715 | 0.48692 | 0.40539 | 0.39280 | 0.56626 | 0.63316 | 0.43006 | 0.57987 | 0.51192 | 0.56590 | 0.47168 |
| | | | Weight decay | 0.76890 | 0.64490 | 0.59879 | 0.54979 | 0.61017 | 0.59380 | 0.69188 | 0.67870 | 0.70871 | 0.70026 | 0.68841 |
| | | | Momentum | 0.00451 | 0.00573 | 0.00385 | 0.00512 | 0.00592 | 0.00430 | 0.00491 | 0.00464 | 0.00474 | 0.00597 | 0.00494 |
| | | 2 | Hidden units | 54 | 64 | 59 | 65 | 42 | 51 | 52 | 55 | 40 | 57 | 48 |
| | | | Learning rate | 0.62897 | 0.54763 | 0.53962 | 0.50163 | 0.60940 | 0.58996 | 0.52708 | 0.52667 | 0.48958 | 0.51287 | 0.43878 |
| | | | Weight decay | 0.52169 | 0.47783 | 0.50376 | 0.50002 | 0.58921 | 0.59686 | 0.46750 | 0.45819 | 0.58481 | 0.56367 | 0.50658 |
| | | | Momentum | 0.00454 | 0.00517 | 0.00357 | 0.00493 | 0.00592 | 0.00472 | 0.00516 | 0.00479 | 0.00454 | 0.00488 | 0.00480 |
| | | 3 | Hidden units | 47 | 44 | 63 | 54 | 53 | 65 | 59 | 51 | 51 | 56 | 56 |
| | | | Learning rate | 0.56236 | 0.60001 | 0.40769 | 0.56111 | 0.60791 | 0.52334 | 0.50628 | 0.46757 | 0.50156 | 0.44181 | 0.44104 |
| | | | Weight decay | 0.47333 | 0.43488 | 0.47475 | 0.54808 | 0.52987 | 0.57111 | 0.47501 | 0.50394 | 0.47855 | 0.44166 | 0.45783 |
| | | | Momentum | 0.00529 | 0.00450 | 0.00619 | 0.00404 | 0.00592 | 0.00542 | 0.00475 | 0.00483 | 0.00520 | 0.00493 | 0.00503 |
| | PCD | 1 | Hidden units | 24 | **38** | 46 | 55 | 47 | 31 | 25 | 25 | 28 | 38 | 25 |
| | | | Learning rate | 0.51945 | **0.59828** | 0.55540 | 0.48854 | 0.61870 | 0.59200 | 0.52847 | 0.52223 | 0.51159 | 0.59228 | 0.56341 |
| | | | Weight decay | 0.73241 | **0.70840** | 0.61023 | 0.63885 | 0.63205 | 0.72091 | 0.64644 | 0.69680 | 0.74135 | 0.63626 | 0.66017 |
| | | | Momentum | 0.00594 | **0.00535** | 0.00537 | 0.00604 | 0.00636 | 0.00531 | 0.00540 | 0.00621 | 0.00517 | 0.00462 | 0.00654 |
| | | 2 | Hidden units | 53 | **60** | 43 | 42 | 47 | 41 | 54 | 50 | 60 | 45 | 52 |
| | | | Learning rate | 0.45762 | **0.45897** | 0.47631 | 0.46645 | 0.65033 | 0.56477 | 0.48949 | 0.42149 | 0.39934 | 0.51284 | 0.49131 |
| | | | Weight decay | 0.49819 | **0.50317** | 0.52903 | 0.50410 | 0.55894 | 0.49775 | 0.49806 | 0.50538 | 0.52402 | 0.49927 | 0.43492 |
| | | | Momentum | 0.00486 | **0.00520** | 0.00497 | 0.00436 | 0.00636 | 0.00464 | 0.00507 | 0.00548 | 0.00440 | 0.00552 | 0.00603 |
| | | 3 | Hidden units | 45 | **46** | 52 | 56 | 47 | 43 | 65 | 45 | 67 | 60 | 50 |
| | | | Learning rate | 0.42481 | **0.48181** | 0.46640 | 0.50393 | 0.58703 | 0.44242 | 0.47473 | 0.48108 | 0.44318 | 0.53141 | 0.51465 |
| | | | Weight decay | 0.43727 | **0.57752** | 0.51211 | 0.52331 | 0.63729 | 0.46758 | 0.44450 | 0.51499 | 0.48335 | 0.52187 | 0.46255 |
| | | | Momentum | 0.00395 | **0.00417** | 0.00566 | 0.00533 | 0.00636 | 0.00505 | 0.00418 | 0.00417 | 0.00595 | 0.00519 | 0.00500 |

Bold values denote the lowest average MSE or values whose Wilcoxon's p-value is above 0.05, i.e., values that are statistically similar

**Table 3.9** Average hyper-parameter values considering Semeion Handwritten Digit dataset

| Layer | Alg | L. | Statistics | IHS | AIWPSO | BA | CS | FA | FPA | BSA | CoBiDE | DE | JADE | RANDOM |
|---|---|---|---|---|---|---|---|---|---|---|---|---|---|---|
| 1 | CD | 1 | Hidden units | 70 | 36 | 51 | 50 | 45 | 60 | 55 | **69** | 73 | 55 | 55 |
| | | | Learning rate | 0.48848 | 0.42688 | 0.48042 | 0.47405 | 0.39179 | 0.41753 | 0.45060 | **0.39458** | 0.45125 | 0.45550 | 0.46821 |
| | | | Weight decay | 0.10000 | 0.30207 | 0.39280 | 0.25304 | 0.45168 | 0.13612 | 0.13876 | **0.10128** | 0.16786 | 0.32846 | 0.10972 |
| | | | Momentum | 0.00455 | 0.00357 | 0.00401 | 0.00499 | 0.00274 | 0.00455 | 0.00424 | **0.00409** | 0.00444 | 0.00518 | 0.00606 |
| | PCD | 1 | Hidden units | 38 | 42 | 45 | 49 | 34 | 53 | 48 | 49 | 35 | 43 | 49 |
| | | | Learning rate | 0.44052 | 0.60338 | 0.48166 | 0.48278 | 0.44034 | 0.61623 | 0.47053 | 0.42232 | 0.41345 | 0.51677 | 0.44187 |
| | | | Weight decay | 0.10010 | 0.18887 | 0.33532 | 0.20518 | 0.47422 | 0.10000 | 0.14014 | 0.10291 | 0.14620 | 0.29313 | 0.11509 |
| | | | Momentum | 0.00439 | 0.00430 | 0.00446 | 0.00449 | 0.00491 | 0.00435 | 0.00505 | 0.00517 | 0.00538 | 0.00471 | 0.00577 |
| 2 | CD | 1 | Hidden units | 16 | 24 | 43 | 33 | 40 | 23 | 18 | 27 | 16 | 27 | 29 |
| | | | Learning rate | 0.64269 | 0.50740 | 0.63046 | 0.48661 | 0.60856 | 0.51439 | 0.63148 | 0.60708 | 0.67221 | 0.51436 | 0.54418 |
| | | | Weight decay | 0.77736 | 0.73480 | 0.70937 | 0.62454 | 0.66910 | 0.71700 | 0.64489 | 0.77167 | 0.78755 | 0.76895 | 0.71001 |
| | | | Momentum | 0.00558 | 0.00509 | 0.00446 | 0.00395 | 0.00647 | 0.00346 | 0.00714 | 0.00491 | 0.00401 | 0.00445 | 0.00555 |
| | | 2 | Hidden units | 49 | 62 | 52 | 41 | 51 | 51 | 48 | 50 | 55 | 45 | 57 |
| | | | Learning rate | 0.47364 | 0.48705 | 0.51975 | 0.56356 | 0.59650 | 0.52145 | 0.51645 | 0.46144 | 0.46614 | 0.59406 | 0.50009 |
| | | | Weight decay | 0.51644 | 0.45940 | 0.51385 | 0.51220 | 0.56529 | 0.47774 | 0.56552 | 0.38143 | 0.35235 | 0.54786 | 0.49230 |
| | | | Momentum | 0.00435 | 0.00559 | 0.00488 | 0.00487 | 0.00666 | 0.00460 | 0.00519 | 0.00420 | 0.00528 | 0.00508 | 0.00409 |
| | PCD | 1 | Hidden units | 30 | 38 | 36 | 46 | 47 | 46 | 26 | 24 | 16 | 28 | 21 |
| | | | Learning rate | 0.51145 | 0.52399 | 0.48580 | 0.61589 | 0.53791 | 0.57940 | 0.54080 | 0.58485 | 0.55226 | 0.55490 | 0.50456 |
| | | | Weight decay | 0.76035 | 0.73038 | 0.66895 | 0.69923 | 0.66610 | 0.76146 | 0.70819 | 0.74702 | 0.70493 | 0.77949 | 0.64618 |
| | | | Momentum | 0.00506 | 0.00546 | 0.00570 | 0.00565 | 0.00417 | 0.00460 | 0.00535 | 0.00354 | 0.00458 | 0.00463 | 0.00516 |
| | | 2 | Hidden units | 49 | 59 | 39 | 50 | 49 | 44 | 46 | 59 | 37 | 55 | 53 |
| | | | Learning rate | 0.40693 | 0.54979 | 0.45286 | 0.51862 | 0.61284 | 0.54677 | 0.53121 | 0.51762 | 0.62388 | 0.67465 | 0.47355 |
| | | | Weight decay | 0.52690 | 0.49358 | 0.52551 | 0.48621 | 0.54843 | 0.46499 | 0.53908 | 0.44290 | 0.44684 | 0.52841 | 0.56107 |
| | | | Momentum | 0.00617 | 0.00656 | 0.00429 | 0.00598 | 0.00417 | 0.00411 | 0.00551 | 0.00576 | 0.00577 | 0.00533 | 0.00530 |

(continued)

**Table 3.9** (continued)

| Layer | Alg | L. | Statistics | IHS | AIWPSO | BA | CS | FA | FPA | BSA | CoBiDE | DE | JADE | RANDOM |
|---|---|---|---|---|---|---|---|---|---|---|---|---|---|---|
| 3 | CD | 1 | Hidden units | 25 | 31 | 35 | 40 | 36 | 21 | 15 | 21 | 13 | 23 | 19 |
| | | | Learning rate | 0.45623 | 0.47945 | 0.44844 | 0.49554 | 0.65543 | 0.58816 | 0.46255 | 0.49091 | 0.67924 | 0.43849 | 0.47585 |
| | | | Weight decay | 0.83627 | 0.71301 | 0.70618 | 0.67923 | 0.72174 | 0.68005 | 0.65421 | 0.70858 | 0.77792 | 0.70635 | 0.73262 |
| | | | Momentum | 0.00534 | 0.00587 | 0.00421 | 0.00549 | 0.00650 | 0.00438 | 0.00502 | 0.00411 | 0.00471 | 0.00575 | 0.00480 |
| | | 2 | Hidden units | 48 | 66 | 57 | 57 | 57 | 55 | 56 | 53 | 46 | 65 | 43 |
| | | | Learning rate | 0.49218 | 0.50469 | 0.51419 | 0.47704 | 0.61155 | 0.51277 | 0.44383 | 0.47918 | 0.40487 | 0.58144 | 0.49837 |
| | | | Weight decay | 0.56203 | 0.64634 | 0.56889 | 0.50804 | 0.61125 | 0.60213 | 0.53538 | 0.42734 | 0.46880 | 0.48684 | 0.51269 |
| | | | Momentum | 0.00515 | 0.00493 | 0.00434 | 0.00419 | 0.00650 | 0.00340 | 0.00399 | 0.00607 | 0.00549 | 0.00560 | 0.00537 |
| | | 3 | Hidden units | 49 | 47 | 47 | 50 | 63 | 54 | 53 | 47 | 60 | 48 | 52 |
| | | | Learning rate | 0.44241 | 0.47279 | 0.46076 | 0.48646 | 0.63856 | 0.45881 | 0.44885 | 0.42203 | 0.44067 | 0.52481 | 0.51374 |
| | | | Weight decay | 0.58616 | 0.41996 | 0.51011 | 0.48673 | 0.65420 | 0.42822 | 0.53627 | 0.48018 | 0.52056 | 0.55745 | 0.57896 |
| | | | Momentum | 0.00420 | 0.00641 | 0.00507 | 0.00636 | 0.00650 | 0.00510 | 0.00398 | 0.00436 | 0.00383 | 0.00396 | 0.00475 |
| | PCD | 1 | Hidden units | 25 | 35 | 34 | 32 | 35 | 13 | 29 | 29 | 13 | 24 | 21 |
| | | | Learning rate | 0.54736 | 0.50874 | 0.44835 | 0.46830 | 0.58631 | 0.61603 | 0.63401 | 0.55630 | 0.55689 | 0.53534 | 0.52184 |
| | | | Weight decay | 0.78853 | 0.75428 | 0.68738 | 0.65386 | 0.70350 | 0.72065 | 0.68351 | 0.76314 | 0.76541 | 0.71391 | 0.66489 |
| | | | Momentum | 0.00391 | 0.00437 | 0.00521 | 0.00560 | 0.00406 | 0.00551 | 0.00545 | 0.00458 | 0.00342 | 0.00440 | 0.00514 |
| | | 2 | Hidden units | 52 | 39 | 53 | 45 | 53 | 51 | 60 | 57 | 55 | 45 | 58 |
| | | | Learning rate | 0.54361 | 0.44707 | 0.51948 | 0.52628 | 0.63192 | 0.49611 | 0.45516 | 0.44575 | 0.51724 | 0.44793 | 0.52341 |
| | | | Weight decay | 0.41558 | 0.50275 | 0.46625 | 0.47226 | 0.59228 | 0.52341 | 0.46947 | 0.53055 | 0.38021 | 0.46047 | 0.57222 |
| | | | Momentum | 0.00463 | 0.00524 | 0.00428 | 0.00471 | 0.00406 | 0.00451 | 0.00453 | 0.00533 | 0.00473 | 0.00380 | 0.00560 |
| | | 3 | Hidden units | 52 | 53 | 55 | 60 | 50 | 51 | 53 | 50 | 43 | 47 | 61 |
| | | | Learning rate | 0.39714 | 0.56683 | 0.58973 | 0.55100 | 0.56347 | 0.50745 | 0.46446 | 0.53937 | 0.44090 | 0.64079 | 0.55210 |
| | | | Weight decay | 0.49576 | 0.54001 | 0.42108 | 0.44537 | 0.63785 | 0.51183 | 0.51329 | 0.58296 | 0.49074 | 0.57689 | 0.47675 |
| | | | Momentum | 0.00502 | 0.00462 | 0.00547 | 0.00427 | 0.00406 | 0.00498 | 0.00474 | 0.00450 | 0.00489 | 0.00481 | 0.00305 |

Bold values denote the lowest average MSE or values whose Wilcoxon's p-value is above 0.05, i.e., values that are statistically similar

**Table 3.10** Average hyper-parameter values considering CalTech 101 Silhouettes dataset

| Layer | Alg | L. | Statistics | IHS | AIWPSO | BA | CS | FA | FPA | BSA | CoBiDE | DE | JADE | RANDOM |
|---|---|---|---|---|---|---|---|---|---|---|---|---|---|---|
| 1 | CD | 1 | Hidden units | **79** | 63 | 63 | 62 | 43 | 80 | 79 | 77 | 74 | 85 | 65 |
| | | | Learning rate | **0.42075** | 0.50650 | 0.53685 | 0.52474 | 0.33131 | 0.50059 | 0.32520 | 0.36895 | 0.44642 | 0.44917 | 0.38533 |
| | | | Weight decay | **0.10000** | 0.14979 | 0.28279 | 0.31002 | 0.36294 | 0.16180 | 0.11270 | 0.11124 | 0.14447 | 0.10524 | 0.11622 |
| | | | Momentum | **0.00395** | 0.00575 | 0.00579 | 0.00480 | 0.00301 | 0.00448 | 0.00581 | 0.00595 | 0.00445 | 0.00615 | 0.00439 |
| | PCD | 1 | Hidden units | 70 | 54 | 52 | 35 | 49 | 72 | 63 | 71 | 74 | 65 | 62 |
| | | | Learning rate | 0.48994 | 0.51606 | 0.53898 | 0.54823 | 0.30069 | 0.43238 | 0.39707 | 0.49553 | 0.44642 | 0.46906 | 0.43330 |
| | | | Weight decay | 0.10063 | 0.18316 | 0.36682 | 0.27416 | 0.18867 | 0.10048 | 0.11245 | 0.10605 | 0.14447 | 0.10451 | 0.11558 |
| | | | Momentum | 0.00556 | 0.00484 | 0.00517 | 0.00496 | 0.00449 | 0.00445 | 0.00425 | 0.00566 | 0.00445 | 0.00411 | 0.00547 |
| 2 | CD | 1 | Hidden units | 52 | 52 | 38 | 46 | 53 | 50 | 39 | 37 | 32 | 57 | 59 |
| | | | Learning rate | 0.43484 | 0.42893 | 0.48042 | 0.47495 | 0.55443 | 0.47328 | 0.49388 | 0.52195 | 0.51051 | 0.53284 | 0.57386 |
| | | | Weight decay | 0.68476 | 0.73549 | 0.62374 | 0.64525 | 0.63411 | 0.67390 | 0.72068 | 0.68482 | 0.65160 | 0.64469 | 0.71296 |
| | | | Momentum | 0.00515 | 0.00535 | 0.00552 | 0.00368 | 0.00600 | 0.00390 | 0.00425 | 0.00344 | 0.00533 | 0.00444 | 0.00454 |
| | | 2 | Hidden units | 47 | 53 | 50 | 41 | 59 | 48 | 43 | 59 | 67 | 50 | 50 |
| | | | Learning rate | 0.51389 | 0.47241 | 0.60258 | 0.48250 | 0.56618 | 0.39817 | 0.57398 | 0.55238 | 0.53212 | 0.50475 | 0.55511 |
| | | | Weight decay | 0.40143 | 0.46011 | 0.56419 | 0.51764 | 0.58570 | 0.50445 | 0.55294 | 0.42686 | 0.47895 | 0.48927 | 0.54928 |
| | | | Momentum | 0.00365 | 0.00453 | 0.00564 | 0.00387 | 0.00600 | 0.00450 | 0.00590 | 0.00431 | 0.00544 | 0.00540 | 0.00597 |
| | PCD | 1 | Hidden units | 45 | 64 | 37 | 48 | 51 | 66 | 38 | 50 | 32 | 64 | 38 |
| | | | Learning rate | 0.55227 | 0.51450 | 0.48681 | 0.46714 | 0.47946 | 0.37420 | 0.49990 | 0.50915 | 0.51051 | 0.52568 | 0.47179 |
| | | | Weight decay | 0.58967 | 0.64592 | 0.60823 | 0.54185 | 0.59370 | 0.64974 | 0.66568 | 0.67036 | 0.65160 | 0.68887 | 0.55182 |
| | | | Momentum | 0.00425 | 0.00459 | 0.00552 | 0.00565 | 0.00554 | 0.00601 | 0.00457 | 0.00371 | 0.00533 | 0.00363 | 0.00481 |
| | | 2 | Hidden units | 64 | 62 | 54 | 51 | 39 | 52 | 45 | 51 | 67 | 47 | 54 |
| | | | Learning rate | 0.59710 | 0.63044 | 0.50150 | 0.46488 | 0.52302 | 0.53368 | 0.52452 | 0.42603 | 0.53212 | 0.53134 | 0.47991 |
| | | | Weight decay | 0.44184 | 0.44811 | 0.49725 | 0.53918 | 0.45025 | 0.48501 | 0.47066 | 0.51208 | 0.47895 | 0.50288 | 0.42359 |
| | | | Momentum | 0.00578 | 0.00584 | 0.00575 | 0.00512 | 0.00554 | 0.00428 | 0.00399 | 0.00565 | 0.00544 | 0.00572 | 0.00495 |

| | | | 43 | 51 | 51 | 34 | 45 | 48 | 31 | 39 | 59 | 37 | 50 |
|---|---|---|---|---|---|---|---|---|---|---|---|---|---|
| **CD** | 1 | Hidden units | 43 | 51 | 51 | 34 | 45 | 48 | 31 | 39 | 59 | 37 | 50 |
| | | Learning rate | 0.52239 | 0.51207 | 0.52097 | 0.48690 | 0.63751 | 0.52362 | 0.51773 | 0.42346 | 0.38359 | 0.47503 | 0.56581 |
| | | Weight decay | 0.71342 | 0.72279 | 0.61374 | 0.62583 | 0.68437 | 0.67406 | 0.69570 | 0.68193 | 0.67821 | 0.71961 | 0.73120 |
| | | Momentum | 0.00431 | 0.00514 | 0.00497 | 0.00553 | 0.00428 | 0.00600 | 0.00446 | 0.00391 | 0.00512 | 0.00546 | 0.00577 |
| | 2 | Hidden units | 49 | 54 | 52 | 41 | 43 | 47 | 49 | 49 | 61 | 49 | 44 |
| | | Learning rate | 0.52924 | 0.51316 | 0.53610 | 0.42102 | 0.49939 | 0.52684 | 0.50960 | 0.57533 | 0.57925 | 0.52413 | 0.54938 |
| | | Weight decay | 0.61271 | 0.48068 | 0.44030 | 0.44930 | 0.57062 | 0.53085 | 0.40449 | 0.46727 | 0.56396 | 0.40061 | 0.44172 |
| | | Momentum | 0.00442 | 0.00495 | 0.00597 | 0.00473 | 0.00428 | 0.00505 | 0.00604 | 0.00419 | 0.00422 | 0.00506 | 0.00605 |
| | 3 | Hidden units | 55 | 52 | 55 | 63 | 46 | 67 | 50 | 59 | 57 | 51 | 55 |
| | | Learning rate | 0.51658 | 0.48736 | 0.45940 | 0.53802 | 0.55106 | 0.53752 | 0.55456 | 0.51279 | 0.55418 | 0.55914 | 0.50365 |
| | | Weight decay | 0.39922 | 0.52719 | 0.58714 | 0.45855 | 0.58377 | 0.58716 | 0.54719 | 0.51086 | 0.42597 | 0.54949 | 0.49739 |
| | | Momentum | 0.00594 | 0.00399 | 0.00633 | 0.00518 | 0.00428 | 0.00457 | 0.00588 | 0.00396 | 0.00575 | 0.00387 | 0.00523 |
| **PCD** | 1 | Hidden units | 56 | 53 | 49 | 51 | 49 | 58 | 51 | 37 | 59 | 48 | 49 |
| | | Learning rate | 0.46399 | 0.40623 | 0.37432 | 0.50307 | 0.48364 | 0.59600 | 0.53473 | 0.44439 | 0.38359 | 0.43948 | 0.49811 |
| | | Weight decay | 0.69432 | 0.63476 | 0.59465 | 0.51970 | 0.59724 | 0.63553 | 0.61455 | 0.60310 | 0.67821 | 0.63087 | 0.65155 |
| | | Momentum | 0.00341 | 0.00550 | 0.00477 | 0.00531 | 0.00451 | 0.00472 | 0.00516 | 0.00504 | 0.00512 | 0.00408 | 0.00444 |
| | 2 | Hidden units | 56 | 51 | 51 | 52 | 59 | 40 | 52 | 49 | 61 | 40 | 51 |
| | | Learning rate | 0.51054 | 0.42227 | 0.52618 | 0.55580 | 0.50243 | 0.42248 | 0.57549 | 0.55893 | 0.57925 | 0.56304 | 0.52403 |
| | | Weight decay | 0.55409 | 0.42987 | 0.47076 | 0.46793 | 0.42596 | 0.43440 | 0.47748 | 0.49206 | 0.56396 | 0.54592 | 0.51342 |
| | | Momentum | 0.00586 | 0.00452 | 0.00493 | 0.00378 | 0.00451 | 0.00623 | 0.00508 | 0.00499 | 0.00422 | 0.00492 | 0.00446 |
| | 3 | Hidden units | 59 | 49 | 52 | 42 | 42 | 55 | 50 | 49 | 57 | 44 | 50 |
| | | Learning rate | 0.49215 | 0.53762 | 0.52214 | 0.59451 | 0.45669 | 0.49241 | 0.58855 | 0.51762 | 0.55418 | 0.48230 | 0.58397 |
| | | Weight decay | 0.53427 | 0.55750 | 0.46411 | 0.45421 | 0.51050 | 0.51246 | 0.47173 | 0.51046 | 0.42597 | 0.52163 | 0.44847 |
| | | Momentum | 0.00522 | 0.00467 | 0.00540 | 0.00457 | 0.00451 | 0.00454 | 0.00527 | 0.00536 | 0.00575 | 0.00434 | 0.00619 |

Bold values denote the lowest average MSE or values whose Wilcoxon's p-value is above 0.05, i.e., values that are statistically similar

than the other methods on the initial iterations. However, it is outperformed by evolution techniques after approximately 15 iterations. Finally, one can also verify that CS is the fastest technique, followed by IHS. On the other hand, AIWPSO is the slowest one.

Regarding future works, we intend to compare meta-heuristic approaches to fine-tuning DBNs to the task of classification.

**Acknowledgments** This study was financed in part by the Coordenação de Aperfeiçoamento de Pessoal de Nível Superior—Brasil (CAPES)—Finance Code 001. The authors also appreciate FAPESP grants #2013/07375-0, #2014/12236-1, #2016/19403-6, #2017/02286-0, #2017/25908-6, #2018/21934-5 and #2019/02205-5, and CNPq grants 307066/2017-7 and 427968/2018-6.

# References

1. Carreira-Perpiñán, M.A., Hinton, G.E.: On contrastive divergence learning. In: Cowell, R.G., Ghahramani, Z. (eds.) Proceedings of the Tenth International Workshop on Artificial Intelligence and Statistics. Society for Artificial Intelligence and Statistics, pp. 33–40 (2005)
2. Civicioglu, P.: Backtracking search optimization algorithm for numerical optimization problems. Appl. Math. Comput. **219**(15), 8121–8144 (2013)
3. Hinton, G.E.: Training products of experts by minimizing contrastive divergence. Neural Comput. **14**(8), 1771–1800 (2002)
4. Hinton, G.E.: A practical guide to training restricted Boltzmann machines. In: Montavon, G., Orr, G., Müller, K.R. (eds.) Neural Networks: Tricks of the Trade. Lecture Notes in Computer Science, vol. 7700, pp. 599–619. Springer, Berlin (2012)
5. Hinton, G.E., Osindero, S., Teh, Y.W.: A fast learning algorithm for deep belief nets. Neural Comput. **18**(7), 1527–1554 (2006)
6. Kennedy, J., Eberhart, R.C.: Swarm Intelligence. Morgan Kaufmann Publishers Inc., San Francisco (2001)
7. Kuremoto, T., Kimura, S., Kobayashi, K., Obayashi, M.: Time series forecasting using restricted Boltzmann machine. In: International Conference on Intelligent Computing, pp. 17–22. Springer, Berlin (2012)
8. LeCun, Y., Bottou, L., Bengio, Y., Haffner, P.: Gradient-based learning applied to document recognition. Proc. IEEE **86**(11), 2278–2324 (1998)
9. Levy, E., David, O.E., Netanyahu, N.S.: Genetic algorithms and deep learning for automatic painter classification. In: Proceedings of the 2014 Annual Conference on Genetic and Evolutionary Computation, pp. 1143–1150. ACM, New York (2014)
10. Liu, K., Zhang, L.M., Sun, Y.W.: Deep Boltzmann machines aided design based on genetic algorithms. In: Applied Mechanics and Materials, vol. 568, pp. 848–851. Trans Tech, Clausthal (2014)
11. Mahdavi, M., Fesanghary, M., Damangir, E.: An improved harmony search algorithm for solving optimization problems. Appl. Math. Comput. **188**(2), 1567–1579 (2007)
12. Nickabadi, A., Ebadzadeh, M.M., Safabakhsh, R.: A novel particle swarm optimization algorithm with adaptive inertia weight. Appl. Soft Comput. **11**, 3658–3670 (2011)
13. Passos, L.A., Papa, J.P.: Fine-tuning infinity restricted Boltzmann machines. In: 2017 30th SIBGRAPI Conference on Graphics, Patterns and Images (SIBGRAPI), pp. 63–70. IEEE, New York (2017)
14. Passos, L.A., Papa, J.P.: On the training algorithms for restricted Boltzmann machine-based models. Ph.D. thesis, Universidade Federal de São Carlos (2018)

15. Passos, L.A., Papa, J.P.: A metaheuristic-driven approach to fine-tune deep Boltzmann machines. Appl. Soft Comput., 105717 (2019, in press). https://www.sciencedirect.com/science/article/abs/pii/S1568494619304983
16. Passos, L.A., Rodrigues, D.R., Papa, J.P.: Fine tuning deep Boltzmann machines through meta-heuristic approaches. In: 2018 IEEE 12th International Symposium on Applied Computational Intelligence and Informatics (SACI), pp. 000,419–000,424. IEEE, New York (2018)
17. Passos, L.A., Rodrigues, D., Papa, J.P.: Quaternion-based backtracking search optimization algorithm. In: 2019 IEEE Congress on Evolutionary Computation. IEEE, New York (2019)
18. Passos, L.A., de Souza Jr, L.A., Mendel, R., Ebigbo, A., Probst, A., Messmann, H., Palm, C., Papa, J.P.: Barrett's esophagus analysis using infinity restricted Boltzmann machines. J. Vis. Commun. Image Represent. **59**, 475–485 (2019)
19. Pereira, C.R., Passos, L.A., Rodrigues, D., Nunes, S.A., Papa, J.P.: JADE-based feature selection for non-technical losses detection. In: VII ECCOMAS Thematic Conference on Computational Vision and Medical Image Processing: VipIMAGE 2019 (2019)
20. Rodrigues, D., Pereira, L.A.M., Almeida, T.N.S., Papa, J.P., Souza, A.N., Ramos, C.O., Yang, X.S.: BCS: a binary cuckoo search algorithm for feature selection. In: IEEE International Symposium on Circuits and Systems, pp. 465–468 (2013)
21. Rodrigues, D., de Rosa, G.H., Passos, L.A., Papa, J.P.: Adaptive improved flower pollination algorithm for global optimization. In: Nature-Inspired Computation in Data Mining and Machine Learning, pp. 1–21. Springer, Berlin (2020)
22. Rosa, G., Papa, J.P., Costa, K., Passos, L.A., Pereira, C., Yang, X.S.: Learning parameters in deep belief networks through firefly algorithm. In: IAPR Workshop on Artificial Neural Networks in Pattern Recognition, pp. 138–149. Springer, Berlin (2016)
23. Salakhutdinov, R., Hinton, G.: Deep Boltzmann machines. In: Artificial Intelligence and Statistics, pp. 448–455 (2009)
24. Smolensky, P.: Parallel distributed processing: explorations in the microstructure of cognition. In: Chap. Information Processing in Dynamical Systems: Foundations of Harmony Theory, pp. 194–281. MIT Press, Cambridge (1986)
25. Storn, R., Price, K.: Differential evolution–a simple and efficient heuristic for global optimization over continuous spaces. J. Glob. Optim. **11**(4), 341–359 (1997)
26. Thornton, C., Hutter, F., Hoos, H.H., Leyton-Brown, K.: Auto-WEKA: combined selection and hyperparameter optimization of classification algorithms. In: Proceedings of the 19th ACM SIGKDD International Conference on Knowledge Discovery and Data Mining, pp. 847–855. ACM, New York (2013)
27. Tieleman, T.: Training restricted Boltzmann machines using approximations to the likelihood gradient. In: Proceedings of the 25th International Conference on Machine Learning (ICML '08), pp. 1064–1071. ACM, New York (2008)
28. Wang, Y., Li, H.X., Huang, T., Li, L.: Differential evolution based on covariance matrix learning and bimodal distribution parameter setting. Appl. Soft Comput. **18**, 232–247 (2014)
29. Whitley, D.: A genetic algorithm tutorial. Stat. Comput. **4**(2), 65–85 (1994)
30. Wilcoxon, F.: Individual comparisons by ranking methods. Biom. Bull. **1**(6), 80–83 (1945)
31. Yang, X.S.: Firefly algorithm, stochastic test functions and design optimisation. Int. J. Bio-Inspired Comput. **2**(2), 78–84 (2010)
32. Yang, X.S., Deb, S.: Cuckoo search via lévy flights. In: Nature and Biologically Inspired Computing (NaBIC 2009). World Congress on, pp. 210–214. IEEE, New York (2009)
33. Yang, X.S., Deb, S.: Engineering optimisation by cuckoo search. Int. J. Math. Model. Numer. Optim. **1**, 330–343 (2010)
34. Yang, X.S., Gandomi, A.H.: Bat algorithm: a novel approach for global engineering optimization. Eng. Comput. **29**(5), 464–483 (2012)

35. Yang, S.S., Karamanoglu, M., He, X.: Flower pollination algorithm: a novel approach for multiobjective optimization. Eng. Optim. **46**(9), 1222–1237 (2014)
36. Yosinski, J., Lipson, H.: Visually debugging restricted Boltzmann machine training with a 3D example. In: Representation Learning Workshop, 29th International Conference on Machine Learning (2012)
37. Zhang, J., Sanderson, A.C.: Jade: adaptive differential evolution with optional external archive. IEEE Trans. Evol. Comput. **13**(5), 945–958 (2009)

# Chapter 4
# Automated Development of DNN Based Spoken Language Systems Using Evolutionary Algorithms

Takahiro Shinozaki, Shinji Watanabe, and Kevin Duh

**Abstract** Spoken language processing is one of the research areas that has contributed significantly to the recent revival in neural network research. For example, speech recognition has been at the forefront of deep learning research, inventing various novel models. Their dramatic performance improvements compared to previous state-of-the-art implementations have resulted in spoken language systems being deployed in a wide range of applications today. However, these systems require intensive tuning of their network designs and the training setups in order to achieve maximal performance. The laborious effort by human experts is becoming a prominent obstacle in system development. In this chapter, we first explain the basic concepts and the neural network-based implementations of spoken language processing systems. Several types of neural network models will be described. We then introduce our effort to automate the tuning of the system meta-parameters using evolutionary algorithms.

## 4.1 Spoken Language Processing Systems

An automatic speech recognition system takes a waveform signal of an utterance and outputs the corresponding text as the recognition result. It functionally corresponds to the human ear. Contrary, a speech synthesis system takes text as input and outputs a waveform signal of the synthesized voice as the output, which corresponds to the human mouth. Depending on the applications, they are used as a stand-alone

T. Shinozaki (✉)
Tokyo Institute of Technology, Yokohama, Kanagawa, Japan
e-mail: shinot@ict.e.titech.ac.jp

S. Watanabe
Johns Hopkins University, Baltimore, MD, USA
e-mail: shinjiw@ieee.org

K. Duh
Johns Hopkins University, Baltimore, MD, USA
e-mail: kevinduh@cs.jhu.edu

© Springer Nature Singapore Pte Ltd. 2020
H. Iba, N. Noman (eds.), *Deep Neural Evolution*, Natural Computing Series,
https://doi.org/10.1007/978-981-15-3685-4_4

application or as a sub-component of other systems such as spoken dialogue systems and spoken translation systems.

### 4.1.1  Principle of Speech Recognition

In speech recognition systems, the input waveform is typically analyzed by short-time Fourier transform by segmenting the waveform with overlapping short windows as shown in Fig. 4.1. The window width and the shift are typically 25 and 10 ms to balance the frequency and time resolutions to capture temporal changes of frequency patterns of sub-phone units. As a result of the analysis, the waveform is converted as a time sequence of fixed-dimensional vectors, where the rate corresponds to the reciprocal of the window shift. The obtained frequency pattern vector may be used as it is, or further analyzed to obtain Mel-frequency cepstral coefficients (MFCCs) [1] or perceptual linear predictive (PLP) [2]. In either the case, the result is a sequence of vectors that contain useful information for speech recognition, where a time position of a vector is referred to as a frame. The process is called *feature extraction*.

Let $O = \langle o_1, o_2, \cdots, o_T \rangle$ be a sequence of acoustic feature vectors of length $T$ extracted from an utterance, and $W = \langle w_1, w_2, \cdots, w_N \rangle$ be a word sequence or a text of length $N$. Speech recognition is formulated as a problem of finding $\hat{W}$ that maximizes the conditional probability $P(W|O)$ as shown in Eq. (4.1), or drawing a sample $\tilde{W}$ from $P(W|O)$ as shown in Eq. (4.2).

$$\hat{W} = \underset{W}{\mathrm{argmax}}\, P(W|O), \tag{4.1}$$

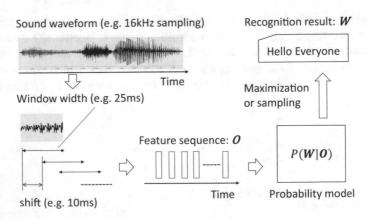

**Fig. 4.1** General framework of speech recognition. A sound signal is first converted to a sequence of feature vectors $O$ by applying a sliding window and frequency analysis, etc. The recognition result is obtained from the posterior distribution $P(W|O)$ of a word sequence $W$ given the feature sequence $O$ by maximization or probability sampling

$$\tilde{W} \sim P\left(W|O\right). \tag{4.2}$$

Finding $\hat{W}$ means outputting the most likely recognition hypothesis, whereas drawing a sample $\tilde{W}$ corresponds to output a hypothesis according to the probability. While the former is more direct to the goal of speech recognition, the latter sometimes have an advantage in the model training.

The conditional probability $P\left(W|O\right)$ may be directly or indirectly modeled. In the latter case, the Bayes' rule is applied as shown in Eq. (4.3).

$$P\left(W|O\right) = \frac{P\left(O|W\right)P\left(W\right)}{P\left(O\right)} \propto P\left(O|W\right)P\left(W\right). \tag{4.3}$$

The models of $P\left(O|W\right)$ and $P\left(W\right)$ are referred to as an *acoustic model* and a *language model*, respectively. The acoustic model describes the generative distribution of the acoustic feature sequence $O$ given the text $W$, whereas the language model describes the distribution of the text $W$. The denominator $P\left(O\right)$ in Eq. (4.3) may be ignored for the maximization or the sampling purposes, since it is a constant in the processes. Hidden Markov model (HMM) has long been used for acoustic modeling in traditional speech recognition systems. The direct modeling of $P\left(W|O\right)$ had been intricate until recently for large vocabulary speech recognition. However, the approach is rapidly developing as end-to-end speech recognition with the progress of deep learning.

### 4.1.2 Hidden Markov Model Based Acoustic Modeling

HMM consists of a finite set of internal states $\{0, 1, \cdots, F\}$, a set of emission distributions $\{P\left(o|s\right)\}$ each of which is associated to a state $s$, and a set of state transition probabilities $\{P\left(s'|s\right)\}$ from a state $s$ to a state $s'$ as shown in Fig. 4.2. The initial state $s = 0$ and the final state $s = F$ represent the beginning and end of the state transitions, and they do not have the emission distribution. HMM gives a model of joint probability of $P\left(O, S\right)$ as shown in Eq. (4.4), where $S = \langle s_0 = 0, s_1, s_2, \cdots, s_T, s_{T+1} = F \rangle$ is a state sequence that starts with the initial state and ends in the final state. By marginalizing over all possible state sequences $S$, the probability of observing the feature sequence is obtained as shown in Eq. (4.5).

$$P_\theta\left(O, S\right) = P_\theta\left(s_{T+1}|s_T\right) \prod_{t=1}^{T} P_\theta\left(s_t|s_{t-1}\right) P_\theta\left(o_t|s_t\right), \tag{4.4}$$

$$P_\theta\left(O\right) = \sum_{S} P_\theta\left(O, S\right), \tag{4.5}$$

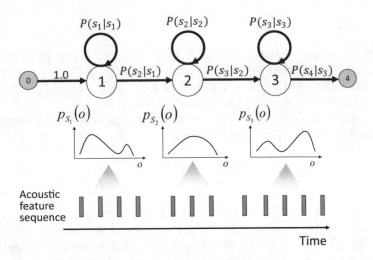

**Fig. 4.2** Hidden Markov model. A circle represents a state, and a directed arc represents a transition. This example HMM has a left-to-right structure with three emission states and an initial and a final states

where $\theta$ indicates a set of all parameters of the emission distributions and the transition probabilities. An acoustic model $P_\theta(O|W)$ is obtained by preparing an HMM for each word sequence $W$ as shown in Eq. (4.6).

$$P_\Theta(O|W) = P_{\theta_W}(O) = \sum_S P_{\theta_W}(O, S), \qquad (4.6)$$

where $\theta_W$ indicates $W$ dependent parameter set, and $\Theta$ is a union of $\theta_W$ for all $W$. Since the number of possible word sequences is exponential to the length of the sequence, separately preparing an HMM for each sequence is intractable both in terms of required memory and parameter estimation from finite training data. Instead, a set of HMMs is prepared to model each phoneme $p$, and an utterance HMM is composed by concatenating the phoneme HMMs according to the pronunciation of the word sequence as shown in Fig. 4.3. The phoneme HMM set is referred to as a mono-phone model.

A limitation of the mono-phone approach is that the same phoneme HMM is used regardless of the surrounding phoneme context in the utterance. Since human voice is generated by modulating the shape of the vocal tract by moving mouth, the change is not instant. Therefore, the spectral pattern of a phoneme is affected by surrounding phonemes. For example, spectral pattern of the same phoneme /ih/ is notably different when it appears in pronunciations of "big" and "bit." Context-dependent phoneme model is used to improve the modeling accuracy, where separate HMMs are prepared for the same phoneme for different preceding and succeeding phoneme contexts. The most popular context-dependent phoneme modeling is tri-phone, where a set of HMMs for a phoneme is prepared for one

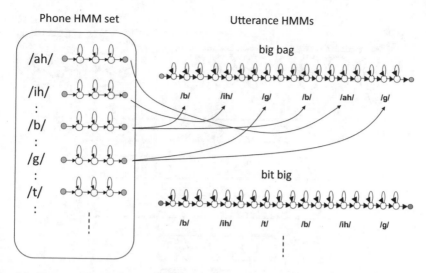

**Fig. 4.3** Phoneme HMM based utterance modeling

preceding and one succeeding phonemes. When the number of phonemes is $N$, the number of tri-phone HMM is $N^3$, which is much larger than $N$ of the monophone model. It causes a problem in the model parameter estimation especially for rare context and phoneme pairs since few or even no samples are available in the training set. To address the problem, clustering is performed for the context-dependent HMM states to control the model complexity by merging the HMM states [3].

The state emission distribution $P(o|s)$ has traditionally been modeled by a mixture of Gaussian distributions (GMM) as shown in Eq. 4.7, where $w_i$ is a mixture weight ($0 < w_i$ and $\sum_i w_i = 1$) and $N(o|\mu_i, \Sigma_i)$ is a Gaussian distribution with mean $\mu_i$ and variance $\Sigma_i$.

$$P(o|s) = \sum_i w_i N(o|\mu_i, \Sigma_i). \tag{4.7}$$

Later, it has been replaced by deep neural networks (DNNs) as shown in Eq. 4.8, where $P(s|o)$ is obtained by the neural network.

$$P(o|s) = \frac{P(s|o)\, p(o)}{P(s)} \propto \frac{P(s|o)}{P(s)}. \tag{4.8}$$

Figure 4.4 shows the whole structure of DNN-HMM. The DNN-HMM often outperforms GMM-HMM with the recognition performance, especially when a larger amount of training data is available. The number of clustered HMM states, neural network structure, and their learning conditions are meta-parameters to be tuned during the system development.

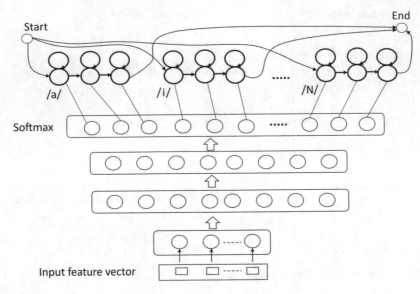

**Fig. 4.4** Example of a DNN-HMM mono-phone model

### 4.1.3 End-to-End Speech Recognition System

For a simple speech recognition task such as vowel recognition from a single feature vector, where $W$ is a set of vowels instead of a variable-length sequence of words and $O$ is a single fixed-dimensional vector rather than the sequence of the vectors, the probability of $P(W|O)$ can be directly modeled by a simple feed-forward neural network with a soft-max output layer as shown in Fig. 4.5. For general cases where $O$ is a feature vector sequence, and $W$ is a word sequence, variable-length input and output need to be handled. Neural networks realize it with some unique architectures such as encoder-decoder network with an attention mechanism [4] and Connectionist Temporal Classification (CTC) [5]. Figure 4.6 shows the architecture of a simple encoder-decoder network without the attention mechanism. It consists of an encoder network and a decoder network. The encoder network accepts a variable-length input and embeds it to a fixed-dimensional vector. The decoder network works by estimating a probability distribution of the next word given the current word $w_t$, from which an output word $w_{t+1}$ is obtained by random sampling following the distribution. Initially, a special word $\langle S \rangle$ that represents the beginning of an utterance is input as $w_0$, and a word $w_1$ is sampled. Then, $w_2$ is obtained using $w_1$ as the input. The process is repeated until a special word $\langle /S \rangle$ is sampled that indicates the end of an utterance. The architecture has generality to handle sequential input and output, and can be used for translation [6] and dialogue systems [7], etc., by simply changing the training data and the input/output representations. In addition, the extended architecture with the attention mechanism can explicitly handle the alignment problem between input and output [8].

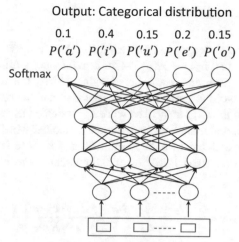

**Fig. 4.5** Frame-wise vowel recognition using a feed-forward neural network. The network directly models $P(W|O)$

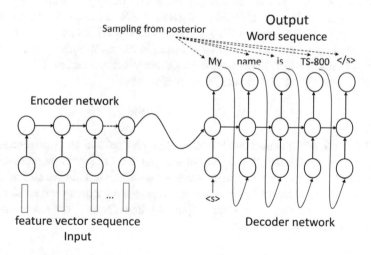

**Fig. 4.6** End-to-end speech recognition system based on a simple encoder-decoder network without an attention mechanism

These systems are referred to as end-to-end systems since they directly model the input/output relationship from $O$ to $W$ by a monolithic neural network in contradistinction to the approaches that construct a system from separately optimized sub-models such as the acoustic and the language models, as discussed in Sect. 4.1.1. The number of hidden layers in the encoder and the decoder networks, the number of neuron units per a hidden layer, the learning conditions, etc., are meta-parameters to be tuned.

### 4.1.4 Evaluation Measures

The results of speech recognition are evaluated by comparing the recognition hypothesis $R = \langle h_1, h_2, \cdots, h_m \rangle$ with a reference word sequence $R = \langle r_1, r_2, \cdots, r_n \rangle$, where $m$ and $n$ are their lengths. Let $h_j$ corresponds to $r_i$ when we make a word by word alignment of the hypothesis and the reference. Figure 4.7 shows an example of the alignment. The word $h_i$ is counted as correctly recognized if it is the same as $r_i$, and mistakenly substituted to another word if it is not. If there is no $h_j$ for $r_i$, it is counted as a deletion error, and if there is no $r_i$ for $h_j$, it is counted as an insertion error. Based on the alignment, word error rate (WER) is defined by Eq. (4.9).

$$\text{WER} = \frac{N_s + N_i + N_d}{n} = \frac{N_s + N_i + N_d}{N_c + N_s + N_d}, \tag{4.9}$$

where $N_c$ is the number of correctly recognized words, and $N_s$, $N_i$, $N_d$ are the numbers of substitution, insertion, and deletion errors. The WER score depends on the alignment, and the lowest score is used as the evaluation score of the recognition hypothesis. The search of the best alignment is efficiently performed by using the dynamic programming [9] algorithm. Smaller WER indicates better recognition performance, and the minimum WER score is 0.0. The WER can take larger values than 1.0 because of the existence of the insertion error. Another measure is word accuracy (WACC), which is obtained by negating WER and adding 1.0 as shown in Eq. (4.10). Larger WACC indicates better performance.

$$\text{WACC} = 1.0 - \text{WER}. \tag{4.10}$$

The WER (or WACC) is evaluated for a development set and an evaluation set. The former score is used during the training of the system for the meta-parameter tuning, and the latter is used as the final performance measure. For dialogue and translation systems where the correct answer is not unique, other measures such as BLEU [10] are used which compare the system output and the reference in a somewhat more relaxed manner in the alignment.

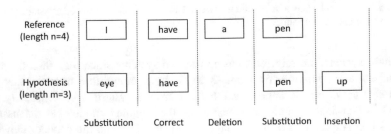

**Fig. 4.7** An example of a word alignment for scoring speech recognition results

---

**Algorithm 1** Genetic algorithm (GA)

---

1: **for** $k = 1$ to $K_0$ **do**
2:     Initialize $x_k$
3: **end for**
4: **while** not convergence **do**
5:     **for** $k = 1$ to $K$ ($K_0$ for the first iteration) **do**
6:         Decode gene $x_k$ to configuration $C_k$
7:         Evaluate configuration $C_k$ to obtain score $y_k = f(C_k)$
8:     **end for**
9:     Generate child genes $\{x_k\}_{k=1}^K$ from current (parent) genes $\{x_k\}_{k=1}^K$ and their scores $\{y_k\}_{k=1}^K$
       by selection, mating, and mutation
10: **end while**
11: **return** Extract the best gene $x^\star$

---

## 4.2   Evolutionary Algorithms

Let $y = f(x)$ be an evaluation function that represents the accuracy of a speech recognition system (or some performance measure of a spoken language processing system) built from tuning meta-parameters represented by $D$-dimensional vector $x$. The process of finding the optimal tuning parameter $x^*$ to maximize the accuracy can be formulated as the following optimization problem:

$$x^* = \underset{x \in \hat{\mathcal{X}}}{\operatorname{argmax}} f(x), \qquad (4.11)$$

where $\hat{\mathcal{X}}$ is a set of candidates for $x$. Because speech recognition systems are extremely complex, there is no analytical form for the solution. We must address this optimization problem without assuming specific knowledge for $f$, i.e., by considering $f$ as a black box. Another important aspect of this problem is that evaluating the function value $f(x)$ is expensive because training a large vocabulary model and computing its development set accuracy can take considerable time. Thus, the key point here is for the black-box optimization to generate an appropriate set of hypotheses $\hat{\mathcal{X}}$ to find the best $x^*$ in the smallest number of the training and evaluation steps ($f(x)$) as possible.

### 4.2.1   Genetic Algorithm

Genetic algorithm (GA) is a search heuristic motivated by the biological evolution process. This algorithm is based on (1) the selection of genes (also called "chromosome representations") according to their scores, pruning inferior genes for the next iteration (generation); (2) mating pairs of genes to form child genes that mix the properties of the parents, and (3) mutation of a part of a gene to produce new gene.

A popular selection method, which we will use in the later experiment, is the *tournament method*. This method first extracts a subset of $M(< K)$ hypotheses $(\hat{\mathcal{X}}_k = \{x_{k'}\}_{k'=1}^{M})$ generated from a total of $K$ genes randomly, and then it selects the best gene $x_{k*}$ in the subset by their scores, i.e.,

$$x_{k*} = \operatorname*{argmax}_{x_{k'} \subset \hat{\mathcal{X}}_k} f(x_{k'}). \tag{4.12}$$

The random subset extraction step can provide variations of genes giving a chance of survival not only to the best gene but also to superior genes, and the best selection step in a subset guarantees the exclusion of inferior genes. This process is repeated $K$ times to obtain a set of survived genes.

For the mating process, a typical method is the *one-point crossover*, which first finds a pair of (parent) genes ($x_{k_1}^{\mathrm{p}}$ and $x_{k_2}^{\mathrm{p}}$) from the selected genes and then swaps the $\{1, \cdots, d\}$ elements to $\{d + 1, \cdots, D\}$ elements of these two vectors to obtain the following new (child) gene pair ($x_{k_1}^{\mathrm{c}}$ and $x_{k_2}^{\mathrm{c}}$):

$$x_{k_1}^{\mathrm{c}} = \begin{bmatrix} x_{k_1,1}^{\mathrm{p}} \\ \vdots \\ x_{k_1,d}^{\mathrm{p}} \\ x_{k_2,d+1}^{\mathrm{p}} \\ \vdots \\ x_{k_2,D}^{\mathrm{p}} \end{bmatrix}, \quad x_{k_2}^{\mathrm{c}} = \begin{bmatrix} x_{k_2,1}^{\mathrm{p}} \\ \vdots \\ x_{k_2,d}^{\mathrm{p}} \\ x_{k_1,d+1}^{\mathrm{p}} \\ \vdots \\ x_{k_1,D}^{\mathrm{p}} \end{bmatrix}. \tag{4.13}$$

The position $d$ is randomly sampled. As the iteration increases, these processes provide appropriate genes that encode optimal DNN configurations.

Algorithm 1 summarizes the GA procedure. The process is repeated until the evaluation score is converged, and the best gene $x^*$ is extracted.

### 4.2.2 Evolution Strategy

Evolution strategy (ES) is a population-based meta-heuristic optimization algorithm that is similar to GA. A difference from GA is that ES represents a gene $x$ by a real-valued vector. Covariance matrix adaptation ES (CMA-ES) [11] is an ES, which is closely related to natural ES [12]. Although both CMA-ES and natural ES have several variations, it has been shown that their core parts are mathematically equivalent [13]. CMA-ES was proposed earlier than natural ES, but the mathematical motivation of natural ES is more concise. Here, we follow the derivation of natural ES as the explanation of CMA-ES.

CMA-ES uses a multivariate Gaussian distribution $N(x|\theta)$ having a parameter set $\theta = \{\mu, \Sigma\}$ to represent a gene distribution, where $\mu$ is a $D$-dimensional mean vector, $\Sigma$ is a $D \times D$-dimensional covariance matrix, and $D$ is the gene size. It seeks a distribution that is concentrated in a region with high values of $f(x)$ such that sampling from the distribution provides superior genes. The search of the distribution is formulated as a maximization problem of the expected value $\mathbb{E}[f(x)|\theta]$ of $f(x)$ under a Gaussian distribution $\mathcal{N}(x|\theta)$ as shown in Eqs. (4.14) and (4.15).

$$\mathbb{E}[f(x)|\theta] = \int f(x) \mathcal{N}(x|\theta)dx, \tag{4.14}$$

$$\hat{\theta} = \underset{\theta}{\mathrm{argmax}} \, \mathbb{E}[f(x)|\theta]. \tag{4.15}$$

To maximize the expectation, the gradient ascent method can be used to iteratively update the current parameter set $\theta_n$ starting from an initial parameter set $\theta_0$, as shown in Eq. (4.16).

$$\hat{\theta}_n = \hat{\theta}_{n-1} + \epsilon \nabla_\theta \mathbb{E}[f(x)|\theta] \,|_{\theta=\hat{\theta}_{n-1}}, \tag{4.16}$$

where $n$ is an iteration index and $\epsilon$ ($> 0$) is a step size. To evaluate the gradient, CMA-ES uses the relation of $\nabla_\theta \log \mathcal{N}(x|\theta) = \frac{1}{\mathcal{N}(x|\theta)} \nabla_\theta \mathcal{N}(x|\theta)$, which is called a "log-trick." By approximating the integration by sampling after applying the log-trick, the gradient is expressed by Eq. (4.19).

$$\nabla_\theta \mathbb{E}[f(x)|\theta] \,|_{\theta=\hat{\theta}_{n-1}} \tag{4.17}$$

$$= \int (f(x) \nabla_\theta \log \mathcal{N}(x|\theta_{n-1})) \mathcal{N}(x|\theta_{n-1})dx \tag{4.18}$$

$$\approx \frac{1}{K} \sum_k^K y_k \nabla_\theta \log \mathcal{N}(x_k|\theta_{n-1}), \tag{4.19}$$

$$x_k \sim \mathcal{N}(x|\theta_{n-1}),$$

where $x_k$ is a gene sampled from the previously estimated distribution $\mathcal{N}(x|\hat{\theta}_{n-1})$, and $y_k$ is the evaluated value of the function $y_k = f(x_k)$. The set of $K$ samples at an iteration step corresponds to a set of individuals at a generation in an evolution. By repeating the generations, it is expected that superior individuals are obtained. Note the formulation is closely related to the reinforcement learning. If we interpret the Gaussian distribution as a policy function taking no input assuming the world is a constant, and regard the gene as an action, it is a special case of the policy gradient based reinforcement learning [14].

Although simple gradient ascent may be directly performed using the obtained gradient, CMA-ES uses the natural gradient $\tilde{\nabla}_\theta \mathbb{E}[f(x)|\theta] = F^{-1} \nabla_\theta \mathbb{E}[f(x)|\theta]$ rather than the original gradient $\nabla_\theta \mathbb{E}[f(x)|\theta]$ to improve the convergence speed, where $F$ is a Fisher information matrix defined by Eq. (4.20).

$$F(\theta) = \int \mathcal{N}(x|\theta) \nabla_\theta \log \mathcal{N}(x|\theta) \nabla_\theta \log \mathcal{N}(x|\theta)^T dx. \tag{4.20}$$

By substituting the concrete Gaussian form for $\mathcal{N}(x|\theta)$, the update formulae for $\hat{\mu}_n$ and $\hat{\Sigma}_n$ are obtained as shown in Eq. (4.21).

$$\begin{cases} \hat{\mu}_n = \hat{\mu}_{n-1} + \epsilon_\mu \sum_{k=1}^K w(y_k)(x_k - \hat{\mu}_{n-1}), \\ \hat{\Sigma}_n = \hat{\Sigma}_{n-1} + \epsilon_\Sigma \sum_{k=1}^K w(y_k) \\ \qquad \cdot \left( (x_k - \hat{\mu}_{n-1})(x_k - \hat{\mu}_{n-1})^\mathsf{T} - \hat{\Sigma}_{n-1} \right), \end{cases} \tag{4.21}$$

where $\mathsf{T}$ is the matrix transpose. Note that, as in [11], $y_k$ in Eq. (4.19) is approximated in Eq. (4.21) as a weight function $w(y_k)$, which is defined as:

$$w(y_k) = \frac{\max\{0, \log(K/2+1) - \log(\mathrm{R}(y_k))\}}{\sum_{k'=1}^K \max\{0, \log(K/2+1) - \log(\mathrm{R}(y_{k'}))\}} - \frac{1}{K}, \tag{4.22}$$

where $\mathrm{R}(y_k)$ is a ranking function that returns the descending order of $y_k$ among $y_{1:K}$ (i.e., $\mathrm{R}(y_k) = 1$ for the highest $y_k$, $\mathrm{R}(y_k) = K$ for the smallest $y_k$, and so forth). This equation only considers the order of $y$, which makes the updates less sensitive to the evaluation measurements (e.g., to prevent different results using word accuracies and the negative sign of error counts).

Algorithm 2 summarizes the CMA-ES optimization procedure, which gradually samples neighboring tuning parameters from the initial values. Because CMA-ES uses a real-valued vector as a gene, it is naturally suited for tuning continuous-valued meta-parameters. To tune discrete-valued meta-parameters, it needs a discretization by some means. The evaluation of $f(x_k)$ can be performed independently for each $k$. Therefore, it is easily adapted to parallel computing environments such as cloud computing services for shorter turnaround times. The number of samples, $K$, is automatically determined from the number of dimensions of $x$ [11], or we can set it manually by considering computer resources.

### 4.2.3   Bayesian Optimization

Even though Bayesian optimization (BO) is motivated differently from ES and GA, in practice, there are several similarities. Especially when it is palatalized, a set of individuals are evaluated at each update stage where a fixed-dimensional vector specifies the configuration of an individual.

While CMA-ES involves a distribution of the tuning parameter $x$ taking the expectation over $x$, BO uses a probabilistic model of the output $y$ to evaluate an acquisition function that evaluates the goodness of $x$. Several acquisition functions

---

**Algorithm 2** CMA-ES

1: Initialization of $\hat{\mu}_0$ and $\hat{\Sigma}_0$, and $y_0^\star = \emptyset$
2: **for** $n = 1$ to $N$ **do**
3:    **for** $k = 1$ to $K$ **do**
4:       Sample $x_k$ from $\mathcal{N}(x | \hat{\mu}_{n-1}, \hat{\Sigma}_{n-1})$
5:       Evaluate $y_k = f(x_k)$
6:    **end for**
7:    Rank $\{y_k\}_{k=1}^K$
8:    Update $\hat{\mu}_n$ and $\hat{\Sigma}_n$
9:    Store $y_n^\star = \max\{y_{1:K}, y_{n-1}^\star\}$ corresponding $x_n^\star$
10: **end for**
11: **return** $\{x_N^\star, y_N^\star\}$

---

have been proposed [15]. Here, we use expected improvement, which is suggested as a practical choice [16]. The expected improvement is defined as:

$$a^{EI}(x_k) = \int \max\{0, y - y_{k-1}^*\} p(y | D_{1:k-1}, x_k) dy, \tag{4.23}$$

where $\max\{0, y - y_{k-1}^*\}$ is an improvement measure based on the best score $y_{k-1}^* = \max_{1 \le k' \le k-1} y_{k'}$ among $k - 1$ previous scores, and $p(y | D_{1:k-1}, x_k)$ is the predictive distribution of $y$ given $x_k$ and the already observed data set $D_{1:k-1} = \{x_{1:k-1}, y_{1:k-1}\}$ modeled by a Gaussian process [17].

BO then performs a deterministic search for the next candidate $\hat{x}_k$ by maximizing the expected improvement over $y$:

$$\hat{x}_k = \underset{x_k}{\operatorname{argmax}}\, a^{EI}(x_k). \tag{4.24}$$

Equation (4.24) selects the $x_k$ that is likely to lead to a high score of $y_k$.

The Gaussian process models the joint probability of the $k$ scores $[y_{1:k-1}^\top, y]^\top$ as a $k$-dimensional multivariate Gaussian with a zero mean vector and a Gram matrix $K$ as covariance matrix:

$$p(y_{1:k-1}, y \mid x_{1:k}) = \mathcal{N}\left(\begin{bmatrix} y_{1:k-1} \\ y \end{bmatrix} \Big| 0, K\right), \tag{4.25}$$

$$K = \begin{bmatrix} G & g(x_k) \\ g(x_k)^\top & g(x_k, x_k) \end{bmatrix}, \tag{4.26}$$

where $g(x, x')$ is a kernel function, $G$ is a Gram matrix with elements $G_{i,j} = g(x_i, x_j)$ for $1 \le i, j \le k - 1$, and $g(x_k) = [g(x_1, x_k), \ldots, g(x_{k-1}, x_k)]^\top$. The predictive distribution of $y$ given $y_{1:k-1}$ is obtained as a univariate Gaussian distribution by using Bayes' theorem:

$$p(y \mid D_{1:k-1}, x_k) = p(y \mid y_{1:k-1}, x_{1:k})$$
$$= \mathcal{N}(y \mid \mu(x_k), \sigma^2(x_k)), \tag{4.27}$$

---
**Algorithm 3** Bayesian optimization (BO)
___

1: Set the domain $\mathcal{X}$ of $\hat{x}_0$, and $\hat{y}_0 = \emptyset$
2: **for** $k = 1$ to $K$ **do**
3:     Compute $\hat{x}_k = argmax_{x_k} a^{EI}(x_k)$
4:     Evaluate $y_k = f(\hat{x}_k)$
5:     Store $y_k^\star = \max\{y_k, y_{k-1}^\star\}$ corresponding $x_k^\star$
6: **end for**
7: **return** $\{x_K^\star, y_K^\star\}$

---

where the mean $\mu(x_k)$ and variance $\sigma^2(x_k)$ are given as:

$$\begin{cases} \mu(x_k) = g(x_k)^\top G^{-1} y_{1:k-1}, \\ \sigma^2(x_k)) = g(x_k, x_k) - g(x_k)^\top G^{-1} g(x_k). \end{cases} \tag{4.28}$$

Based on this predictive distribution, we can analytically evaluate the expected improvement $a^{EI}(x_k)$ by substituting Eq. (4.27) into (4.23), and numerically obtain $\hat{x}_k$ by Eq. (4.24).

The basic algorithm of BO is shown in Algorithm 3. While one needs to set initial values for $x$ for CMA-ES, one needs to set the domain of $x$ for BO. Parallelization can be performed when computing the expected improvement function $a^{EI}(x_k)$ with Monte Carlo sampling. However, the greedy search resulting from BO often selects tuning parameters on the edges of the parameter domains, which leads to extremely long function evaluations when the dimension of $x$ is large. We have observed that these actually make the evaluation difficult in our experiments.

## 4.3   Multi-Objective Optimization with Pareto Optimality

In Sect. 4.2, we explained meta-parameter optimization methods for single objectives, such as the recognition accuracy. Sometimes, other objectives are also important in real applications. For example, smaller DNN size is preferable because it affects the computational costs for both training and decoding. In this section, we explain multi-objective CMA-ES with Pareto optimality.

### 4.3.1   Pareto Optimality

Without loss of generality, assume that we wish to maximize $J$ objectives with respect to $x$ jointly, which are defined as:

$$F(x) \triangleq [f_1(x), f_2(x), \ldots, f_J(x)]. \tag{4.29}$$

Because objectives may conflict, we adopt a concept of optimality known as Pareto optimality [18]. For jointly optimizing multiple objectives, it needs to satisfy the following terms:

$$
\begin{cases}
f_j(\mathbf{x}_k) \geq f_j(\mathbf{x}_{k'}) \ \forall\ j = 1, .., J \\
f_j(\mathbf{x}_k) > f_j(\mathbf{x}_{k'}) \ \exists\ j = 1, .., J.
\end{cases}
\tag{4.30}
$$

Then, we say that $\mathbf{x}_k$ *dominates* $\mathbf{x}_{k'}$ and write $F(\mathbf{x}_k) \rhd F(\mathbf{x}_{k'})$. Given a set of candidate solutions, $\mathbf{x}_k$ is *Pareto optimal* iff no other $\mathbf{x}_{k'}$ exists such that $F(\mathbf{x}_{k'}) \rhd F(\mathbf{x}_k)$.

Pareto optimality formalizes the intuition that a solution is good if no other solution outperforms (dominates) it in all objectives. Given a set of candidates, there are generally multiple Pareto-optimal solutions; this is known as the Pareto frontier. Note that an alternative approach is to combine multiple objectives into a single objective via a weighted linear combination:

$$
\sum_j \beta_j f_j(\mathbf{x}),
\tag{4.31}
$$

where $\sum_j \beta_j = 1$ and $\beta_j > 0$. The advantage of the Pareto definition is that weights $\beta_j$ need not be specified and it is more general, i.e., the optimal solution obtained by any setting of $\beta_j$ is guaranteed to be included in the Pareto frontier. Every $\{\mathbf{x}_{1:K}\}$ can be ranked by using the Pareto frontier, which can adapt to meta-heuristics.

### 4.3.2   CMA-ES with Pareto Optimality

We realize multi-objective CMA-ES for a low WER and small model size by modifying the rank function $R(y_k)$ used in Eq. (4.22). Given a set of solutions $\{\mathbf{x}_k\}$, we first assign rank $= 1$ to those on the Pareto frontier. Then, we exclude these rank 1 solutions and compute the Pareto frontier again for the remaining solutions, assigning them rank 2. This process is iterated until no $\{\mathbf{x}_k\}$ remain, and we ultimately obtain a ranking of all solutions according to multiple objectives. The remainder of CMA-ES remains unchanged; by this modification, future generations are drawn to optimize multiple objectives rather than a single objective. With some bookkeeping, this ranking can be computed efficiently in $O(J \cdot K^2)$ [19].

Algorithm 4 summarizes the CMA-ES optimization procedure with Pareto optimality, which is used to rank the multiple objectives $F(\mathbf{x}_k)$. The obtained rank is used to update the mean vector and covariance matrix of CMA-ES. CMA-ES gradually samples neighboring tuning parameters from the initial values and finally provides a subset of solutions, $\{\mathbf{x}, F(\mathbf{x})\}$, that lie on the Pareto frontier (rank 1) of all stored $N \times K$ samples.

---

**Algorithm 4** Multi-objective CMA-ES

---

1: Initialization of $\hat{\mu}_0$ and $\hat{\Sigma}_0$
2: **for** $n = 1$ to $N$ **do**
3:    **for** $k = 1$ to $K$ **do**
4:       Sample $x_k$ from $\mathcal{N}(x|\hat{\mu}_{n-1}, \hat{\Sigma}_{n-1})$
5:       Evaluate $J$ objectives $F(x_k) \triangleq [f_1(x_k), f_2(x_k), \ldots, f_J(x_k)]$
6:    **end for**
7:    Rank $\{F(x_k)\}_{k=1}^{K}$ according to Pareto optimality
8:    Update $\hat{\mu}_n$ and $\hat{\Sigma}_n$
9: **end for**
10: **return** subset of solutions $\{x, F(x)\}$ that lie on the Pareto frontier (rank 1) of all stored $N \times K$ samples

---

### 4.3.3   Alternative Multi-Objective Methods

There is a rich literature of multi-objective methods for genetic algorithms. Refer to [20, 21] for a survey of techniques. One class of methods utilizes Pareto optimality in estimating the fitness $F(x)$ of each solution. Examples include the widely used NSGA-II [19], and the Pareto CMA-ES method we described in Sect. 4.3.2 adopts a very similar approach.

There are also multi-objective genetic algorithms that do not utilize the concept of Pareto fitness. For example, VEGA [22] divides the selection of offspring population into seperate groups based on different objectives, then allow crossover operations across groups. HGLA [23] runs a genetic algorithm on a linear combination of objectives; the combination weights are not fixed but evolved simultaneously with the solutions. All these methods should be applicable to the problem of automatic optimization of the DNN meta-parameters, but we are not aware of any large-scale empirical evaluation.

For Bayesian optimization, [24] proposed an acquisition function which chooses the $x$ to maximally reduce the entropy of the posterior distribution over the Pareto set. This has been evaluated for automatic optimization of speed and accuracy of DNNs on the MNIST image classification, with promising results. There are also methods based on using a combination of multiple objectives to a single objective, e.g. [25].

## 4.4   Experimental Setups

### 4.4.1   General Setups

We applied the evolutionary algorithms to tune large vocabulary speech recognition systems [26]. Figure 4.8 shows the overall tuning process. The experiments were performed using the Kaldi speech recognition toolkit with speech data from the

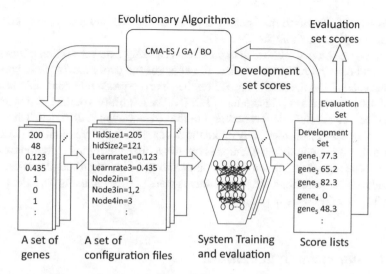

**Fig. 4.8** Evolutionary tuning process of ASR systems

corpus of spontaneous Japanese (CSJ) [27], which is a popular Japanese speech dataset. We performed two separate experiments with training sets having different amounts of data: one consists of 240 h of academic presentations, whereas the other is a 100-h subset. A common development set consisting of 10 academic presentations was used in GA, CMA-ES, and BO to evaluate the individuals for the black-box optimization. The official evaluation set defined in CSJ consisting of 10 academic presentations totalling 110 min was used as the evaluation set.

Acoustic models were trained by first creating a GMM-HMM by maximum likelihood estimation and then building a DNN-HMM by pre-training and fine-tuning using alignments generated by the GMM-HMM. For the performance evaluation of the system, the DNN-HMM was used as the final model. The language model was a 3-gram model trained on CSJ with academic and other types of presentations, which amounted to 7.5 million words in total. The vocabulary size was 72 k. Speech recognition was performed using the OpenFST WFST decoder [28]. As an initial configuration, we borrowed the settings from the Kaldi recipe for the Switchboard corpus (i.e., egs/swbd/s5b). We chose the recipe because this task was similar, while the language was different and because it was manually well tuned and publicly available.

For the experiments, TSUBAME 2.5 supercomputer[1] was used. A maximum of 44 NVIDIA K20X GPGPUs was used in parallel through the message-passing

---

[1]https://www.gsic.titech.ac.jp/en/tsubame.

interface (MPI). We used the Spearmint package[2] for BO and the Python version of Hansen's implementation[3] for CMA-ES.

Further, we ran two additional experiments utilizing a newer version of the Kaldi toolkit and the CSJ recipe to confirm the effect of the evolution.[4] One is based on the nnet1 script and the other is based on the chain script. While nnet1 adopts basic neural network structure, chain adopts TDNN. The definitions of the training and the evaluation sets are the same as before, but the development is different. The reason is that the recipe scripts internally make the development set by holding-out a subset of the training set, and the new recipe script has a different implementation from the old one. The new development set amounted to 6.5 h having 4000 utterances from 39 academic presentations. The experiments were performed using TSUBAME 3.0 using 30 P100 GPGPUs in parallel.

### 4.4.2 Automatic Optimizations

In the evolution experiments, feature types, DNN structures, and learning parameters were optimized. The first and second columns of Table 4.1 describe these variables. We specify three base feature types (feat_type) for the GMM-HMM and DNN-HMM models: MFCC,PLP, and filter bank (FBANK). The dimensions of these features were 13, 13, and 36, respectively. The GMM-HMMs were first trained directly using the specified base features and their delta [29] and delta-delta. Then, they were re-trained using 40-dimensional LDA [30]-compressed and MLLT [31]-transformed features that were obtained from composite features made by concatenating 9 frames of the base features, and fMLLR [31]-based speaker adaptive training was performed. The DNN-HMMs were trained using features that were expanded again from the fMLLR features, splicing 5 pre- and post-context frames. The other settings were the same as those used in the Kaldi recipe.

CMA-ES uses genes represented as real-valued vectors, mappings from a real scalar value to a required type are necessary, depending on the parameters. For the mapping, we used $ceil(10^x)$ for converting positive continuous values to integers (e.g., splice). Similarly, we used $10^x$ for positive real values (e.g., learning rates), and $mod(ceil(abs(x)*3),3)$ for a multiple choice (feature type). For example, if a value of feature type (feat_value) in a gene is $-1.7$, it is mapped to 0, and indicates MFCC. If it is 1.4, it is mapped to 2, which corresponds to PLP in our implementation. The third column of the tables presents the baseline settings, which was also used as an initial meta-parameter configuration. The MFCC-based baseline system with the 240-h training set and K20X GPGPU took 12 h for the RBM pre-training and 70 h for fine-tuning.

---

[2]https://github.com/JasperSnoek/spearmint.

[3]https://www.lri.fr/~hansen/cmaes_inmatlab.html.

[4]We ran main experiments in 2015, and the additional experiments in 2018.

**Table 4.1** Meta-parameters subject to optimization and their automatically tuned results for the system using 240-h training data

| Name of meta-parameters | Description | Baseline | Values obtained by evolution using 240 h training data | | | | | |
|---|---|---|---|---|---|---|---|---|
| | | | gen1 | gen2 | gen3 | gen4 | gen5 | gen6 |
| feat_type | MFCC, FBANK, or PLP | MFCC | MFCC | MFCC | MFCC | MFCC | MFCC | MFCC |
| splice | Segment length for DNN | 5 | 6 | 9 | 10 | 17 | 21 | 18 |
| nn_depth | Number of hidden layers | 6 | 7 | 6 | 6 | 6 | 5 | 7 |
| hid_dim | Units per layer | 2048 | 1755 | 1907 | 2575 | 1905 | 2904 | 3304 |
| param_stddev_first | Init parameters in 1st RBM | 1.0E−1 | 1.1E−1 | 1.3E−1 | 1.1E−1 | 1.2E−1 | 0.7E−1 | 0.6E−1 |
| param_stddev | Init parameters in other RBMs | 1.0E−1 | 1.0E−1 | 1.3E−1 | 1.0E−1 | 2.3E−1 | 1.9E−1 | 1.6E−1 |
| rbm_lrate | RBM learning rate | 4.0E−1 | 5.2E−1 | 5.7E−1 | 4.1E−1 | 4.7E−1 | 3.6E−1 | 3.6E−1 |
| rbm_lrate_low | Lower RBM learning rate | 1.0E−2 | 1.3E−2 | 1.1E−2 | 0.8E−2 | 0.7E−2 | 0.8E−2 | 1.1E−2 |
| rbm_l2penalty | RBM Lasso regularization | 2.0E−4 | 2.1E−4 | 2.2E−4 | 1.2E−4 | 1.6E−4 | 1.9E−4 | 1.5E−4 |
| learn_rate | Learning rate for fine tuning | 8.0E−3 | 7.3E−3 | 6.5E−3 | 7.8E−3 | 4.4E−3 | 5.3E−3 | 3.7E−3 |
| momentum | Momentum for fine tuning | 1.0E−5 | 0.9E−5 | 0.9E−5 | 0.4E−5 | 0.9E−5 | 0.4E−5 | 0.7E−5 |

CMA-ES with Pareto (CMA-ES+P) was used for the automatic tuning

The population sizes of the black-box optimizations were 20 for the 100-h training set and 44 for the 240-h training set. The WERs used for the optimizations were evaluated using the development set. For the evaluations of each individual, a limit was introduced for the training time at each generation. If a system did not finish the training within 2.5 and 4 days for the 100-h training set and the 240-h training set, respectively, the training was interrupted and the last model in the iterative back-propagation training at that time was used as the final model. The GA-based optimization was performed based on WER and DNN size, and it is referred to as GA(WER, Size) in the following experiments. Ten initial genes ($= N_0$) were manually prepared. Basically, gene A wins over gene B if its WER is lower than that of B. However, gene A having a higher WER wins over gene B if the difference of the WER is less than 0.2% and the DNN size of gene A is less than 90% of that of gene B. The tournament size $M$ was three. For the mutation process, Gaussian noise with zero mean and 0.05 standard deviation was uniformly added to the gene. For CMA-ES, two types of experiments were performed. One was the single-objective experiment based on WER, and the other was the multi-objective experiment based on WER and DNN size using the Pareto optimality. In the following, the former is referred to as CMA-ES, and the latter is referred to as CMA-ES+P. In both cases, the initial mean vector of the multivariate Gaussian was set equal to the baseline settings. For CMA-ES+P, the maximum WER thresholds were set so that they included the top 1/2 and 1/3 of the populations at each generation for the trainings using the 100- and 240-h data sets, respectively. The BO-based tuning was performed using WER as the objective. The search range of the meta-parameters was set from 20 to 600% of the baseline configuration.

For the additional experiments using the newer version of Kaldi, we reduced the number of meta-parameters; our motivation is to evaluate the evolution in more detail under a variety of conditions. For the experiment using nnet1, the optimized meta-parameters were splice, nn_depth, hid_dim, learn_rate and momentum. These were a subset of meta-parameters, deemed to be most important in modern architectures, in Table 4.1. As an initial configuration of the evolution, we borrowed values again from the SWB recipe. For the additional experiment using chain, we used the initial value used in the CSJ recipe.

In these evolution experiments, TDNNs were trained using lattice-free MMI [32] without the weight averaging based parallel processing.[5] The initial TDNN structure was slightly modified from its original version to make the meta-parameter setting a little simpler for a variable number of layers as shown in Fig. 4.9. While in the original structure, layers 2 to 4 had different sub-sampling structures than other layers, all the layers had the same sub-sampling structure in our experiment. Note if necessary, it is possible to allow different structures for each layer by preparing separate meta-parameters for them. In total, 7 meta-parameters shown in Table 4.5 were optimized. Unlike the currently released nnet1 script in the CSJ recipe where

---

[5]We disabled the default option of the parallel training to make the experiments tractable in our environment as it requires a large number of GPUs.

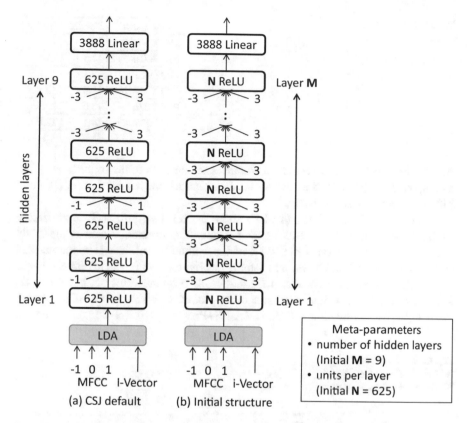

**Fig. 4.9** TDNN model structures for chain based systems. (**a**) is the original structure used in CSJ recipe and (**b**) is the one used as an initial configuration in our evolution experiments. The arrows with numbers at the hidden layers indicate the time splicing index

our evolution results had been integrated, the tuning of chain so far is based on the human effort by the Kaldi community, and this is the first evolution based optimization. The training set was the 240-h data set. The initial nnet1 and chain systems spent 14 and 18 h, respectively, using a P100 GPGPU. If a system did not finish the training within 24 h in the evolution processes, the training was interrupted and the last model at that time was used as the final model. The population size was 30.

## 4.5 Results

Table 4.2 shows the WERs and DNN sizes for systems with the default configuration using the 100- and 240-h training sets with one of the three types of features. Among the features, MFCC was the default in the Switchboard recipe, and it

**Table 4.2** WER of base
systems

| Training data | Dev set | Eval set |
|---|---|---|
| MFCC 100 h | 14.4 | 13.1 |
| PLP 100 h | 14.5 | 13.1 |
| FBANK 100 h | 15.1 | 13.8 |
| MFCC 240 h | 13.5 | 12.5 |
| PLP 240 h | 13.6 | 12.5 |
| FBANK 240 h | 14.1 | 13.0 |

yielded the lowest WERs for the development set for both of the training sets. The
corresponding WERs for the evaluation set were 13.1 and 12.5% for the 100- and
240-h training sets, respectively.

Figures 4.10, 4.11, 4.12, and  4.13 show the results when each optimization
method was used with the 100-h training data. The horizontal axis is the DNN
size, and the vertical axis is the WER of the evaluation set. The baseline marked
on the figure is the MFCC-based system. Ideally, we want systems on the lower
side of the plot when WER based single-objective optimizations (CMA-ES, BO)
were performed, and on lower-left side of the plot when WER and model size based
multi-objective optimizations (GA, CMA-ES+P) were performed. Figure 4.10 is a

**Fig. 4.10** Results of
GA(WER, Size) when the
100-h training data were used

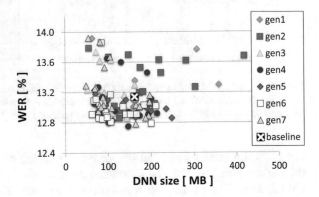

**Fig. 4.11** Results of
CMA-ES when the 100-h
training data were used

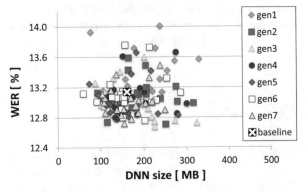

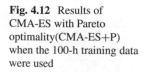 **Fig. 4.12** Results of CMA-ES with Pareto optimality(CMA-ES+P) when the 100-h training data were used

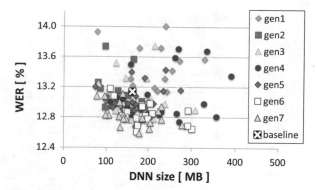

**Fig. 4.13** Results of BO when the 100-h training data were used

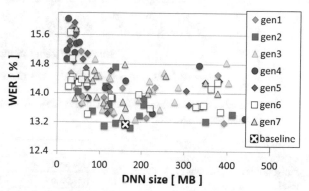

scatter plot of the GA(WER, Size). The distribution is oriented to the left side of the plot with the progress of generations, but the WER reduction was relatively small. Figure 4.11 presents the results of the single-objective CMA-ES. The distribution shifted towards lower WERs and lower DNN file sizes from the baseline with the progress of generations. The reason that it trended to a lower DNN size was probably due to the time limit imposed on the DNN training. In the evolution process, the ratio of individuals that hit the limit was approximately 35%. If an individual has a large DNN size, then it is likely that it hits the limit. Then, the WER is evaluated using a DNN at that time before the back-propagation converges, which is a disadvantage for that individual. Figure 4.12 presents the results of the multi-objective CMA-ES+P. The result is similar to that produced by using CMA-ES, but the distribution is oriented more to the lower-left side of the plot.

Figure 4.13 presents the results using BO for the optimization. In this case, the initial configuration is not directly specified, but the ranges of the meta-parameters are specified. We found that specifying a proper range was actually not straightforward and required knowledge of the problem. That is, if the ranges are too wide, then the initial samples are coarsely distributed in the space, and it is likely that the systems have lower performance. Meanwhile, if the ranges are too narrow, then it is likely that the optimal configuration is not included in the search space. Consequently, the improvement by BO was smaller than that by the CMA-

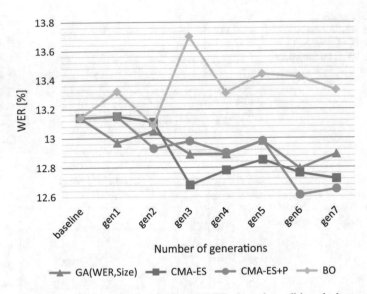

Fig. 4.14 Number of generations and evaluation set WER. At each condition, the best system was chosen by using the development set

**Table 4.3** WER and DNN size of the best system when the 100-h training data was used

|  | WER [%] | | |
| Opt. method | Dev | Eval | DNN size [MB] |
| Baseline | 14.4 | 13.1 | 161.8 |
| GA(WER, Size) | 14.1 | 13.0 | 234.5 |
| CMA-ES | 14.0 | 12.7 | 225.5 |
| CMA-ES+P | 14.0 | 12.7 | 202.4 |
| BO | 14.2 | 13.1 | 110.6 |

ES. Carefully setting the ranges might solve the problem but would again assume expert human knowledge.

Figure 4.14 shows the WER of the evaluation set based on the best systems chosen by using the development set at each generation. CMA-ES evolved more efficiently than GA(WER, Size) and BO. Table 4.3 shows the evaluation results of the best systems chosen by the development set WER through all the generations. The evaluation set WERs by CMA-ES and CMA-ES+P were both 12.7%.[6] However, a smaller DNN model size was obtained by using CMA-ES with Pareto. The DNN model size by CMA-ES was 225.5 Mb, whereas it was 202.4 Mb when CMA-ES+P was used, which was 89.8% of the former. The selected feature type was all MFCC except for the 7th generation, which was PLP.

---

[6]In the table, we scored the evaluation set WERs of systems that gave the lowest development set WER through all the generations. Therefore, they were not necessarily the same as the minimum of the generation wise evaluation set WERs shown in Fig. 4.14.

Figure 4.15 shows the results of CMA-ES+P using the 240-h training data. Approximately 70% of the individuals completed the training within the limit of 4 days. This figure shows that the distributions shifted towards lower WERs and lower DNN file sizes with the progress of generations.

Figure 4.16 shows the WERs of the best systems selected at each generation based on the development set WER when the 240-h training set was used. Although the development set error rate monotonically decreased with the number of the generation, the evaluation set error rate appeared to be saturated after the fourth

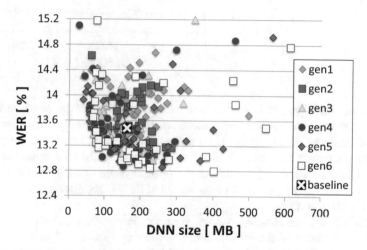

**Fig. 4.15** The DNN model size and the development set WER when the 240-h training set was used with CMA-ES+P. The results of the *n*-th generation are denoted as "gen *n*"

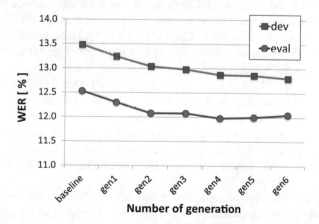

**Fig. 4.16** The development and evaluation set WERs of the best systems at each generation when the 240-h training set was used with CMA-ES+P. The systems were chosen by the development set WER. In the figure "dev" and "eval" indicate the results of the development and the evaluation sets, respectively

**Fig. 4.17** Pareto frontier
derived from the results from
the initial to the 6th
generation using the 240-h
training data. In the figure
"dev" and "eval" indicate the
results of the development
and the evaluation sets,
respectively

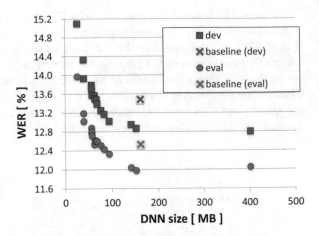

generation, which might have resulted from overfitting to the development set
because we used the same development set for all the generations. The lowest
WER of the development set was obtained at the 6th generation. The corresponding
evaluation set error rate was 12.1%. The difference in the evaluation set WERs
between the baseline (12.5%) and the optimized system (12.1%) was 0.48%, and
this was statistically significant under the MAPSSWE significance test [33]. The
relative WER was 3.8%.

If desired, we can choose a system from the Pareto frontier that best matches
the required balance of the WER and the model size. Figure 4.17 shows the Pareto
frontier derived from the results from the initial to the 6th generation using the 240-
h training data. This figure shows that if we choose a system with approximately
the same WER as the initial model, then we can obtain a reduced model size that is
only 41% of the baseline. That is, the model size was reduced by 59%. The decoding
time of the evaluation set by the reduced model was 79.5 min, which was 85.4% of
the 93.5 min by the baseline. Similarly, the training time of the reduced model was
54.3% of that of the baseline model.

Columns 4 to 9 of Table 4.1 show the meta-parameter configurations obtained as
the result of evolution using the 240-h training set. These are the configurations that
yielded the lowest development set WERs at each generation. When we analyze the
obtained meta-parameters, although the changes were not monotonic for most of the
meta-parameters, we found that splice size was increased by more than three times
from the initial model. We also note that the learning rate decreased by more than
half from the initial condition.

As a supplementary experiment, sequential training [34] was performed using the
best model at the 4th generation as an initial model. Because the sequential training
is computationally intensive, it took an additional 7 days. After the training, the
WER was further reduced, and a WER of 10.9% was obtained for the evaluation
set. This value was lower than the WER of 11.2% obtained with sequential training
using the baseline as the initial model. The difference was statistically significant,
which confirms the effectiveness of the proposed method.

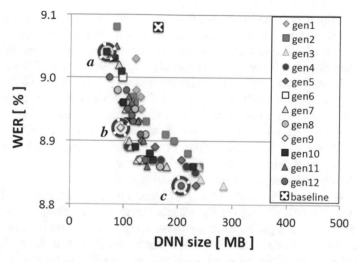

**Fig. 4.18** Evolution result of the nnet1 based system. The CMA-ES with Pareto based evolution (CMA-ES+P) was applied to nnet1 of the newer version of Kaldi with reduced tuning meta-parameters. The baseline is the initial model of the evolution. Only individuals on the Pareto frontier at each generation are plotted for visibility

Figure 4.18 shows the result of the additional experiments of nnet1 using the newer version of Kaldi with the reduced meta-parameters. The figure plots the development set WER and DNN model size. The evolution was performed by CMA-ES with Pareto (CMA-ES+P) and the process was repeated for 12 generations. Approximately 77% of the individuals had completed the training within the 24-h limit. In the figure, only the results of genes on the Pareto frontier at each generation were plotted for visibility. The gene marked as "*a*" gave the smallest DNN size, while the gene marked as "*c*" gave the lowest WER (There were three genes with the smallest WER and *c* was the one with the smallest DNN size.). Gene "*b*" gave both smaller DNN size and smaller WER than the initial system. Table 4.4 describes properties of these representative genes. In this experiment, the improvement in the evaluation set WER from the baseline initial configuration was minor even when

**Table 4.4** Summary of three representative genes in the additional nnet1 experiment with CMA-ES+P

| Gene | Generation | WER [%] | | DNN size [MB] | Decoding time [min] |
| --- | --- | --- | --- | --- | --- |
| | | Dev | Eval | | |
| Baseline | 0 | 9.1 | 11.9 | 161.0 | 90.5 |
| *a* | 10 | 9.0 | 12.0 | 66.5 | 70.9 |
| *b* | 9 | 8.9 | 11.8 | 93.6 | 80.4 |
| *c* | 12 | 8.8 | 11.8 | 207.3 | 99.4 |

Gene *a* gave the smallest model size, and gene *c* gave the lowest development set WER. Gene *b* balances the model size and WER reductions. See Fig. 4.18 for their positions

choosing the gene with the lowest WER in the development set. We conjecture this was probably because the initial meta-parameters were already close to optimal in terms of WER. The reduction of the number of meta-parameters might also have limited the room for improvement though we chose the ones that we thought important based on our previous experiments. However, the evolution had an effect of reducing the DNN size. When the gene "*b*" was chosen, it gave slightly lower WER on the evaluation set and largely reduced DNN size of 93.6 (MB), which was 58% of the initial model of 161.0 (MB). If the gene "*a*" was chosen, the WER of the evaluation set slightly increased from 11.9 to 12.0%, but the model size reduced to 66.5 (MB), which was only 40% of the initial model. Accordingly, the decoding time of the evaluation set was reduced from 90.5 to 70.9 min.

Figure 4.19 shows the result of the evolution based optimization of the chain script. Approximately 63% of the individuals completed the training within the 24-h limit. In this case, larger improvement than nnet1 was obtained both in reducing the WER and the model size. Figure 4.20 shows the WERs of the best systems selected at each generation based on the development set WER. While there was a little random behavior in the evaluation set WER, overall, a consistent trend of WER reduction was observed both in the development and the evaluation set. Table 4.5 shows corresponding changes of the meta-parameters. Different from the changes

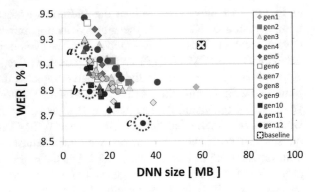

**Fig. 4.19** Evolution result of the chain based system. The CMA-ES with Pareto based evolution (CMA-ES+P) was applied to the chain script of the newer version of Kaldi. Only individuals on the Pareto frontier at each generation are plotted for visibility

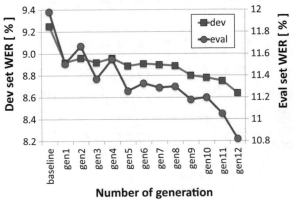

**Fig. 4.20** The development and evaluation set WERs of the best TDNN-based systems at each generation with CMA-ES+P. The systems were chosen by the development set WER. The results of the development and the evaluation sets are indicated by "dev" and "eval," respectively

**Table 4.5** Meta-parameters subject to optimization and their automatically tuned results for the TDNN-based system

| Description of meta-parameters | Baseline | Values obtained by evolution using 240 h training data | | | | | | | | | | | |
|---|---|---|---|---|---|---|---|---|---|---|---|---|---|
| | | gen1 | gen2 | gen3 | gen4 | gen5 | gen6 | gen7 | gen8 | gen9 | gen10 | gen11 | gen12 |
| Number of hidden layers | 9 | 8 | 12 | 9 | 8 | 11 | 9 | 8 | 9 | 10 | 9 | 12 | 10 |
| Units per layer | 625 | 641 | 308 | 396 | 518 | 320 | 354 | 384 | 349 | 461 | 311 | 282 | 427 |
| Learning rate during the initial iteration | 1.0E−3 | 1.1E−3 | 1.0E−3 | 1.5E−3 | 1.5E−3 | 1.2E−3 | 1.9E−3 | 1.4E−3 | 1.8E−3 | 3.0E−3 | 1.3E−3 | 7.0E−3 | 1.1E−2 |
| Learning rate during the final iteration | 1.0E−4 | 0.7E−4 | 1.4E−4 | 1.4E−4 | 0.8E−4 | 1.1E−4 | 0.9E−4 | 0.9E−4 | 0.6E−4 | 0.6E−4 | 0.7E−4 | 0.4E−4 | 0.9E−4 |
| Weight of cross-entropy cost | 1.0E−1 | 1.1E−1 | 0.7E−1 | 1.6E−1 | 1.7E−1 | 1.3E−1 | 1.4E−1 | 1.3E−1 | 1.4E−1 | 0.9E−1 | 1.5E−1 | 1.2E−1 | 1.4E−1 |
| Coefficient for l2 norm | 5.0E−5 | 3.5E−5 | 5.5E−5 | 4.4E−5 | 3.2E−5 | 3.0E−5 | 2.1E−5 | 2.7E−5 | 1.7E−1 | 1.5E−1 | 2.9E−5 | 0.5E−5 | 0.7E−5 |
| Coefficient for leaky hmm | 1.0E−1 | 1.5E−1 | 1.1E−1 | 1.5E−1 | 1.0E−1 | 1.9E−1 | 2.1E−1 | 1.6E−1 | 3.0E−1 | 4.7E−1 | 1.0E−1 | 4.0E−1 | 4.0E−1 |

CMA-ES with Pareto (CMA-ES+P) was used for the automatic tuning. A gene was selected at each generation that gave the best development set WER

**Table 4.6** Summary of three representative genes in the additional chain experiment

| Gene | Generation | WER [%] | | DNN size [MB] | Decoding time [min] |
|------|-----------|---------|---------|---------------|---------------------|
|      |           | Dev | Eval | | |
| CSJ default |  | 9.3 | 11.7 | 53.7 | 22.6 |
| Baseline | 0 | 9.3 | 12.0 | 59.7 | 23.0 |
| a | 11 | 9.2 | 11.9 | 9.1 | 13.1 |
| b | 12 | 8.9 | 11.5 | 11.5 | 13.4 |
| c | 12 | 8.6 | 10.8 | 34.5 | 16.1 |

Gene a gave the smallest model size, and gene c gave the lowest development set WER. Gene b balances the model size and WER reductions. See Fig. 4.19 for their positions

of the WERs, it is seen that none of their changes was monotonic revealing their complex mutual interactions. A remarkable change after 12 generations was the large reduction of units in the hidden layers (units per layer) from 625 of the baseline to 427.

In Fig. 4.19, three representative genes are marked as in the nnet1 results. Table 4.6 describes their details. The evaluation set WER of the gene b was 11.5% and it was 0.5% lower than the baseline initial structure. While the improvement was only 0.2% when compared to the CSJ default (11.7%), the model size reduction was significant from 53.7 (MB) to 11.5 (MB). When the gene c was used, evaluation set WER was 10.8% and the relative reduction was 7.8 and 9.7% compared to the CSJ default and the baseline initial configuration, respectively. Their differences were both statistically significant by the MAPSSWE test. Moreover, the model size was reduced to 57.7% of the original size. The decoding time of 22.6 min of the CSJ default settings was reduced to 16.1 min.

## 4.6 Conclusion

In this chapter, we have introduced the basic principles of spoken language processing, focusing on speech recognition. We have performed an automatic optimization of the meta-parameters by using evolutionary algorithms without human expert elaboration. In the experiments using the 100-h training set, multi-objective GA, CMA-ES, CMA-ES with Pareto (CMA-ES+P) and BO were compared. Both of the CMA-ES methods and GA yielded lower WERs than the baseline. Among them, CMA-ES and CMA-ES+P provided lower WERs than GA. By using CMA-ES+P to jointly minimize the WER and the DNN model size, a smaller DNN size than single-objective CMA-ES was obtained while keeping the WER. CMA-ES was more convenient for optimizing speech recognition systems than BO, which requires the ranges of the meta-parameters to be specified. Moreover, we ran additional experiments using the newer version of the Kaldi toolkit and demonstrated the consistent effectiveness of the CMA-ES+P based approach. Especially, the tuned chain system was significantly superior to the default system both in WER and the model size. Other than experiments introduced here, we have also applied CMA-ES to language modeling and neural machine translation and have achieved automatic performance improvements [35, 36].

When the meta-parameter tuning is applied to the neural network training, there is a double structure of learning; one is the estimation of the neural network connection weights, and the other is the meta-parameter tuning of the network structure and the learning conditions. Currently, the tuning process only uses the performance score, and the learned network weight parameters are all discarded at each generation. Future work includes improving the optimization efficiency by introducing a mechanism to transmit knowledge learned by ancestors to descendants.

# References

1. Davis, S.B., Mermelstein, P.: Comparison of parametric representations for monosyllabic word recognition in continuously spoken sentences. IEEE Trans. Acoust. Speech Signal Process. **28**(4), 357–366 (1980)
2. Hermansky, H.: Perceptual linear predictive (PLP) analysis of speech. J. Acoust. Soc. Am. **87**(4), 1738–1752 (1990)
3. Odell, J.J.: The use of context in large vocabulary speech recognition, Ph.D. Thesis, Cambridge University (1995)
4. Chorowski, J.K., Bahdanau, D., Serdyuk, D., Cho, K., Bengio, Y.: Attention-based models for speech recognition. In: Advances in Neural Information Processing Systems (NeurIPS), pp. 577–585 (2015)
5. Graves, A., Mohamed, A.-R., Hinton, G.: Speech recognition with deep recurrent neural networks. In: 2013 IEEE International Conference on Acoustics, Speech and Signal Processing (ICASSP), pp. 6645–6649. IEEE, Piscataway (2013)
6. Sutskever, I., Vinyals, O., Le, Q.V.: Sequence to sequence learning with neural networks. In: Advances in Neural Information Processing Systems (NeurIPS), pp. 3104–3112 (2014)
7. Vinyals, O., Le, Q.: A neural conversational model. Preprint. arXiv:1506.05869 (2015)
8. Bahdanau, D., Cho, K., Bengio, Y.: Neural machine translation by jointly learning to align and translate. Preprint. arXiv:1409.0473 (2014)
9. Bellman, R.E., Dreyfus, S.E.: Applied Dynamic Programming. Princeton University Press, Princeton (1962)
10. Papineni, K., Roukos, S., Ward, T., Zhu, W.-J.: BLEU: a method for automatic evaluation of machine translation. In: Proceedings of the 40th Annual Meeting on Association for Computational Linguistics, Stroudsburg, ACL '02, pp. 311–318. Association for Computational Linguistics, Stroudsburg (2002)
11. Hansen, N., Müller, S.D., Koumoutsakos, P.: Reducing the time complexity of the derandomized evolution strategy with covariance matrix adaptation (CMA-ES). Evol. Comput. **11**(1), 1–18 (2003)
12. Wierstra, D., Schaul, T., Glasmachers, T., Sun, Y., Peters, J., Schmidhuber, J.: Natural evolution strategies. J. Mach. Learn. Res. **15**(1), 949–980 (2014)
13. Akimoto, Y., Nagata, Y., Ono, I., Kobayashi, S.: Bidirectional relation between CMA evolution strategies and natural evolution strategies. In: Proceedings of Parallel Problem Solving from Nature (PPSN), pp. 154–163 (2010)

14. Sutton, R.S., McAllester, D., Singh, S., Mansour, Y.: Policy gradient methods for reinforcement learning with function approximation. In: Proceedings of the 12th International Conference on Neural Information Processing Systems, NIPS'99, pp. 1057–1063 (1999)
15. Brochu, E., Cora, V.M., De Freitas, N.: A tutorial on Bayesian optimization of expensive cost functions, with application to active user modeling and hierarchical reinforcement learning. Preprint. arXiv:1012.2599 (2010)
16. Snoek, J., Larochelle, H., Adams, R.P.: Practical Bayesian optimization of machine learning algorithms. In: Advances in Neural Information Processing Systems 25 (2012)
17. Rasmussen, C.E., Williams, C.K.I.: Gaussian Processes for Machine Learning. MIT Press, Cambridge (2006)
18. Miettinen, K.: Nonlinear Multiobjective Optimization. Springer, Berlin (1998)
19. Deb, K., Pratap, A., Agarwal, S., Meyarivan, T.: A fast and elitist multiobjective genetic algorithm: NSGA-II. IEEE Trans. Evol. Comput. 6(2), 182–197 (2002)
20. Deb, K., Kalyanmoy, D.: Multi-Objective Optimization Using Evolutionary Algorithms. John Wiley & Sons, Inc., New York (2001)
21. Zitzler, E., Thiele, L.: Multiobjective evolutionary algorithms: a comparative case study and the strength Pareto approach. IEEE Trans. Evol. Comput. 3(4), 257–271 (1999)
22. David Schaffer, J.: Multiple objective optimization with vector evaluated genetic algorithms. In: Proceedings of the 1st International Conference on Genetic Algorithms, Hillsdale, pp. 93–100. L. Erlbaum Associates Inc., Mahwah (1985)
23. Hajela, P., Lin, C.Y.: Genetic search strategies in multicriterion optimal design. Struct. Optim. 4(2), 99–107 (1992)
24. Hernandez-Lobato, D., Hernandez-Lobato, J., Shah, A., Adams, R.: Predictive entropy search for multi-objective Bayesian optimization. In: Balcan, M.F., Weinberger, K.Q. (eds.) Proceedings of The 33rd International Conference on Machine Learning, New York, 20–22 Jun. Proceedings of Machine Learning Research, vol. 48, pp. 1492–1501 (2016)
25. Knowles, J.: ParEGO: a hybrid algorithm with on-line landscape approximation for expensive multiobjective optimization problems. IEEE Trans. Evol. Comput. 10(1), 50–66 (2006)
26. Moriya, T., Tanaka, T., Shinozaki, T., Watanabe, S., Duh, K.: Evolution-strategy-based automation of system development for high-performance speech recognition. IEEE/ACM Trans. Audio Speech Lang. Process. 27(1), 77–88 (2019)
27. Furui, S., Maekawa, K., Isahara, H.: A Japanese national project on spontaneous speech corpus and processing technology. In: Proceedings of ASR'00, pp. 244–248 (2000)
28. Allauzen, C., Riley, M., Schalkwyk, J., Skut, W., Mohri, M.: OpenFST: a general and efficient weighted finite-state transducer library. In: Implementation and Application of Automata, pp. 11–23. Sprinter, Berlin (2007)
29. Furui, S.: Speaker independent isolated word recognition using dynamic features of speech spectrum. IEEE Trans. Acoustics Speech Signal Process. 34, 52–59 (1986)
30. Haeb-Umbach, R., Ney, H.: Linear discriminant analysis for improved large vocabulary continuous speech recognition. In: Proceedings of International Conference on Acoustics, Speech, and Signal Processing, vol. 1, pp. 13–16 (1992)
31. Gales, M.J.F.: Maximum likelihood linear transformations for HMM-based speech recognition. Comput. Speech Lang. 12, 75–98 (1998)
32. Povey, D., Peddinti, V., Galvez, D., Ghahremani, P., Manohar, V., Na, X., Wang, Y., Khudanpur, S.: Purely sequence-trained neural networks for ASR based on lattice-free MMI. In: Interspeech, pp. 2751–2755 (2016)
33. Gillick, L., Cox, S.: Some statistical issues in the comparison of speech recognition algorithms. In: Proceedings of International Conference on Acoustics, Speech, and Signal Processing, pp. 532–535 (1989)
34. Vesely, K., Ghoshal, A., Burget, L., Povey, D.: Sequence-discriminative training of deep neural networks. In: Proceedings of Interspeech, pp. 2345–2349 (2013)

35. Tanaka, T., Moriya, T., Shinozaki, T., Watanabe, S., Hori, T., Duh, K.: Automated structure discovery and parameter tuning of neural network language model based on evolution strategy. In: Proceedings of the 2016 IEEE Workshop on Spoken Language Technology, pp. 665–671 (2016)
36. Qin, H., Shinozaki, T., Duh, K.: Evolution strategy based automatic tuning of neural machine translation systems. In: Proceeding of International Workshop on Spoken Language Translation (IWSLT), pp. 120–128 (2017)

# Chapter 5
# Search Heuristics for the Optimization of DBN for Time Series Forecasting

**Takashi Kuremoto, Takaomi Hirata, Masanao Obayashi, Kunikazu Kobayashi, and Shingo Mabu**

**Abstract** A deep belief net (DBN) with multi-stacked restricted Boltzmann machines (RBMs) was proposed by Hinton and Salakhutdinov for reducing the dimensionality of data in 2006. Comparing to the conventional methods, such as the principal component analysis (PCA), the superior performance of DBN received the most attention by the researchers of pattern recognition, and it even brought out a new era of artificial intelligence (AI) with a keyword "deep learning" (DL). Deep neural networks (DNN) such as DBN, deep auto-encoders (DAE), and convolutional neural networks (CNN) have been successfully applied to the fields of dimensionality reduction, image processing, pattern recognition, etc., nevertheless, there are more AI disciplines in which they could be applied such as computational cognition, behavior decision, forecasting, and others. Furthermore, the architectures of conventional deep models are usually handcrafted, i.e., the optimization of the structure of DNN is still a problem. In this chapter, we mainly introduce how DBNs were firstly adopted to time series forecasting systems by our original studies, and two kinds of heuristic optimization methods for structuring DBNs are discussed: particle swarm optimization (PSO), a well-known method in swarm intelligence; and random search (RS), which is a simpler and useful algorithm for high dimensional hyper-parameter exploration.

## 5.1 Introduction

Human-level software such as game control [1], Alpha-Go [2, 3] were developed and attracted people all over the world in recent years. Deep learning and reinforcement learning [4] were the key techniques of these remarkable inventions.

T. Kuremoto (✉) · T. Hirata · M. Obayashi · S. Mabu
Yamaguchi University, Ube, Japan
e-mail: wu@yamaguchi-u.ac.jp; m.obayas@yamaguchi-u.ac.jp; mabu@yamaguchi-u.ac.jp

K. Kobayashi
Aichi Prefectural University, Nagakute, Japan
e-mail: kobayashi@ist.aichi-pu.ac.jp

© Springer Nature Singapore Pte Ltd. 2020
H. Iba, N. Noman (eds.), *Deep Neural Evolution*, Natural Computing Series,
https://doi.org/10.1007/978-981-15-3685-4_5

131

Deep learning (DL), which indicates a kind of machine learning methods using multi-layered artificial neural networks, is studied from the middle of the 2000s. Hinton et al. proposed a stacked auto-encoder (SAE) which used multiple restricted Boltzmann machines (RBM), a kind of stochastic neural networks, to compose a deep belief net (DBN) successfully applied to dimensionality reduce and high dimensional data classification [5–7]. Since deep conventional neural networks (DCNN) [8] won the champion of image recognition competition ILSVRC 2012 [9], DL has become the most popular method of artificial intelligence (AI) research. However, a problem of deep learning is how to configure an optimal deep learning model for different tasks. The number of units and the number of layers affect the performance of models seriously, and during learning process, i.e., the optimization of parameters including weights of connections between units and suitable learning rates are also very important. So the structure of deep neural networks (DNN) has been designed by grid and manual search, and even was called "a work of art." In [10], Bergstra and Bengio proposed a random search method which is more efficient. In [11], Ba and Frey proposed a dropout method for optimizing DNNs and it is popularly used now. Recently, Nowakowski, Dorogyy, and Doroga-Ivaniuk proposed to use a genetic algorithm (GA) to construct and optimizing DNNs [12]. In our previous study [13–15], particle swarm optimization (PSO) [16], a well-known swarm intelligent method proposed by Kennedy and Eberhart in 1995, was adopted to find the optimal number of units in different restricted Boltzmann machines (RBM) which were used in DBN for time series forecasting. However, the main problem of PSO-based DBN architecture optimization is the large computation cost for the reason that each DBN needs to be trained with learning of RBMs and fine-tuning of DBN whereas one particle presents one DBN. In our experiment with a personal computer (3.2 Hz×1 CPU, 8 GB×2 RAM) for 5000 data training, it needs almost 1 week [13, 14]. Instead of PSO, we applied Bergstra and Bengio's random search (RS) to decide the number of units and layers of RBMs in DBNs which were composed by RBMs or RBMs and multi-layered perceptron (MLP). The RS algorithm is simpler than PSO and more available to obtain the optimal hyper-parameters than conventional grid search (GS), i.e., manual exploration. For the same problem in the experiment, RS spent only about half of the time that PSO took for the training of DBN [17–19]. For other state-of-the-art DL models, such as VGG16 [20], ResNet [21], the architectures were also designed by trial and error until 2 years ago [22, 23]. In [22], Wang et al. firstly evolved CNN architecture by PSO in 2017. Recently, Fernandes Junior and Yen proposed a novel PSO algorithm which has a variable size of particles for CNN architecture optimization [23].

   In this chapter, it is firstly, and consistently to explain how a DBN is optimized its architecture by PSO or RS according to our a series of original works [13–15, 17–19]. Experimental results using chaotic time series such as Lorenz chaos [24] and Hénon map [25], and benchmark CATS data [26, 27] showed the effectiveness of the optimized DBNs.

## 5.2 DBNs with RBMs and MLP

A deep belief net (DBN) proposed by Hinton et al. [5–7] inspired the study of deep learning. DBN is composed by multiple restricted Boltzmann machines (RBM) which have the ability of representation for the high dimensional data, such as images and audio processing. In [5], Hinton and Salakhutdinov showed the powerful compression and reconstruction functions of DBN with 4 RBMs as encoder and 4 RBMs as decoder, comparing to the conventional principal component analysis (PCA).

### 5.2.1 RBM

RBM is a variation of a Boltzmann machine, a stochastic neural network, but with 2 layers: visible layer and hidden layer (See Fig. 5.1). The connection between units (neurons) in a RBM is restricted to those in different layers, and the weights of those synaptic connections are the same in both directions, i.e., $w_{ij} = w_{ji}$. The output of each unit in the visible layer and the hidden layer is 0 or 1,

$$p(h_j = 1|\mathbf{v}) = \frac{1}{exp\left(-b_j - \sum_i v_i w_{ij}\right)}, \tag{5.1}$$

$$p(v_i = 1|\mathbf{h}) = \frac{1}{exp\left(-b_i - \sum_j h_i w_{ji}\right)}, \tag{5.2}$$

So the likelihood functions of layers are as follows.

$$p(\mathbf{h}|\mathbf{v}) = \prod_j p(h_j|\mathbf{v}), \tag{5.3}$$

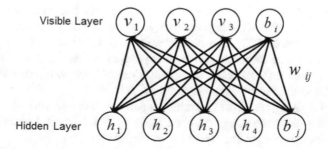

**Fig. 5.1** A restricted Boltzmann machine (RBM)

$$p(\mathbf{v}|\mathbf{h}) = \prod_i p(v_j|\mathbf{h}). \tag{5.4}$$

Network energy of RBM is given by

$$E(\mathbf{v}, \mathbf{h}) = -\sum_i b_i v_i - \sum_j b_j h_j - \sum_{ij} v_i h_j w_{ij}. \tag{5.5}$$

To represent an input vector, Markov chain Monte Carlo (MCMC) method is used for RBM. The state of hidden layer $\mathbf{h}$ can be obtained by sampling with $p(\mathbf{h}|\mathbf{v})$. Then according to the probability $p^{(1)}(\mathbf{v}|\mathbf{h})$ the new state of the visible layer $\mathbf{v}^{(1)} \in \{0, 1\}$ can be obtained by the stochastic sampling. After repeating this process with $k = 1, 2, \ldots, \infty$ steps, i.e., a MCMC process, the state of RBM convergences to stable.

### 5.2.2 Training of RBM and DBN

Parameters $(w_{ij}, b_i, b_j) \equiv \theta$ of RBM are optimized by the gradient of the log likelihood function.

$$\theta \leftarrow \theta + \alpha \frac{\partial p(\mathbf{v}, \mathbf{h})}{\partial \theta}, \tag{5.6}$$

where $0 \le \alpha \le 1$ is a learning rate.

In fact, the learning rule of RBM is as follows.

$$\Delta w_{ij} = p^{(k-1)}(v_i = 1)p^{k-1}(h_j = 1) - p^{(k)}(v_i = 1)p^{(k)}(h_j = 1), \tag{5.7}$$

$$\Delta b_j = p^{k-1}(h_j = 1) - p^{(k)}(h_j = 1), \tag{5.8}$$

$$\Delta b_i = p^{k-1}(v_i = 1) - p^{(k)}(v_i = 1), \tag{5.9}$$

where $k = 1, 2, \ldots, K$ is the step of MCMC. In practice, when the number of training iterations (epochs) is large enough, $K = 1$ also works well, and this kind of contrastive divergence (CD) algorithm is called CD-1. Furthermore, instead of the original data input to the visible units, it is possible to "initialize a Markov chain at the state in which it ended for the previous model," which training algorithm is called persistent contrastive divergence (PCD) [28, 29].

For DBN training, each RBM is trained separately using the above learning rules as pre-training, and then using error back-propagation (BP) [30] as fine-tuning to modify the parameters obtained from the pre-training.

Now, let $v_i^L \in \{0, 1\}$ and $h_j^L \in \{0, 1\}$ be the visible unit $i$ and hidden unit $j$'s state in layer $L$, and the connection between $i$ and $j$ with a weight $w_{ij}^L = w_{ji}^L \in [0,1]$, and a bias $b_i^L = b_j^L$, the training rule of BP for these parameters' modification is given as follows.

$$\Delta w_{ij}^L = -\epsilon \left( \sum_i \frac{\partial E}{\partial w_{ji}^{L+1}} w_{ji}^{L+1} \right) \left( 1 - h_j^L \right) v_i^L, \qquad (5.10)$$

$$\Delta b_i^L = -\epsilon \left( \sum_i \frac{\partial E}{\partial w_{ji}^{L+1}} w_{ji}^{L+1} \right) \left( 1 - h_j^L \right), \qquad (5.11)$$

where $E = \sum_{t=1}^{T} (y_t - \hat{y}_t)^2$ is the squared errors of output of the network.

### 5.2.3  DBNs used in Time Series Forecasting

There have been thousands of studies of time series forecasting approached by artificial neural networks [31, 32]. The principle of neural forecasting systems is illustrated in Fig. 5.2 For a time series data $y(t), t = 1, 2, \ldots, T, n \le T$ data are used as the input of a neural network: $\mathbf{x}(t) = (y(t), y(t - \tau), y(t - 2\tau), \ldots, y(t - n\tau))$, where $n\tau \le T$, $\tau$ is an integer that indicates time lag, and $n$ is the dimension of the input. The output of the network is a scalar $\hat{y}(t + \tau)$. Using $N$ samples $(\mathbf{x}(t), y(t + \tau)), (\mathbf{x}(t + 1), y(t + 1 + \tau)), \ldots, (\mathbf{x}(t + N - 1), y(t + N - 1 + \tau))$,

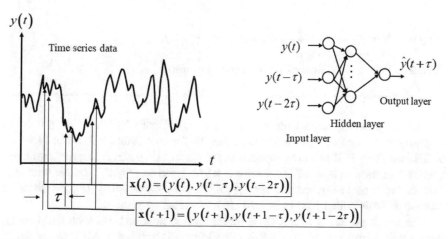

**Fig. 5.2** Time series forecasting by artificial neural networks

the neural network can be trained by gradient methods of loss (error) functions, or cross entropy functions.

For example, a feed-forward neural network (See Fig. 5.2) with an input layer, a hidden layer, and an output layer, named multi-layered perceptron (MLP) [30] can be trained by BP algorithm as follows.

### Algorithm I: Error Back-Propagation (BP Algorithm)

Step 1    Input data $\mathbf{x}(t) = (y(t), y(t - \tau), \ldots, y(t - n\tau))$ to MLP.

Step 2    Predict a future data $y(t + \tau)$ according to Eqs. (5.12) and (5.13).

$$f(y + \tau) = \frac{1}{1 + exp\left(-\sum_{j=1}^{K+1} w_j f(z_j)\right)}, \qquad (5.12)$$

$$f(z_j) = \frac{1}{1 + exp\left(-\sum_{i=0}^{n} v_i y(t - i\tau)\right)}, \qquad (5.13)$$

where $v_i$ and $w_j$ are the connection weights of input units with hidden units, and hidden units with output unit.

Step 3    Calculate the modification of connection weights, $\Delta w_j$, $\Delta v_{ji}$ according to Eqs. (5.14) and (5.15).

$$\Delta w_j = -\epsilon (y(t + \tau) - \hat{y}(t + \tau)) y(t + \tau)(1 - y(t + \tau)) z_j, \qquad (5.14)$$

$$\Delta v_{ji} = -\epsilon (y(t + \tau) - \hat{y}(t + \tau)) y(t + \tau)(1 - y(t + \tau)) z_j w_j (1 - z_j) x_i. \qquad (5.15)$$

Step 4    Modify the connections, $w_j \leftarrow w_j + \Delta w_j$,
          $v_{ji} \leftarrow v_{ji} + \Delta v_{ji}$ .

Step 5    For the next time step $t + 1$, return to step 1.

For the powerful representation ability of DBN with RBMs, it is utilized to capture the feature of time series data then to forecast future values. In [13–15], DBNs with 2–4 RBMs were proposed in the field of time series forecasting firstly, (Fig. 5.3 gives a sample of DBN with 2 RBMs) and in [17–19], DBNs with 0–3 RBMs and a multi-layered perceptron (MLP) were constructed (Fig. 5.4 shows a sample of DBN with 1 RBM and 1 MLP) as forecasting systems.

The training of DBN with RBMs and MLP is the same as DBN with RBMs only, which was introduced in Sect. 5.2.2, but adding BP training to MLP as shown in Algorithm I.

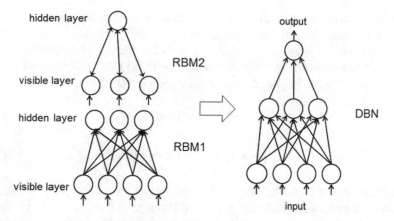

**Fig. 5.3**  A DBN with 2 RBMs [14]

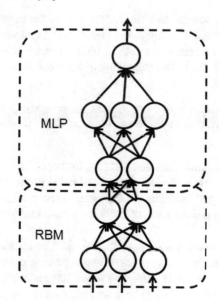

**Fig. 5.4**  A DBN with 2 RBMs and MLP [17]

## 5.3    Design DBN with PSO or Random Search

### 5.3.1    Design DBN with PSO

To design the optimal structure of DBN, Kennedy and Eberhart's particle swarm optimization (PSO) [16] is adopted in [13–15].

PSO is a well-known evolutionary computation method for optimization problems. Let candidate solutions (particles) $\mathbf{x}_i^k \in \Re^n$ be positions in the hyper-parameter space at time $k$, $i = 1, 2, \ldots, m$ is the number of particles. They are evaluated and improved by the cost function $\min_{\mathbf{x}} f(\mathbf{x})$ using its moment $\mathbf{x}_i^k + \mathbf{v}_i^k$, its historic best position $\mathbf{pbest}_i^k$, and the best particle $\mathbf{gbest}^k$ in the swarm.

$$\mathbf{x}_i^{k+1} = \mathbf{x}_i^k + \mathbf{v}_i^k \tag{5.16}$$

$$\mathbf{v}_i^{k+1} = w\mathbf{v}_i^k + c_1 r_1 \left( \mathbf{pbest}_i^k - \mathbf{x}_i^k \right) + c_2 r_2 \left( \mathbf{gbest}^k - \mathbf{x}_i^k \right), \tag{5.17}$$

where the initial values of $\mathbf{x}$, $\mathbf{v}$ are random numbers, $k = 1, 2, \ldots, K$ is the iteration number, damping coefficient $w \geq 0$, parameters $c_1, c_2 \geq 0$, $r_1, r_2 \in [0,1]$ are random numbers. Repeat Eqs. (5.16) and (5.17), the $\mathbf{gbest}^k$ is modified and converges to the optimal solution.

For DBN structure design, the number of layers, the number of units, and the learning rate of different RBMs are set as elements of input vector of PSO. The cost function is the mean squared error (MSE) between the teacher data $y(t + \tau)$ and the output of DBN $\hat{y}(t + \tau)$. The procedure of PSO for DBN architecture design is shown in Algorithm II.

### Algorithm II: Design a DBN with PSO

Step 1    Select hyper-parameters, such as the number of RBMs, the number of units in different RBMs, the learning rate of RBM pre-training, the learning rate of fine-tuning of DBN, as the exploration space of particles of PSO.

Step 2    Decide the population size of particles $P$ and limitation of iteration number $I$.

Step 3    Initialize the start position $\mathbf{x}_i^{k=0}$ and $\mathbf{v}_i^{k=0}$ using random integer or real numbers in exploration ranges of hyper-parameter space.

Step 4    Evaluate each particle using the mean squared error (MSE) between the prediction value $\hat{y}(t + \tau)$ and teacher data $y(t + \tau)$, and find the best position $\mathbf{pbest}_i^k$ of a particle from its history, and the best particle position of the swarm $\mathbf{gbest}^k$.

Step 5    Renew positions $\mathbf{x}_i^{k+1}$ and velocities $\mathbf{v}_i^{k+1}$ of particles by Eqs. (5.16) and (5.17), respectively. Notice that integer elements in these vectors need to be rounded off for Step 4.

Step 6    Finish: if the evaluation function (prediction MSE of training data) converged, or $k = I$; else return to Step 4.

## 5.3.2  Design DBN with RS

To find an optimal architecture of neural network, grid search (GS) is conventionally used. The number of layers, the number of units on each units and the learning rate, etc., are considered as elements in a hyper-parameter vector space. In GS, the value of each element is tuned according to the reduction of loss function of the network, whereas values of other elements are fixed in the previous values, and this process is iterated for all elements of the hyper-parameter space. For the tuning intervals of GS are usually fixed, the optimal parameter in the continuous space may be leaked (See Fig. 5.5).

As an alternative to GS, Bergstra and Bengio proposed a random search (RS) method in [10]. As shown in Fig. 5.6, various values of elements of hyper-parameters can be selected as candidates, and repeating this process enough times, the optimal architecture of neural networks is available to be found more than GS. In [17–19], RS was used to find the number of RBMs, the number of units, learning rates of RBMs, and learning rate of fine-tuning to construct a DBN with multiple RBMs and a MLP instead of PSO. The algorithm of RS is very simple which is shown in Algorithm III.

---

**Algorithm III: Design a DBN with RS**

---

Step 1    Select hyper-parameters, such as the number of RBMs, the number of units in different RBMs, the learning rate of RBM pre-training, the learning rate of fine-tuning of DBN, as the exploration space of RS.

Step 2    Decide the iteration number $I$ of exploration.

Step 3    Initialize the hyper-parameter vector $x^{k=0}$ using random real numbers (in limited intervals of different elements).

Step 4    Store the current mean squared error (MSE) between the prediction value $\hat{y}(t + \tau)$ and teacher data $y(t + \tau)$.

Step 5    Compare the stored MSEs to find the smallest one and using the corresponding hyper-parameter $x^k$ as the optimal structure of DBN.

Step 6    Finish: if the evaluation function (prediction MSE of training data) converged, or $k = I$;
          else return to Step 3.

---

**Fig. 5.5** Grid search

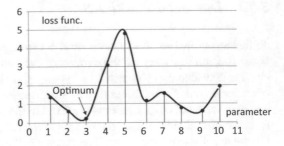

**Fig. 5.6** Random search

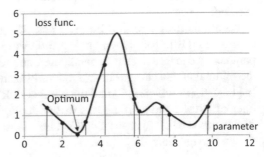

## 5.4 Experimental Comparison of PSO and RS

Time series forecasting experiments with different DBNs structured by PSO and RS are reported in this section. Three time series data were utilized in the experiments: Lorenz chaos [24], Hénon map [25], and CATS benchmark [26, 27].

Chaos is a complex nonlinear phenomena in which dynamical states are sensitive to initial conditions, and impossible for long-term prediction in general. In our experiments, for the chaotic time series forecasting (Lorenz chaos and Hénon map), one-ahead forecasting was performed. Meanwhile, for CATS data, long-term forecasting, which uses the output of predicted values as the next input to DBN, was executed.

### 5.4.1 One-ahead Prediction of Lorenz Chaos

Lorenz chaos [24] is given by 3-dimension nonlinear formulae as follows.

$$\begin{cases} \frac{dx}{dt} = \sigma x + \sigma y; \\ \frac{dy}{dt} = -xz + rx - y; \\ \frac{dz}{dt} = xy - bz. \end{cases} \quad (5.18)$$

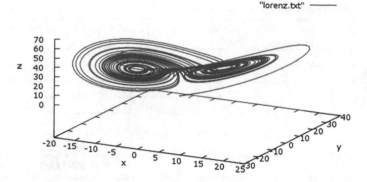

**Fig. 5.7** Lorenz chaos [24]

Let parameter $\sigma = 10, b = 28, r = 8/3, \Delta t = 0.01$, an attractor, i.e., the well-known "butterfly effect," can be obtained as in Fig. 5.7.

The chaotic time series of $x - axis$ of Lorenz chaos were used in the forecasting experiment. The length of time series was 1000, and within it, 600 data were used as training samples, 200 for validation, and 200 unknown data were used as a test set.

The exploration ranges of the number of RBMs were 1–4 for DBN with RBMs and 0–4 for DBN with RBMs and MLP. The ranges of number of units for input and hidden layers of DBNs were 2–20. For learning rates of RBMs and MLP were 0.1–0.00001, and the limitations of training iterations (epochs) were 5000 for RBM's pre-training and 10,000 for fine-tuning using BP algorithm.

### 5.4.1.1   DBN Decided by PSO for Lorenz Chaos

Four neural networks were optimized by PSO algorithm (See Algorithm II) for Lorenz chaos forecasting: a MLP, a DBN with 2 RBMs [13, 14], a DBN with 1 RBM and 1 MLP [15], and a DBN with 2 RBMs and 1 MLP [15]. The number of particles (population) was 10, the number of iterations was 15, and the exploration ranges of the number of units in each layers were [2–20] except the output layer which had 1 unit, respectively. The interval of time lag $\tau = 1$, the range of learning rates of RBM and BP were [$10^{-1}$, $10^{-5}$]. The upper number of training iterations (epochs) were 5000 for RBM's pre-training and 10,000 for fine-tuning using BP algorithm. The exploration results of PSO are shown in Table 5.1.

The training and one-ahead forecasting results by the 2-RBM DBN decided by PSO are shown in Figs. 5.8 and 5.9, respectively. Prediction accuracy (MSE) in detail is shown in Table 5.2, where the model composed by 1 RBM and 1 MLP had the lowest training and forecasting error (in bold values).

**Table 5.1** Hyper-parameters optimized by PSO for Lorenz chaos

| Model | Structure | Learning rate of RBM1 | Learning rate of RBM2 | Learning rate of BP |
|---|---|---|---|---|
| MLP | 5-2-1 | – | – | 0.1 |
| 2 RBMs | 7-7-1 | 0.093 | 0.074 | 0.092 |
| 1 RBM + MLP | 5-9-2-1 | 0.1 | – | 0.1 |
| 2 RBMs + MLP | 7-12-19-2-1 | 0.09 | 0.22 | 0.089 |

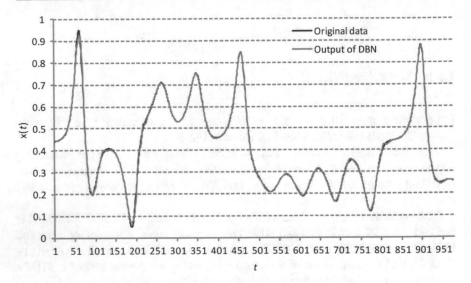

**Fig. 5.8** Training result of Lorenz chaos using a 2-RBM DBN with PSO structuring

### 5.4.1.2 DBN Decided by RS for Lorenz Chaos

Different from the experiment reported in Sect. 5.4.1.1, the number of RBMs and MLP were also adopted in the hyper-parameter space in the case of RS for DBN architecture decision. So the four models described in Sect. 5.4.1.1 were able to be investigated in one optimization process. Other conditions used in this experiment were the same as in the case of PSO for Lorenz chaotic time series forecasting. The explored hyper-parameters are shown in Table 5.3 which means that the optimal architecture of DBN was composed by one RBM (5-11) and MLP (11-2-1), i.e., 5-11-2-1, and the learning rates of RBM (CD-1 learning) and BP learning were 0.042 and 0.091, respectively.

The prediction accuracies of a DBN decided by RS for Lorenz chaos are shown in Table 5.4. The training MSE ($= 0.94 \times 10^{-5}$) was lower than PSO method ($= 1.0 \times 10^{-5}$), but RS's forecasting MSE was higher than PSO (RS $= 0.5 \times 10^{-5}$, PSO $= 0.3 \times 10^{-5}$).

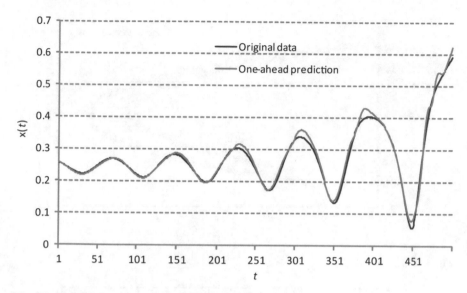

**Fig. 5.9** One-ahead prediction result of Lorenz chaos using a 2-RBM DBN with PSO structuring

**Table 5.2** Prediction accuracies (MSE) for Lorenz chaos by different models optimized by PSO

| Model | Training error | Forecasting error |
|---|---|---|
| MLP | 2.7 | 1.0 |
| 2 RBMs | 3.2 | 1.3 |
| **1 RBM +MLP** | **1.0** | **0.3** |
| 2 RBMs + MLP | 1.6 | 0.7 |

Unit: $\times 10^{-5}$

**Table 5.3** Hyper-parameters optimized by RS for Lorenz chaos

| Model | Structure | Learning rate of RBM | Learning rate of BP |
|---|---|---|---|
| 1 RBM + MLP | 5-11-2-1 | 0.42 | 0.1 |

**Table 5.4** Prediction accuracies (MSE) of Lorenz chaos by the optimal model given by RS

| Model | Training error | Forecasting error |
|---|---|---|
| 1 RBM + MLP | 0.94 | 0.5 |

Unit: $\times 10^{-5}$

## 5.4.2   One-ahead Prediction of Hénon Map

Hénon [25] is given by 2-dimension nonlinear formulae as follows.

$$\begin{cases} x(t+1) = 1 - ax(t)^2 + y(t); \\ y(t+1) = bx(t). \end{cases} \qquad (5.19)$$

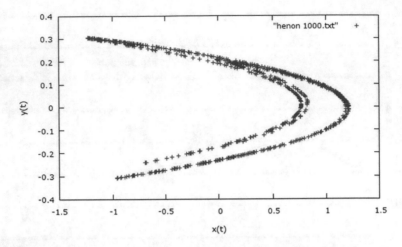

**Fig. 5.10** Hénon map [24]

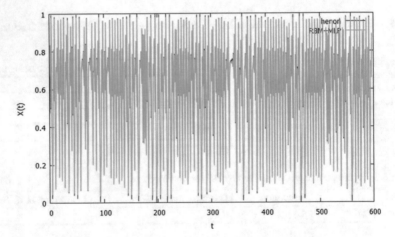

**Fig. 5.11** Training result of Hénon map using a DBN with 2 RBMs and MLP by RS structuring

Let parameter $a = 1.4, b = 0.3$, the attractor can be obtained as in Fig. 5.10. The chaotic time series of $x - axis$ of Hénon map were used. The length of time series was 1000, and within it, 600 data were used as training samples, 200 for validation, and 200 unknown data were used as a test set. The training and one-ahead forecasting results using a DBN with 2 RBMs and MLP structured by RS are shown in Fig. 5.11.

In experiment conditions for Hénon map time series forecasting with PSO/RS structure optimization process were as same as in the case of Lorenz chaos (Sect. 5.4.1) Hyper-parameters of DBN optimized by PSO and RS are shown in Tables 5.5 and 5.6. From Table 5.6, as same as for Lorenz chaos, DBN with 1 RBM and 1 MLP was chosen for Hénon map by RS, meanwhile, DBN with 2 RBMs and

**Table 5.5** Hyper-parameters optimized by PSO for Hénon map

| Model | Structure | Learning rate of RBM1 | Learning rate of RBM2 | Learning rate of BP |
|-------|-----------|-----------------------|-----------------------|---------------------|
| MLP | 2-16-1 | – | – | 0.1 |
| 2 RBMs | 2-16-1 | 0.025 | 0.082 | 0.041 |
| 1 RBM + MLP | 2-9-9-1 | 0.1 | – | 0.1 |
| 2 RBMs + MLP | 2-11-20-15-1 | 0.1 | 0.1 | 0.085 |

**Table 5.6** Hyper-parameters optimized by RS for Lorenz chaos

| Model | Structure | Learning rate of RBM | Learning rate of BP |
|-------|-----------|----------------------|---------------------|
| 1 RBM + MLP | 14-18-14-1 | 0.0001 | 0.0895 |

**Table 5.7** Comparison of accuracies (MSE) between PSO and RS using Hénon map

| Model | Training PSO/RS | Forecasting PSO/RS |
|-------|-----------------|--------------------|
| MLP | 288/– | 79/– |
| 2 RBMs | 3912/– | 918/– |
| 1 RBM + MLP | 135/**5.52** | 36/**9.09** |
| 2 RBMs + MLP | **130**/– | **31**/– |

Unit: $\times 10^{-5}$

1 MLP was chosen by PSO. The one-ahead prediction accuracies (MSEs) of DBNs structured by PSO and RS using Hénon map data are listed in Table 5.7. The highest training accuracy and forecasting accuracy were achieved by the DBN with 1 RBM and MLP found by RS, which $MSE$ are $5.52 \times 10^{-5}$ and $9.09 \times 10^{-5}$ (in bold values).

### 5.4.3 Long-term Prediction of CATS Benchmark

Since the middle of the 1980s, there have been thousands of publications of time series forecasting using the method of artificial neural networks (ANN). In 2004 International Joint Conference on Neural Networks (IJCNN'04), Lendasse et al. organized a time series forecasting competition with a benchmark data CATS [26], and the results were reported in [27]. CATS includes 5000 time points data, which are separated into 5 blocks, and the last 20 data in each block are hidden by the organizer (See Fig. 5.12).

The evaluation of the competition is the prediction accuracies which are the average MSEs of the five blocks named $E_1$, and the average MSEs of the first four blocks $E_2$. So for the unknown 20 data, long-term prediction is necessary, i.e., using the output of the predictor as input recurrently.

$$E_1 = \frac{\sum_{t=981}^{1000}(y_t - \hat{y}_t)^2}{100} + \frac{\sum_{t=1981}^{2000}(y_t - \hat{y}_t)^2}{100} + \frac{\sum_{t=2981}^{3000}(y_t - \hat{y}_t)^2}{100} \quad (5.20)$$

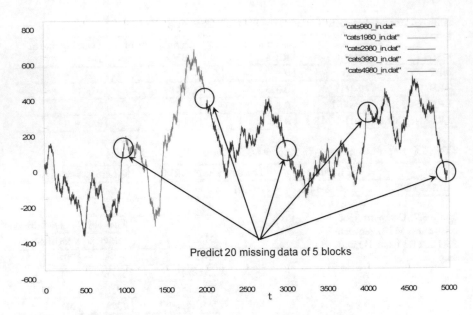

**Fig. 5.12** CATS benchmark [26, 27]

**Table 5.8** Comparison of accuracies (MSE) between PSO and Random Search (RS) using CATS benchmark (∗)

| Model | E1 | E2 |
|---|---|---|
| The best of IJCNN'04 [27] | 408 | 222 |
| DBN with 2 RBMs using PSO [13, 14] | 1215 | 979 |
| DBN with 1 RBM and MLP using RS | 257 | 252 |

∗ long-term forecasting

$$+\frac{\sum_{t=3981}^{4000}(y_t - \hat{y}_t)^2}{100} + \frac{\sum_{t=4981}^{5000}(y_t - \hat{y}_t)^2}{100} \tag{5.21}$$

$$E_2 = \frac{\sum_{t=981}^{1000}(y_t - \hat{y}_t)^2}{80} + \frac{\sum_{t=1981}^{2000}(y_t - \hat{y}_t)^2}{80} \tag{5.22}$$

$$+\frac{\sum_{t=2981}^{3000}(y_t - \hat{y}_t)^2}{80} + \frac{\sum_{t=3981}^{4000}(y_t - \hat{y}_t)^2}{80}. \tag{5.23}$$

The prediction results by conventional methods and DBNs with PSO and RS are shown in Table 5.8. Comparing to the best of IJCNN'04 $E_1 = 408$ which used a Kalman smoother method [27], the DBN structured by RS had a remarkable advance with $E_1 = 257$. In the case of $E_2$, the DBN had a similar accuracy to the best of IJCNN'04, which used ensemble models [27]. Hyper-parameters of DBNs optimized by PSO and RS for different block data are shown in Table 5.9. It seems that RS had better prediction accuracies than PSO from Table 5.8; however, a fact is that MLP was not composed in the DBN optimized by PSO in [13, 14]. In

**Table 5.9** Hyper-parameters optimized by PSO and RS for CATS benchmark

| Model | Structure PSO/RS | Learning rate of RBM(s) PSO/RS | Learning rate of BP PSO/RS |
|---|---|---|---|
| First block | 20-18-1/18-20-14-1 | 0.47, 0.99/0.083 | 0.96/0.082 |
| Second block | 11-3-1/16-19-15-1 | 0.85, 0.95/0.0046 | 0.99/0.076 |
| Third block | 8-13-1/18-18-13-1 | 0.90, 0.77/0.046 | 0.87/0.088 |
| Fourth block | 20-7-1/18-14-19-1 | 0.92, 0.88/0.12 | 0.53/0.083 |
| Fifth block | 20-14-1/19-19-17-1 | 0.17, 0.19/0.079 | 0.39/0.075 |

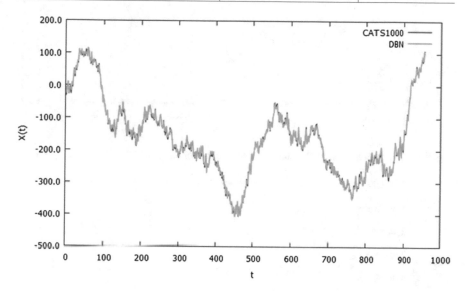

**Fig. 5.13** The training result of a DBN for the 1st block of CATS structured by RS

Figs. 5.13 and 5.14, the training result and long-term prediction (20 future unknown data) of a DBN (18-20-14-1) for the first block of CATS are shown, respectively. The DBN was given by RS, which has 1 RBM and 1 MLP. The number of input units of RBM was 18, and output 20. The number of MLP units was 20, and hidden units were 14, output 1. The learning rates of RBM and MLP were 0.083 and 0.082, respectively. Details of DBNs for different blocks of CATS are shown in Table 5.9. The efficiencies of PSO and RS can be observed by the change of prediction errors (MSE) during training iteration. For example, the change of MSE of the DBN optimized in RS learning is shown in Fig. 5.15, and the evolution of 10 particles in PSO algorithm during training can be confirmed by Fig. 5.16. The change of the number of input layers and the number of hidden layers of RBMs was yielded by the best particle of swarm during PSO exploration (See Figs. 5.17 and 5.18, respectively).

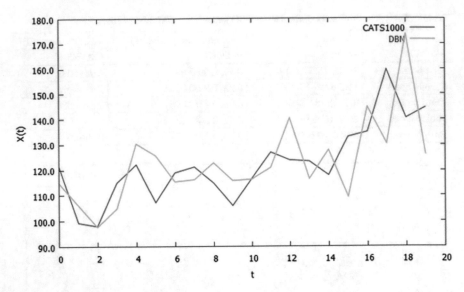

**Fig. 5.14** The long-term predict result of a DBN for the 1st block of CATS structured by RS

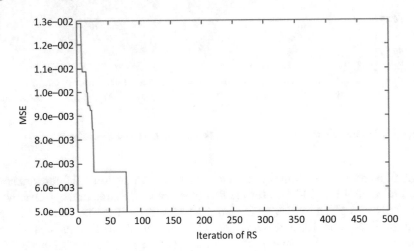

**Fig. 5.15** The change of MSE in RS learning for a DBN with 1 RBM and 1 MLP

## 5.5 Advanced Forecasting Systems Using DBNs

The study of time series forecasting has a long history because it plays an important role in social activities such as finance, economy, industry, etc., and natural sciences. The most famous linear prediction model is the autoregressive integrated moving average (ARIMA) which was proposed in the 1970s [33]. Artificial neural networks (ANN) became novel efficient predictors as nonlinear models since the 1980s

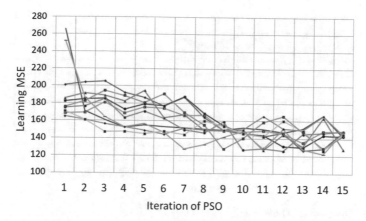

**Fig. 5.16**  The evolution of particles in PSO learning for a DBN with 2 RBMs

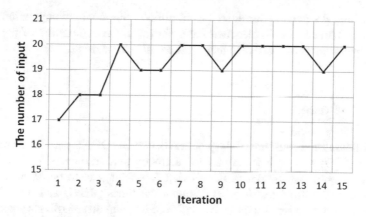

**Fig. 5.17**  The number of input units

[26, 27, 31]. Considering the linear and nonlinear features of time series data, Zhang proposed a hybrid forecasting method which using ARIMA to forecast the unknown data in the future, then using a trained ANN which can forecast the residual of predicted data by ARIMA. In [17, 34], Zhang's hybrid model was modified by using a DBN, which was structured by RS introduced in Sect. 5.3, to forecast the unknown data in future, and then using ARIMA to forecast residuals. It is confirmed by the comparison experiments that the hybrid models ARIMA+DBN and DBN+ARIMA were priority to using ARIMA or DBN independently. For real time series forecasting, a DBN with reinforcement learning method "stochastic gradient ascent (SGA)" [35] is proposed recently [19, 36]. SGA was used as fine-tuning method instead of the conventional BP method. In forecasting experiments, the number of Sunspot, sea level pressure, atmospheric CO2 concentration, etc. provided by Aalto University [37], Hyndman [38] were utilized. All of these advanced DBNs were designed by the RS algorithm, i.e., Algorithm III, and the

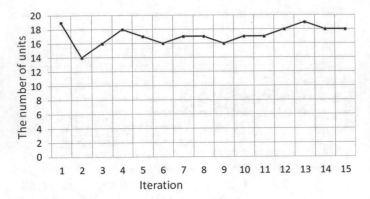

**Fig. 5.18** The number of hidden units

reason is that RS may be considered a simplified PSO in which population size is one, and the process is simpler than PSO. However, the comparison of precisions and computational costs between them still remains to be done.

## 5.6 Conclusion

To find the optimal structure of deep neural networks (DNN), particle swarm optimization (PSO) and random search (RS) methods were introduced in this chapter. As a novel nonlinear model of time series forecasting, deep belief nets (DBN) composed by multiple restricted Boltzmann machines (RBM) and multi-layered perceptron (MLP) were structured by PSO and RS, and these DBNs were superior to conventional methods according to chaotic time series and benchmark data forecasting experiment results. However, as we know, the evolutionary computation including swarm intelligence methods is developed rapidly, meanwhile PSO and RS utilized in the papers mentioned here were the original versions. So it is very interesting to adopt advanced optimization methods to decide the optimal structure of DNNs.

## References

1. Mnih, V., et al.: Human-level control through deep reinforcement learning. Nature **518**, 529–533 (2015)
2. Sivler, D., et al.: Mastering the game of go with deep neural networks and tree search. Nature **529**, 484–489 (2016)
3. Sivler, D., et al.: Mastering the game of go without human knowledge. Nature **550**, 354–359 (2017)

4. Williams, R.J., Simple statistical gradient following algorithms for connectionist reinforcement learning. Mach. Learn. **8**, 229–256 (1992)
5. Hinton, G.E., Salakhutdinov, R.R.: Reducing the dimensionality of data with neural networks. Science **313**, 504–507 (2006)
6. Hinton, G.E., Osindero, S., The, Y.W.: A faster learning algorithm for deep belief nets. Neural Comput. **18**(7), 1527–1544 (2006)
7. Roux, N.L., Bengio, Y.: Representational power of restricted Boltzmann machines and deep belief networks. Neural Comput. **20**(6), 1631–1649 (2008)
8. Krizhevsky, A., Sutskever, I., Hinton, G.E.: ImageNet classification with deep convolutional neural networks. In: Proceedings of the 25th International Conference on Neural Information Processing Systems (NIPS'12), vol.1, pp. 1097–1105 (2012)
9. IMAGENET Large Scale Visual Recognition Challenge (2012). http://www.image-net.org/challenges/LSVRC/2012/
10. Bergstra, J., Bengio, Y.: Random search for hyper-parameter optimization. J. Mach. Learn. Res. **13**, 281–305 (2012)
11. Ba, J., Frey, B.: Adaptive dropout for training deep neural networks. In: Advances in Neural Information Processing Systems (NIPS2013), vol. 26, pp. 3084–3092 (2013)
12. Nowakowski, G., Dorogyy, Y., Doroga-Ivaniuk, O.: Neural network structure optimization algorithm. J. Autom. Mob. Robot. Intell. Syst. **12**(1), 5–13 (2018)
13. Kuremoto, T., Kimura, S., Kobayashi, K., Obayashi M.: Time series forecasting using restricted Boltzmann machine. In: Proceedings of the 8th International Conference on Intelligent Computing (ICIC 2012). Communications in Computer and Information Science (CCIS), vol. 304, pp. 17–22. Springer, Berlin (2012)
14. Kuremoto, T., Kimura, S., Kobayashi, K., Obayashi, M.: Time series forecasting using a deep belief network with restricted Boltzmann machines. Neurocomputing **137**(5), 47–56 (2014)
15. Kuremoto, T., Hirata, T., Obayashi, M., Mabu, S., Kobayashi, K.: Forecast chaotic time series data by DBNs. In: Proceedings of the 7th International Congress on Image and Signal Processing (CISP 2014), pp. 1304–1309 (2014)
16. Kennedy, J., Eberhart, R.C.: Particle swarm optimization. In: IEEE International Conference on Neural Networks, pp. 1942–1948 (1995)
17. Hirata, T., Kuremoto, T., Obayashi, M., Mabu, S.: Time series prediction using DBN and ARIMA. In: Proceedings of International Conference on Computer Application Technologies (CCATS 2015), pp. 24–29 (2015)
18. Hirata, T., Kuremoto, T., Obayashi, M., Mabu, S., Kobayashi, K.: Deep belief network using reinforcement learning and its applications to time series forecasting. In: Proceedings of International Conference on Neural Information Processing (ICONIP' 16). Lecture Notes in Computer Science (LNCS), vol. 9949, pp. 30–37. Springer, Berlin (2016)
19. Hirata, T., Kuremoto, T., Obayashi, M., Mabu, S., Kobayashi, K.: Forecasting real time series data using deep belief net and reinforcement learning. J. Robot. Netw. Artif. Life **4**(4), 260–264 (2018). https://doi.org/10.2991/jrnal.2018.4.4.1
20. Simonyan, K., Zisserman, A.: Very deep convolutional networks for large-scale image recognition. arXiv preprint:1409.15566, CISP2014
21. He, K., Zhang, X., Ren, S., Sun, J.: Deep residual learning for image recognition. In: 2016 IEEE Conference on Computer Vision and Pattern Recognition (CVPR), pp. 770–778 (2016)
22. Wang, B., Sun, Y., Xue, B., Zhang, M.: Evolving deep convolutional neural networks by variable-length particle swarm optimization for image classification. In: 2018 IEEE Congress on Evolutionary Computation (CEC), pp. 1514–1521 (2018)
23. Fernandes Jr., F.E., Yen, G.G.: Particle swarm optimization of deep neural networks architectures for image classification. Swarm Evol. Comput. **49**, 62–73 (2019)
24. Lorenz, E.N.: Deterministic nonperiodic flow. J. Atmos. Sci. **20**, 130–141 (1963)
25. Henon, M.: A two-dimensional mapping with a strange attractor. Commun. Math. Phys. **50**(1), 69–77 (1976)
26. Lendasse, A., Oja, E., Simula, O., Verleysen, M.: Time series prediction competition: the CATS benchmark. In: Proceedings of international joint conference on neural networks (IJCNN'04), pp. 1615–1620 (2004)

27. Lendasse, A., Oja, E., Simula, O., Verleysen, M.: Time series prediction competition: the CATS benchmark. Neurocomputing **70**, 2325–2329 (2007)
28. Tieleman, T.: Training restricted Boltzmann machines using approximations to the likelihood gradient. In: Proceedings of the 25 International Conference on Machine Learning (ICML '08), pp. 1064–1071 (2008)
29. Kuremoto, T., Tokuda, S., Obayashi, M., Mabu, S., Kobayashi, K.: An experimental comparison of deep belief nets with different learning methods. In: Proceedings of 2017 RISP International Workshop on Nonlinear Circuits, Communications and Signal Processing (NCSP 2017), pp. 637–640 (2017)
30. Rumelhart, D.E., Hinton, G.E., Williams, R.J.: Learning representation by back-propagating errors. Nature **232**(9), 533–536 (1986)
31. NN3: http://www.neural-forecasting-competition.com/NN3/index.htm
32. Kuremoto, T., Obayashi, M., Kobayashi, M.: Neural forecasting systems. In: Weber, C., Elshaw, M., Mayer, N.M. (eds.) Reinforcement Learning, Theory and Applications, Chapter 1, pp. 1–20 (2008). InTech
33. Box, G.E.P., Pierce, D.A.: Distribution of residual autocorrelations in autoregressive-integrated moving average time series models. J. Am. Stat. Assoc. **65**(332), 1509–1526 (1970)
34. Hirata, T., Kuremoto, T., Obayashi, M., Mabu, S., Kobayashi, K.: A novel approach to time series forecasting using deep learning and linear model. IEEJ Trans. Electron. Inf. Syst. **136**(3), 248–356 (2016, in Japanese)
35. Kimura, H., Kobayashi, S.: Reinforcement learning for continuous action using stochastic gradient ascent. In: Proceedings of 5th Intelligent Autonomous Systems (IAS-5), pp. 288–295 (1998)
36. Kuremoto, T., Hirata, T., Obayashi, M., Mabu, S., Kobayashi, K.: Training deep neural networks with reinforcement learning for time series forecasting. In: Time Series Analysis - Data, Methods, and Applications, InTechOpen (2019)
37. Aalto University Applications of Machine Learning Group Datasets. ⟨http://research.ics.aalto.fi/eiml/datasets.shtml⟩ (01-01-17)
38. Hyndman, R.J.: Time Series Data Library (TSDL) (2013). ⟨http://robjhyndman.com/TSDL/⟩ (01-01-13)

# Part III
# Structure Optimization

# Chapter 6
# Particle Swarm Optimization for Evolving Deep Convolutional Neural Networks for Image Classification: Single- and Multi-Objective Approaches

Bin Wang, Bing Xue, and Mengjie Zhang

**Abstract** Convolutional neural networks (CNNs) are one of the most effective deep learning methods to solve image classification problems, but the design of the CNN architectures is mainly done manually, which is very time consuming and requires expertise in both problem domains and CNNs. In this chapter, we will describe an approach to the use of particle swarm optimization (PSO) for automatically searching for and learning the optimal CNN architectures. We will provide an encoding strategy inspired by computer networks to encode CNN layers and to allow the proposed method to learn variable-length CNN architectures by focusing only on the single objective of maximizing the classification accuracy. A surrogate dataset will be used to speed up the evolutionary learning process. We will also include a multi-objective way for PSO to evolve CNN architectures in the chapter. The PSO-based algorithms are examined and compared with state-of-the-art algorithms on a number of widely used image classification benchmark datasets. The experimental results show that the proposed algorithms are strong competitors to the state-of-the-art algorithms in terms of classification error. A major advantage of the proposed methods is the *automated design* of CNN architectures without requiring human intervention and good performance of the learned CNNs.

## 6.1 Introduction

Image classification has been attracting increasing interest both from the academic and industrial researchers due to the exponential growth of images in terms of both the number and the resolution, and the meaningful information extracted from images. Convolutional neural networks (CNNs) have been investigated to solve the

B. Wang · B. Xue · M. Zhang (✉)
School of Engineering and Computer Science, Victoria University of Wellington, Wellington, New Zealand
e-mail: bin.wang@ecs.vuw.ac.nz; bing.xue@ecs.vuw.ac.nz; mengjie.zhang@ecs.vuw.ac.nz

© Springer Nature Singapore Pte Ltd. 2020
H. Iba, N. Noman (eds.), *Deep Neural Evolution*, Natural Computing Series,
https://doi.org/10.1007/978-981-15-3685-4_6

155

image classification tasks and are considered a state-of-the-art approach to image classification. However, designing the architectures of CNNs for specific tasks can be extremely complex, which can be seen from some existing efforts done by researchers, such as LeNet [20, 21], AlexNet [17], VGGNet [27], and GoogLeNet [34]. In addition, one cannot expect to get the optimal performance by applying the same architecture on various tasks, and the CNN architecture needs to be adjusted for each specific task, which will bring tremendous work as there are a large number of types of image classification tasks in industry.

In order to solve the complex problem of the CNN architecture design, evolutionary computation (EC) has recently been leveraged to automatically design the architecture without much human effort involved. Interested researchers have done excellent work on the automated design of the CNN architectures by using genetic programming (GP) [31] and genetic algorithms (GAs) [30, 33], such as large-scale evolution of image classifiers (LEIC) method [24] recently proposed by Google, which have shown that EC can be used in learning CNN architectures that are competitive with the state-of-the-art algorithms designed by humans. However, the learning process for large data is too slow due to the high computational cost for most of the methods and it might not be practical for industrial use.

A lot of work has been done in order to improve using EC to evolve a CNN architecture, such as the recent proposed EvoCNN using GAs [32]. One of the improvements in EvoCNN is that during the fitness evaluation, instead of training the model for 25,600 steps in LEIC, it only trains each individual by 10 epochs, which dramatically speeds up the learning process. The rationale behind EvoCNN using 10 epochs is that the researchers believe that training 10 epochs can obtain the major trend towards a good CNN architecture, which would be decisive to the final performance of a model, having been verified by their experiments. However, not a lot of research has been done by using other EC methods to evolve the architectures of CNNs, so we would like to introduce another EC method for evolving the architectures of CNNs without any human interference. Since particle swarm optimization (PSO) has the advantages of easy implementation, lower computational cost, and fewer parameters to adjust, a PSO method will be introduced in this chapter. As the fixed-length encoding of the particle in traditional PSO is a big challenge for evolving the architectures of CNN since the optimal CNN architecture varies for different tasks, a pseudo-variable-length PSO with a flexible encoding scheme inspired by computer networks will be introduced to break the fixed-length constraint. In order to reduce the computational cost, a small subset, which samples from the training dataset, will be used as a surrogate dataset to speed up the evolutionary process of the PSO method.

The classification accuracy has been significantly improved on hard problems in recent years because the rapid development of hardware capacity makes it possible to train very deep CNNs. A couple of years ago, VGG [27], which was deemed as a very deep CNN, only had 19 layers, but the recently proposed ResNet [10] and DenseNet [12] were capable of effectively training CNNs of more than 100 layers, which dramatically reduced the classification error rate. However, it is hard to deploy CNNs in real-life applications mainly because a trade-off between the classification accuracy and inference latency needs to be made, which is hard to be decided by application developers. Taking DenseNet as an example, although several DenseNet architectures are manually evaluated [12], there are two obstacles for applying DenseNet in real-life applications: Firstly, the hyperparameters may not be optimized, and for different tasks, the optimal model is not fixed, so before integrating DenseNet into applications, the hyperparameters have to be fine-tuned, which can be very complicated; Secondly, since an optimal DenseNet for a specific task may be extremely deep, the inference latency can be too long in some real-life applications such as web applications or mobile applications given limited hardware resource. This means that the classification accuracy may need to be compromised by reducing the complexity of DenseNet in order to reduce the inference latency to an acceptable amount of time. This chapter will also include a multi-objective PSO method to evolve CNN architectures, which will produce a Pareto set of solutions, so the real-world end users can choose one according to the computing resource and the classification requirement.

### 6.1.1   Goals

1. The first overall goal of this chapter is demonstrate how to develop an effective and efficient PSO method to automatically discover good architectures of CNNs. The specific objectives of this goal are to:

    (a) Design a new particle encoding scheme that has the ability of effectively encoding a CNN architecture, and develop a new PSO algorithm based on the new encoding strategy.
    (b) Design a method to break the constraint of the fixed-length encoding of traditional PSO in order to learn variable-length architectures of CNNs. We will introduce a new layer called disabled layer to attain a pseudo-variable-length particle.
    (c) Develop a fitness evaluation method using a surrogate dataset instead of the whole dataset to significantly speed up the evolutionary process.

2. The second overall goal is to depict a multi-objective particle swarm optimization (MOPSO) method to balance the trade-off between the classification accuracy and the inference latency, which is named MOCNN. MOCNN automatically tunes the hyperparameters of CNNs and deploys the trained model for better industrial use. To be more specific, an MOPSO method will be developed to search for a Pareto front of models. Therefore, industrial users can obtain the most suitable model for their specific problem based on their image classification task and the target devices. The specific objectives of this goal are listed below:

(a) As DenseNet achieved the competitive classification accuracy comparing the state-of-the-art methods, in order to reduce the search space, this work will focus on optimizing the hyperparameters of each dense block by pre-defining the number of dense blocks, such as the growth rate of each dense block, and the number of layers of each dense block. An encoding strategy will be introduced to encode the dense blocks;

(b) There are two major factors—classification accuracy and computational cost, which are decisive to the performance of the neural network. An MOPSO application will be developed to optimize the hyperparameters of dense blocks by jointly considering the classification accuracy and the computational cost. The two specific objectives are classification accuracy and FLOPs (floating point operations) where FLOPs can reflect the computational cost of both training and inference;

## 6.2 Background

### 6.2.1 CNN Architecture

Figure 6.1 exhibits a general architecture of a CNN with two convolutional (Conv) layers, two pooling layers, and two fully-connected layers—one hidden layer and one output layer at the end [14]. It is well-known that when designing a deep CNN architecture, the number of Conv layers, pooling layers, and fully-connected layers before the last output layer have to be properly defined along with their positions and configurations. Different types of layers have different configurations as follows: filter size, stride size, and feature maps are the main attributes of the configuration for the Conv layer; Kernel size, stride size, and pooling type—max-pooling or average pooling—are the important parameters for the configuration of the pooling layer; and the number of neurons is the key attribute of the fully-connected layer.

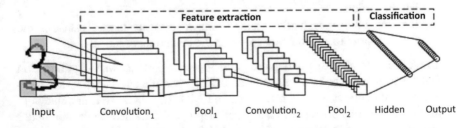

**Fig. 6.1** A general architecture of the convolutional neural network [14]

## 6.2.2   Internet Protocol Address

An Internet Protocol address (*IP address*) is a numerical label assigned to each device connected to a computer network that uses the Internet Protocol for communication [23]. In order to identify the network of an IP address, the subnet is introduced, which is often described in *CIDR* (Classless Inter-Domain Routing) style [7] by combining the starting IP address and the length of the *subnet mask* together. Both the IP address used for identifying the host and its corresponding subnet used for distinguishing different networks are carried by a network interface. For example, a standard IP address of a 32-bit number could be 192.168.1.251, and a standard subnet carries the starting IP address and the length of the subnet mask could be 192.168.1.0/8, which indicates the IP address in the subnet starts from 192.168.1.0 and the length of the subnet mask is 8 defining an IP range from 192.168.1.0 to 192.168.1.255.

Although the outlook of the IP address is a sequence of decimal numbers delimited by full stops, the binary string under the hood is actually used for the network identification, which inspires the new PSO encoding scheme. As there are several attributes in the configuration of each type of CNN layers, each of which is an integer value within a range, each value of the attribute can be smoothly converted to a binary string, and several binary strings, each of which represents the value of an attribute, can be concatenated to a large binary string to represent the whole configuration of a specific layer. It is obvious that the binary string suits the requirement of encoding CNN layers to particles. However, in this way, a huge number converted from the binary string will have to be used as one dimension of the particle vector, which may result in a horrendous searching time in PSO. On the other hand, in the IP structure, instead of utilizing one huge integer to mark the identification (ID) of a device in a large network, in order to make the IP address readable and memorable, it divides a huge ID number into several decimal values less than 256, each of which is stored in one byte of the IP address. In this way, the binary string can be divided into several bytes, and each byte comprises one dimension of the particle vector. The convergence of PSO can be facilitated by splitting one dimension of a large number to several dimensions of small numbers because in each round of the particle updates, all of the dimensions can be concurrently learned and the search space of one split dimension is much smaller. In this chapter, the new particle encoding scheme will use this idea to gain the flexibility of encoding various types of layers into a particle, and drastically cut down the learning process, which will be described in the next section.

## 6.2.3   DenseNet

A DenseNet is composed of several dense blocks, which are connected by a convolutional layer followed by a pooling layer, and before the first dense block, the input is

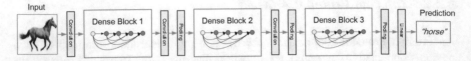

**Fig. 6.2** DenseNet architecture (image taken from [12])

filtered by a convolutional layer. An example of a DenseNet comprising three dense blocks is outlined in Fig. 6.2. Apart from the dense blocks, the hyperparameters of the other layers are fixed. The hyperparameters for the convolutional layer before the first block are problem-specific based on the image size in order to reduce the image size of the input feature maps passed to the first block; while the hyperparameters of the layers between blocks are problem-agnostic, which are a $1 \times 1$ convolutional layer and a $2 \times 2$ average pooling layer. However, the hyperparameters of dense blocks vary depending on specific image classification tasks, which are the number of layers in the dense block and the *growth rate* of the dense block. The *growth rate* is the number of output feature maps for each convolutional layer in the dense block. The output $x_l$ is calculated according to Formula (6.1), where $[x_0, x_1, \ldots, x_{l-1}]$ refers to the concatenation of the feature maps obtained from layer $0, 1, \ldots, l-1$, and $H_l$ represents a composite function of three consecutive operations, which are batch normalization (BN) [13], a rectified linear unit (ReLU) [9], and $3 \times 3$ convolution (Conv).

$$x_l = H_l([x_0, x_1, \ldots, x_{l-1}]) \tag{6.1}$$

---

**Algorithm 1** OMOPSO

---

1: $P, A \leftarrow$ Initialize swarm, initialize empty $\epsilon$-archive;
2: $g \leftarrow$ Set the current generation $g$ to 0;
3: $L \leftarrow$ Select leaders from $P$;
4: Send $L$ to $A$;
5: *crowding(L)*
6: **while** $g < g_{max}$ **do**
7:     **for** *particle* **in** $P$ **do**
8:         Select leader, updating, mutation and evaluation
9:         Update *pbest*
10:     **end for**
11:     $L \leftarrow$ Update leaders
12:     Send $L$ to $A$;
13:     *crowding(L)*
14:     $g \leftarrow g + 1$
15: **end while**
16: Report results in $A$

---

## 6.2.4 OMOPSO

OMOPSO [26] is a multi-objective optimization approach based on Pareto dominance. The crowding distance is established as a second discrimination criterion in addition to Pareto dominance to select the leaders. To calculate the crowding distance of a particular solution, it calculates the average distance of two surrounding points on either side of the particular solution along each of the objectives. The pseudocode of OMOPSO is written in Algorithm 1. There are a few items in the algorithm that need to be pointed out. First of all, there are two archives used by the algorithm: the first archive stores the current leaders that are used for performing the updating, and the other one carries the final solutions. The final solutions are the non-dominant solutions according to $\epsilon$-Pareto dominance [19]. In addition, when the maximum number of leaders is exceeded, the crowding distance [6, 22] is used to filter out the leaders in order to keep the number of leaders within the maximum number limit. The solution set comprised of the selected leaders is represented by $L$ in Algorithm 1, and the final solution set is called $\epsilon$-archive represented by $A$ in Algorithm 1. Thirdly, for each particle, when selecting a leader for the updating of OMOPSO, the binary tournament based on the crowding distance is applied. Finally, PSO update is performed for the leaders. Then, the particles are divided into three parts of equal size, and three mutation schemes are applied on the three parts, respectively. The first part has no mutation at all, the second part has uniform mutation (i.e., the variability range allowed for each decision variable is kept constant over generations), and the third part has non-uniform mutation (i.e., the variability range allowed for each decision variable decreases over time).

## 6.3 The Methods

This section provides two main methods in details—one with single objective and the other with multiple objectives.

## 6.3.1 Single-Objective PSO Method

In this sub-section, the new IP-based PSO (IPPSO) method for evolving deep CNNs, which focuses solely on a single objective of classification accuracy, will be presented in detail.

#### 6.3.1.1 Algorithm Overview

---

**Algorithm 2** Framework of IPPSO

$P \leftarrow$ Initialize the population with the proposed particle encoding strategy;
$P_{id} \leftarrow empty$;
$P_{gd} \leftarrow empty$;
**while** termination criterion is not satisfied **do**
    update velocity and position of each particle shown in Algorithm 4;
    evaluate the fitness value of each particle;
    update $P_{id}$ and $P_{gd}$;
**end while**

---

Algorithm 2 outlines the framework of the single-objective PSO method. There are mainly four steps, which are initializing the population by using the particle encoding strategy, updating the position and velocity, fitness evaluation, and checking whether the termination criterion is met.

#### 6.3.1.2 Particle Encoding Strategy to Encode CNN Layers

The IPPSO encoding strategy is inspired by how the Network IP address works. Although the CNN architecture is comprised of three types of layers—convolutional layer, pooling layer, and fully-connected layer, and the encoded information of different types of layers varies in terms of both the number of parameters and the range in each parameter shown in Table 6.1, a Network IP address with a fixed length of enough capacity can be designed to accommodate all the types of CNN

**Table 6.1** The parameters of different types of CNN layers—convolutional, pooling, fully-connected, and disabled layer with an example in the example column

| Layer type | Parameter | Range | # of Bits | Example value |
|---|---|---|---|---|
| Conv | Filter size | [1,8] | 3 | 2(001) |
| | # of feature maps | [1,128] | 7 | 32(000 1111) |
| | Stride size | [1,4] | 2 | 2(01) |
| | *Summary* | | 12 | 001 000 1111 01 |
| Pooling | Kernel size | [1,4] | 2 | 2(01) |
| | Stride size | [1,4] | 2 | 2(01) |
| | Type: 1 (maximal), 2 (average) | [1,2] | 1 | 2(1) |
| | Place holder | [1,128] | 6 | 32(00 1111) |
| | *Summary* | | 11 | 01 01 0 00 1111 |
| Fully-connected | # of Neurons | [1,2048] | 11 | 1024(011 11111111) |
| | *Summary* | | 11 | 011 11111111 |
| Disabled | Place holder | [1,2048] | 11 | 1024(011 11111111) |
| | *Summary* | | 11 | 011 11111111 |

layers, and then the Network IP can be divided into numerous subsets, each of which can be used to define a specific type of CNN layers.

First of all, the length of the binary string under the IP-based encoding scheme needs to be designed. With regard to Conv layers, firstly, there are three key parameters—filter size, number of feature maps, and stride size listed in the column of parameter in Table 6.1, which are the fundamental factors affecting the performance of CNNs. Secondly, based on the size of benchmark datasets, the range of the parameters is set to [1,8], [1,128], and [1,4] for the aforementioned three parameters, respectively, shown in the column of range in table 6.1. Thirdly, taking a CNN architecture with the filter size of 2, number of feature maps of 7, and stride size of 2 as an example, the decimal values can be converted to the binary strings of 001, 000 1111, and 01, where the binary string converted from the decimal value is filled with 0s until the length reaches the corresponding number of bits, illustrated in the column of Example Value in Table 6.1. Lastly, the total number of bits of 12 and the sample binary string of 001 000 1111 01 by concatenating the binary strings of the three parameters are displayed in the summary row of Conv layer in Table 6.1. In terms of pooling layers and fully-connected layers, the total number of bits and the sample binary string can be obtained by following the same process of Conv layers, which are listed in the summary rows of pooling and fully-connected layers in Table 6.1. As the largest number of bits to represent a layer is 12 as shown in Table 6.1 and the unit of an IP address is one byte—8 bits, there will be 2 bytes required to accommodate the 12 bits IP address.

In addition, the subnets for all types of CNN layers need to be defined according to the number of bits of each layer illustrated in Table 6.1 and CIDR style will be used to represent the subnet. As there are three types of CNN layers, we need to define three subnets with enough capacity to represent all the types of layers. Starting with the Conv layer, 0.0 is designed as the starting IP address of the subnet; in addition, the total length of the designed 2-byte IP address is 16 and the total number of bits required by the Conv layer is 12, so the subnet mask length is 4 calculated by subtracting the total number of bits from the length of the IP address, which brings the subnet representation to 0.0/4 with the range from 0.0 to 15.255. Regarding the pooling layer, the starting IP address is 16.0 obtained by adding 1 to the last IP address of the Conv layer, and the subnet mask length is 5 calculated in the same way as that of the Conv layer, which results in 16.0/5 with the range from 16.0 to 23.255 as the subnet representation of the pooling layer. Similarly, the subnet 24.0/5 with the range from 24.0 to 31.255 is designed as the subnet of the fully-connected layer. In order to make the subnets clear, all of the subnets are depicted in Table 6.2.

### 6.3.1.3   Pseudo-Variable-Length Representation by Introducing Disabled Layers

As the particle length of PSO is fixed after initialization, in order to cope with the variable length of the architectures of CNNs, an effective way of disabling some

**Table 6.2** Four subnets distributed to the three types of CNN layers and the disabled layer

| Layer type | Subnet (CIDR) | IP range |
|---|---|---|
| Convolutional layer | 0.0/4 | 0.0–15.255 |
| Fully-connected layer | 16.0/5 | 16.0–23.255 |
| Pooling layer | 24.0/5 | 24.0–31.255 |
| Disabled layer | 32.0/5 | 32.0–39.255 |

**Table 6.3** An example of IP addresses—one for each type of CNN layers

| Layer type | Binary (filled to 2 bytes) | IP address |
|---|---|---|
| Convolutional layer | (0000)001 000 1111 01 | 2.61 |
| Pooling layer | (00000)01 01 0 00 1111 | 18.143 |
| Fully-connected layer | (00000)011 11111111 | 27.255 |
| Disabled layer | (00000)01111111111 | 35.255 |

of the layers in the encoded particle vector will be used to achieve this purpose. Therefore, another layer type called the disabled layer and the corresponding subnet named the disabled subnet are introduced. To achieve a comparable probability for the disabled layer, the least total number of bits of 11 among all three types of CNN layers is set as the number of bits of the disabled layer, so the disabled subnet comes to 32.0/5 with the range from 32.0 to 39.255, shown in Table 6.2, where each layer will be encoded into an IP address of 2 bytes. Table 6.3 shows how the example in Table 6.1 is encoded into IP addresses by combining all the binary string of each parameter of a specific layer into one binary string, filling the combined binary string with zeros until reaching the length of 2 bytes, applying the subnet mask on the binary string, and converting the final binary string to an IP address with one byte as a unit delimited by full stops. For instance, the sample binary string of the Conv layer in Table 6.1 is 001 000 1111 01, which is filled to 0000 001 000 1111 01 to reach the length of 2 bytes. Then, 2-byte binary string—0000 0010 and 0011 1101, can be obtained by applying the subnet mask, in which the starting IP address of the subnet is added to the binary string. Finally, the IP address of 2.61 is achieved by converting the first byte to the decimal value of 2 and the second byte to 61.

### 6.3.1.4 An Example of the Encoding Strategy

After converting each layer into a 2-byte IP address, the position and velocity of PSO can be defined. However, there are a few parameters that need to be mentioned first—*max_length* (maximum number of CNN layers), *max_fully_connected* (maximum fully-connected layers with the constraint of at least one fully-connected layer) listed in Table 6.4. The encoded data type of the position and the velocity will be a byte array with a fixed length of *max_length* × 2 and each byte will be deemed as one dimension of the particle.

**Table 6.4** Parameter list

| Parameter name | Parameter meaning | Value |
|---|---|---|
| max_length | Maximum length of CNN layers | 9 |
| max_fully_connected | Maximum fully-connected layers given at least there is one fully-connected layer | 3 |
| N | Population size | 30 |
| k | The training epoch number before evaluating the trained CNN | 10 |
| num_of_batch | The batch size for evaluating the CNN | 200 |
| c1 | Acceleration coefficient array for $P_{id}$ | [1.49618,1.49618] |
| c2 | Acceleration coefficient array for $P_{gd}$ | [1.49618,1.49618] |
| w | Inertia weight for updating velocity | 0.7298 |
| $v_{max}$ | Maximum velocity | [4,25.6](0.1 × search range) |

(a)

| 2.61(C) | 18.143(P) | 2.61(C) | 35.255(D) | 27.255(F) |
|---|---|---|---|---|

(b)

| 2 | 61 | 18 | 143 | 2 | 61 | 35 | 255 | 27 | 255 |
|---|---|---|---|---|---|---|---|---|---|

**Fig. 6.3** An example IP Address and its corresponding particle vector. (**a**) An example of IP addresses in a particle containing 5 CNN layers. (**b**) An example of a particle vector with 5 CNN layers encoded

Here is an example of a particle vector to explain how the CNN architecture is encoded and how it copes with variable length of CNN architecture. Assume max_length is 5, a sequence of IP addresses representing a CNN architecture with the maximum number of 5 layers can be encoded into 5 IP addresses in Fig. 6.3a by using the sample IP addresses in Table 6.3, where C represents a Conv layer, P represents a pooling layer, F represents a fully-connected layer, and D represents a disabled layer. The corresponding particle vector with the dimension of 10 is shown in Fig. 6.3b. Since there is one disabled layer in the example, the actual number of layers is 4. However, after a few PSO updates, the seventh dimension and the eighth dimension of the particle vector may become 18 and 143, respectively, which turns the third IP address representing a disabled layers to a pooling layer, so the updated particle carries a CNN architecture of 5 layers. Conversely, after a few updates, the fifth dimension and the sixth dimension of the particle vector may become 35 and 255, respectively, which makes the third IP address fall into the disabled subnet, so the actual number of layers is 3. To conclude, as shown in this example, the particle with IPPSO encoding scheme is capable of representing variable-length architectures of CNNs—3, 4, and 5 in this example.

### 6.3.1.5 Population Initialization

In terms of the population initialization, after the size of the population is set up, individuals are randomly created until reaching the population size. For each individual, an empty vector is initialized first, and each element in it will be used to store a Network Interface containing the IP address and subnet information. The first element will always be a Conv layer. From the second to ($max\_length -$ $max\_fully\_connected$) layer, each element can be filled with a Conv layer, pooling layer, or disabled layer. From ($max\_length - max\_fully\_connected$) to ($max\_length - 1$) layer, it can be filled with any of the four types of layers until the first fully-connected is added, and after that only fully-connected layers or disabled layers are allowed. The last element will always be a fully-connected layer with the size the same as the number of classes. In addition, each layer will be generated with the random settings—a random IP address in a valid subnet.

### 6.3.1.6 Fitness Evaluation

---
**Algorithm 3** Fitness evaluation

---
**Input:** The population $P$, the training epoch number $k$, the training set $D_{train}$, the fitness
  evaluation dataset $D_{fitness}$, the batch size $batch\_size$;
**Output:** The population with fitness $P$;
  **for** individual $ind$ **in** $P$ **do**
    $i \leftarrow 1$;
    **while** $i <= k$ **do**
      *Train the connection weights of the CNN*
      *represented by individual ind*;
    **end while**
    $accy\_list \leftarrow$ Batch-evaluate the trained model on the dataset $D_{fitness}$ with the batch size
    $batch\_size$ and store the accuracy for each batch;
    $mean \leftarrow$ Calculate the mean value of $acc\_list$
    $fitness \leftarrow mean$;
    $P \leftarrow$ Update the fitness of the individual $ind$ in the population $P$;
  **end for**
  **return** $P$;

---

In order to speed up the CNN training process in fitness evaluation, a small subset is randomly sampled from the whole training dataset, which is used as the surrogate dataset. The surrogate dataset is then split into the training set $D_{train}$ and the fitness evaluation dataset $D_{fitness}$, shown in Algorithm 3. Before performing the fitness evaluation, a proper weight initialization method has to be chosen, and Xavier weight initialization [8] is chosen as it has been proved as an effective way, and has been implemented in most of deep learning frameworks. With regard to the fitness evaluation (shown in Algorithm 3), each individual is decoded to a CNN architecture with its settings, which will be trained for $k$ epochs on the first part of the training dataset. Then the partially trained CNN will be batch-evaluated on the second part of the training dataset, which will produce a series of accuracies. Finally, we calculate the mean value of the accuracies for each individual, which will be stored as the individual fitness.

### 6.3.1.7   Update Particle with Velocity Clamping

$$v_{id}(t+1) = w * v_{id}(t) + c_1 * r_1 * (P_{id} - x_{id}(t)) +$$
$$c_2 * r_2 * (P_{gd} - x_{id}(t)) \tag{6.2}$$

$$x_{id}(t+1) = x_{id}(t) + v_{id}(t+1) \tag{6.3}$$

---

**Algorithm 4** Update particle with velocity clamping

---

**Input:** particle individual vector $ind$, acceleration coefficient array for $P_{id}$ $c_1$, acceleration coefficient array for $P_{gd}$ $c_2$, inertia weight $w$, max velocity array $v_{max}$;
**Output:** updated individual vector $ind$;
  **for** element $interface$ **in** $ind$ **do**
    $i \leftarrow 0$;
    **for** $i$ < number of bytes of IP address in $interface$ **do**
      $x \leftarrow$ the $i$th byte of the IP address in the $interface$;
      $(r_1, r_2)) \leftarrow$ uniformly generate $r_1, r_2$ between [0, 1];
      $v_{new} \leftarrow$ Update velocity based on Equation 6.4;
      $v_{new} \leftarrow$ Apply velocity clamping using $v_{max}$;
      $x_{new} \leftarrow x + v_{new}$
      **if** $x_{new}$ > 255 **then**
        $x_{new} \leftarrow x_{new} - 255$;
      **end if**
    **end for**
  **end for**
  $fitness \leftarrow$ evaluate the updated individual $ind$;
  $(P_{id}, P_{gd}) \leftarrow$ Update $pbest$ and $gbest$ by comparing their $fitness$;
  **return** $ind$;

---

In Algorithm 4, as each layer is encoded into an interface with 2 bytes in the particle vector, and we want to control the acceleration coefficients for each byte, the two acceleration coefficients implemented as two float arrays with the size of 2 are required shown in Eq. 6.4. $v$ and $x$ are decimal values of the $i$th byte of the 2-byte IP address and its corresponding velocity, $P_{id}$ and $P_{gd}$ are decimal values of the $i$th byte of the IP address of the local best and the global bet, respectively, and $w, r_1, r_2$ are the same as traditional PSO in Eq. 6.2. The major difference is how the acceleration coefficients are implemented—$c_1[i]$ and $c_2[i]$ are the acceleration coefficients for the $i$th byte of the IP address, where $i$ is 1 or 2 in the case of 2-byte IP encoding, comparing to a singular value for each of the acceleration coefficients in traditional PSO. The reason of separating the acceleration coefficients for each byte of the IP address is that different parameters may fall into different bytes of the IP address, and the ability to explore a specific parameter more than others may be needed when fine-tuning the learning process.

After the coefficients defined, the velocity is initialized as a vector of all zeros. Then, we go through each byte in the particle and update the velocity and position by using the corresponding coefficients for that byte. The updated position may be an array of real value, but our proposed byte array must contain integer values, so each byte of the updated position needs to be rounded off to achieve a byte array. Since there are some constraints for each interface in the particle vector according to its position in the particle vector, e.g. the second interface can only be a Conv layer, a pooling layer, or a disabled layer, the new interface needs to be replaced by an interface with a random IP address in a valid subnet if the new interface does not fall in a valid subnet. After all the bytes being updated, the new particle is evaluated, and the fitness value is compared with the local best and the global best in order to update the two bests if needed.

$$v_{new} = w \times v + c_1[i] \times r_1 \times (P_{id} - x) + c_2[i] \times r_2 \times (P_{gd} - x) \tag{6.4}$$

#### 6.3.1.8 Best Individual Selection and Decoding

The global best of PSO will be reported as the best individual. In terms of the decoding, a list of network interfaces, stored in every 2 bytes from left to right in the particle vector of the global best, can be extracted from a particle vector. According to the subnets in Table 6.2, the type of layer can be distinguished, and then based on Table 6.1, the IP address can be decoded into different sets of binary string, which indicates the parameter values of the layer. After decoding all the interfaces in the global best, the final CNN architecture can be attained by connecting all of the decoded layers in the same order as that of the interfaces in the particle vector.

### 6.3.2 Multi-Objective PSO Method

This sub-section describes a multi-objective method for evolving CNNs.

#### 6.3.2.1 Algorithm Overview

The framework of the multi-objective PSO method named MOCNN has three steps. The first step is to initialize the population based on the proposed particle encoding strategy. At the second step, the multi-objective PSO algorithm called OMOPSO [26] is applied to optimize the two objectives, which are the classification accuracy and the FLOPs. Lastly, the non-dominant solutions in the Pareto set are retrieved,

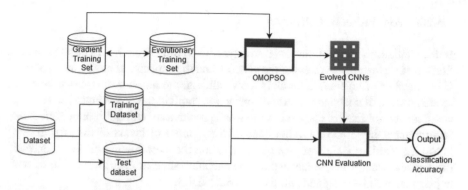

**Fig. 6.4** The flowchart of the experimental process

from which the actual user of the CNNs can choose one based on the usage requirements.

Figure 6.4 shows the overall structure of the system. The dataset is split into a training set and a test set, and the training set is further divided into a *gradient training set* and a *evolutionary training set*. The gradient training set and the evolutionary training set are passed to the proposed OMOPSO method for the objective evaluation. The proposed OMOPSO method produces non-dominant solutions, which are the optimized CNN architectures. Depending on the trade-off between the classification accuracy and the hardware resource capability, one of the non-dominant solutions can be selected for actual use. The CNN evaluation needs to be fine-tuned for the selected CNN architecture, and the whole training set and the test set are used to obtain the final classification accuracy.

### 6.3.2.2 Particle Encoding Strategy

In DenseNet, the hyperparameters, which need to be optimized, are the number of bocks, the number of layers in each block, and the growth rate of each block. For each of the block, a vector with the length of two can represent the number of layers and the growth rate in the block. Once the number of blocks is defined, the number of layers and the growth rate in each block can be encoded into a vector with the fixed length of 2 × the number of blocks. Figure 6.5 shows an example of the vector, which carries the hyperparameters of DenseNets with 3 blocks.

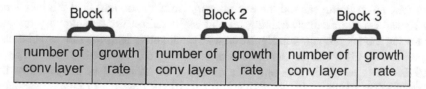

**Fig. 6.5** An example of a particle vector

### 6.3.2.3 Population Initialization

Before initializing the population, the range of each dimension has to be worked out first based on the effectiveness of the network and the capacity of hardware resource. If the number of layers in a block is too small, e.g. the number of layers is smaller than 2, there will not be any shortcut connections built in the dense block, and a very small number of feature maps, i.e. a too small growth rate, will not produce effective feature maps either. On the other hand, if the number of layers or the growth rate is too big, the hardware resource required to run the experiment will likely exceed the actual capacity of the hardware. The specific range of each dimension of our experiment will be designed and listed in Sect. 6.4.3.

The initial population is randomly generated based on the range of each dimension, whose pseudocode is composed in Algorithm 5. To be more specific, when randomly generating an individual, a random value is generated according to the range of each dimension from the first dimension until the last dimension; by repeating the individual generation process until the population size is satisfied, the whole initial population with a fixed population size will be successfully generated.

---

**Algorithm 5** Population initialization

**Input:** particle dimension $d$, population size $p_s$, a list of dimension value range $r$;
1: $P \leftarrow$ Empty population set;
2: $i \leftarrow 0$;
3: **while** $i < p_s$ **do**
4:      $ind \leftarrow$ Empty particle;
5:      **while** $j < d$ **do**
6:          $ind[j] \leftarrow$ Generate a random number within the range $r[j]$;
7:      **end while**
8:      $P \leftarrow$ Append $ind$ to $P$;
9: **end while**
10: **return** $P$;

---

### 6.3.2.4 Objective Evaluation

As MOCNN simultaneously optimizes the classification accuracy and the FLOPs, in the objective evaluation of MOCNN, both of them are calculated and returned as the objectives of the individual shown in Algorithm 6. When obtaining the classification accuracy, before training the individual representing a DenseNet with its specific hyperparameters, the training dataset is divided into two parts, which are the gradient training set and the evolutionary training set, and then the individual is trained on the gradient training set and evaluated on the evolutionary training set using a back propagation algorithm with an adaptive learning rate called Adam optimization [15] with the default settings, which are $\alpha = 0.001$, $\beta_1 = 0.9$, $\beta_2 = 0.999$, and $\epsilon = 10^{-8}$. The optimization target of MOCNN is to maximize the classification accuracy; in regard to the computational cost, the FLOPs is calculated for the individual, which is used as the second objective, and MOCNN will try to minimize the number of FLOPs. Since the number of FLOPs is fixed once the CNN architecture is defined, the calculation of FLOPs can be performed at any stage regardless of the CNN training process.

---

**Algorithm 6** Objective evaluation

---

**Input:** individual $ind$, maximum epochs $e_{max}$, accuracy list of trained CNNs $acc_{saved}$;
  1: **if** $ind$ **in** $acc_{saved}$ **then**
  2:     $acc_{best}$ ← retrieve the accuracy of $ind$ from $acc_{saved}$;
  3: **else**
  4:     $acc_{best}, e_{best}, e$ ← 0, 1, 0;
  5:     **while** $e < e_{max}$ **do**
  6:         Apply Adam optimization [15] to train $ind$ on the gradient training set $d_t$;
  7:         $acc_t$ ← evaluate the trained $ind$ on the evolutionary training set $d_t$;
  8:         **if** $acc_t > acc_{best}$ **then**
  9:             $acc_{best}, e_{best}$ ← $acc_t, e$;
 10:         **else if** $e - e_{best} > 10$ **then**
 11:             **break**
 12:         **end if**
 13:     **end while**
 14:     $acc_{saved}$ ← Append $ind$ and $acc_{best}$ to $acc_{saved}$;
 15: **end if**
 16: $flops$ ← calculated FLOPs of $ind$;
 17: $ind$ ← update the accuracy and FLOPs of $ind$ by $acc_{best}$ and $flops$;
 18: **return** $ind$;

---

Since training CNNs takes much longer time than that of calculating FLOPs, a couple of methods have been implemented to reduce the computational cost of getting the classification accuracy. First of all, an early stop criterion of terminating the training process when the accuracy does not increase in the next 10 epochs is adopted to potentially reduce the epochs of the training process, which as a result, decreases the training time. It worked particularly effective to search for CNN architectures because the complexity of different individuals may vary significantly, which may require a various number of epochs to completely train different individuals. For example, as the CNN architecture can be as simple as one or two layers with a very small number of feature maps, the number of epochs needed to train the CNN can be very small. When the CNN architecture is as complicated as one containing hundreds of layers with a really large number of feature maps in each layer, it will require many more epochs to completely train the complicated CNN. Therefore, it is hard to define a fixed number of epochs to be used by the objective evaluation to train CNNs with various complexities. Instead, MOCNN sets a maximum number of epochs, which is large enough to fully train the most complicated CNNs in our search space, and utilizes the early stop criterion to stop the training process at an earlier stage in order to save the computational cost. In addition, since each individual will be evaluated by the objective evaluation in each generation, there may be a large number of CNNs evaluated across the whole evolutionary process, among which there may be individuals representing the same CNN architecture duplicately trained and evaluated. For the purpose to prevent the same CNN from the duplicate training, the classification accuracy obtained for each individual in the objective evaluation is stored in the memory, which is persisted just before the program finishes, and loaded at the beginning of the program. In the objective evaluation, before training the individual, a search for the individual in the stored classification accuracy is performed first, and the training procedure will be executed only when the search result is empty.

Adam optimization [15] is chosen as the backpropagation algorithm, and the whole training dataset is used to evaluate the CNNs. As to our best knowledge, two other methods of objective evaluation were found being used in the area of using EC method to automatically design CNN architectures. The first method used in [24] is to use stochastic gradient descent (SGD) [3] with a scheduled learning rate, e.g. 0.1 as the learning rate before 150 epochs, and the learning rate divided by 10 at the epoch of 150 and 200, respectively. From the settings of SGD for training VGGNet [27], ResNet [10], and DenseNet [12], it can be observed that the SGD settings are quite different, which means that a set of SGD settings may be good for a specific type of CNNs, but may not work well for other types of CNNs. Therefore, it is very hard to perform a fair comparison between two various CNNs that need SGDs with different settings to optimize, which results in the preference of a specific set of CNNs in the EC algorithm. The second method is to train the CNN for a small number of epochs used in [2, 36, 37, 39]. It speeds up the training process by restraining the number of training epochs, which relies on the assumption that the CNN architecture with a good performance at the beginning would perform well in the end, but to our best effort, a strong evidence has not been found to prove the assumption in either theoretical or empirical study. As a result, the evolutionary process may prefer the CNN architectures that perform well at the beginning without any guarantee of achieving a good classification accuracy in the end, but it is the classification accuracy in the end that should be used to select CNN architectures. Both of these two methods may introduce some bias toward a specific set of CNN architectures. However, by using the Adam optimization to train the CNNs on the whole training dataset, it could mitigate or even eliminate the bias of the aforementioned two methods because the learning rate will be automatically adapted during the training process based on the CNN architecture and the dataset, and the training process will stop until the convergence of the Adam optimization. So, the objective evaluation method in MOCNN is expected to be able to reduce the bias.

## 6.4   Experiment Design

### 6.4.1   Benchmark Datasets

To examine the performance of the IPPSO method, three datasets are chosen from the widely used image classification benchmark datasets. They are the MNIST Basic (MB) [18], the MNIST with Rotated Digits plus Background Images (MRDBI), [18] and the Convex Sets (CS) [18].

The first two benchmark datasets are two of the MNIST [21] variants for classifying 10 hand-written digits (i.e., 0–9). There are a couple of reasons for using MNIST variants instead of MNIST. Firstly, as the classification accuracy of MNIST has achieved 97%, in order to challenge the algorithm, different noises (e.g., random backgrounds, rotations) are added into these MNIST variants from the MNIST to improve the complexity of the dataset. Secondly, there are 12,000 training images and 50,000 test images in these variants, which further challenges the classification algorithms due to the much less training data but more test data. The third benchmark dataset is for recognizing the shapes of objects (i.e., convex or not), which contains 8,000 training images and 50,000 test images. Since it is a two-class classification problem comparing to 10 classes of MNIST dataset, and the images contain shapes rather than digits, it is chosen as a supplement benchmark to the two MNIST variants in order to thoroughly test the performance of IPPSO.

Each image in these benchmarks is with the size 28 × 28. Another reason for choosing these three benchmark datasets is that different algorithms have reported their promising results, so it is convenient for comparing the performance of IPPSO with these existing algorithms.

However, for examining the performance of MOCNN, based on the computational cost of the algorithm that needs to be evaluated and the hardware resource to run the experiment, the CIFAR-10 dataset will be chosen as the benchmark dataset. It consists of 60,000 color images with the size of 32 × 32 in 10 classes, and each class contains 6000 images. The whole dataset is divided into the training dataset of 50,000 images and the test dataset of 10,000 images [16].

### 6.4.2   Peer Competitors

In the experiments of IPPSO, state-of-the-art algorithms, that have reported promising classification errors on the chosen benchmarks, are collected as the peer competitors of IPPSO. To be specific, the peer competitors on the three benchmarks are CAE-2 [25], TIRBM [28], PGBM+DN1 [29], ScatNet-2 [4], RandNet-2 [5], PCANet-2 (softmax) [5], LDANet-2 [5], SVM+RBF [18], SVM+Poly [18], NNet [18], SAA-3 [18], and DBN-3 [18], which are from the literature [5] recently

**Table 6.5** Parameter list

| Parameter | Value |
|---|---|
| *Objective evaluation* | |
| Initial learning rate | 0.1 |
| Batch size | 128 |
| Maximum epochs | 300 |
| *Particle encoding* | |
| Number of blocks | 4 |
| The range of growth rate in all four blocks | 8 to 32 |
| The range of number of layers in the first block | 4 to 6 |
| The range of number of layers in the second block | 4 to 12 |
| The range of number of layers in the third block | 4 to 24 |
| The range of number of layers in the fourth block | 4 to 16 |
| *OMOPSO* | |
| $\epsilon$ values in the format of [accuracy, FLOPs] | [0.01, 0.05] |

published and the provider of the benchmarks.[1] However, for MOCNN, we will focus on the quality of the Pareto set, and will compare the non-dominated solution with the best classification accuracy with DenseNet.

### 6.4.3 Parameter Settings

All the parameter settings of IPPSO are set based on the conventions in the communities of PSO [35] and deep learning [11] which are listed in Table 6.4.

IPPSO is implemented in TensorFlow [1], and each copy of the code runs on a computer equipped with two GPU cards with the identical model number GTX1080. Due to the stochastic nature of IPPSO, 30 independent runs are performed on each benchmark dataset, and the mean results are used for the comparisons unless otherwise specified.

As the multi-objective PSO method consists of two parts, which are the multi-objective EC algorithm called OMOPSO and the process of training deep CNNs in the objective evaluation, the parameters listed in Table 6.5 are set according to the conventions of the communities of EC and deep learning with the consideration of the computational cost and complexity of the search space in MOCNN method.

---

[1]http://www.iro.umontreal.ca/lisa/twiki/bin/view.cgi/Public/DeepVsShallowComparisonICML 2007.

However, several parameters are specific to MOCNN method, which will be discussed in details in the following paragraphs.

First of all, since the proposed particle encoding strategy is exclusively designed for MOCNN, the parameters are customized for effectively and efficiently running our MOCNN experiment. As the purpose of MOCNN is to explore the Pareto front of the multi-objective problem of deep CNNs, MOCNN is not focusing on setting a new benchmark of the classification accuracy. DenseNet-121, which is the least complex DenseNet reported in the DenseNet paper [12], is chosen as the most complex CNN to be searched by MOCNN due to our limited memory, computational capacity of our GPU resource and time constraint. Although DenseNet-121 was not the best DenseNet reported in its paper, the classification accuracy was only slightly worse than the more complicated DenseNets, and the computational cost of training DenseNet-121 is quite high, so the least complex DenseNet is set as the maximum complexity given that the training process needs to be performed 400 (20 individuals $\times$ 20 generations) times in the evolutionary process. As a result, the number of blocks is fixed to 4; 32, which is the growth rate of DenseNet-121, is set as the maximum value of the growth rate; and the maximum number of layers for the first, second, third, and fourth block is configured as 6, 12, 24, and 16, respectively, which is the same as that of DenseNet-121. In terms of the lower bound of the parameters, if there are too few layers in a block, the dense connection will not work effectively, and if the growth rate is too small, it will cause the issue of a very limited number of extracted features, which will not provide enough useful features for the classification algorithm. Therefore, 4 and 8 are chosen as the lower bounds of the number of layers in each block and the growth rate, respectively.

In addition, the maximum epochs used to train the CNNs in objective evaluation is set to 300 based on the number of epochs used to train the most complex CNN in the search space. To be more specific, 100, 200, and 300 epochs were examined for training DenseNet-121 to see whether DenseNet-121 could be fully trained. It turned out training DenseNet-121 for 300 epochs can guarantee the convergence on the CIFAR-10 dataset used as the benchmark dataset in our experiment.

Furthermore, as the $\epsilon$ value defines the number of non-dominant solutions, which is demonstrated in Sect. 6.2.4. A few $\epsilon$ values are investigated for each of the objectives. A smaller value of $\epsilon$ produces fewer non-dominant solutions, while more non-dominant solutions are obtained by increasing the value of $\epsilon$. However, $\epsilon$ value does not affect the evolutionary process of MOCNN, so the $\epsilon$ value is configured purely based on the number of non-dominant solutions that are preferred to be displayed in the final result, where the actual industrial users of the proposed method can choose the best solution by considering the classification accuracy and the computational cost.

Finally, the population size and the maximum generation need to be designed for the experiment. 20 and 50 are chosen from the widely used population sizes based on the convention of the EC community and the high computational cost of our experiment. The reason for running two experiments with different population sizes is to explore how population size will affect the results of MOCNN method. Due to the time constraint, 400 to 500 evaluations are used, which may take 2 weeks.

Therefore, the experiment with 20 individuals will run for 20 generations and the other one with 50 individuals will run for 10 generations. In order to better refer these two experiments, the experiment with 20 individuals and 20 generations is called *EXP-20-20*, and *EXP-50-10* represents the experiment with 50 individuals and 10 generations.

## 6.5   Results and Analysis

### 6.5.1   Results and Analysis of Single-Objective PSO Method

#### 6.5.1.1   Overall Performance

Experimental results on all the three benchmark datasets are shown in Table 6.6 where the last three rows denote the mean classification errors, the best classification errors, and the standard deviations of the classification errors obtained by IPPSO from the 30 runs, and the other rows show the best classification errors reported by peer competitors.[2] In order to conveniently investigate the comparisons, the terms "(+)" and "(−)" are provided to indicate whether the result generated by IPPSO is better or worse than the best result obtained by the corresponding peer competitor. The term "—" means there is no available result reported from the provider or cannot be counted.

It is clearly shown in Table 6.6 that by comparing the mean classification errors of IPPSO with the best performance of the peer competitors, IPPSO performs the second best on the MB dataset, which is only a little bit worse than LDANet-2. IPPSO is the best on the MDRBI dataset, which is the most complicated dataset among these three, and the fifth best on the CS dataset, which is not ideal but very competitive.

#### 6.5.1.2   Evolved CNN Architectures

Although IPPSO is performed on each benchmark with 30 independent runs, only one is chosen on each benchmark for this description purpose shown in Table 6.7. Since disabled layers have been removed during the decoding process, they do not show up in the learned CNN architectures. Therefore, it turns out that IPPSO is able to learn a variable-length CNN architecture, which can be obviously seen from the listed architectures—6 CNN layers for the MB and CS benchmark and 8 CNN layers for the MDRBI benchmark.

---

[2]Most deep learning methods only report the best result.

**Table 6.6** The classification errors of IPPSO against the peer competitors on the MB, MDRBI, and CS benchmark datasets

| Classifier | MB | MDRBI | CS |
|---|---|---|---|
| CAE-2 | 2.48(+) | 45.23(+) | — |
| TIRBM | — | 35.50(+) | — |
| PGBM + DN-1 | — | 36.76(+) | — |
| ScatNet-2 | 1.27(+) | 50.48(+) | 6.50(−) |
| RandNet-2 | 1.25(+) | 43.69(+) | 5.45(−) |
| PCANet-2 (softmax) | 1.40(+) | 35.86(+) | 4.19(−) |
| LDANet-2 | 1.05(−) | 38.54(+) | 7.22(−) |
| SVM + RBF | 3.03(+) | 55.18(+) | 19.13(+) |
| SVM + Poly | 3.69(+) | 56.41(+) | 19.82(+) |
| NNet | 4.69(+) | 62.16(+) | 32.25(+) |
| SAA-3 | 3.46(+) | 51.93(+) | 18.41(+) |
| DBN-3 | 3.11(+) | 47.39(+) | 18.63(+) |
| IPPSO(mean) | 1.21 | 34.50 | 12.06 |
| IPPSO(best) | 1.13 | 33 | 8.48 |
| IPPSO(standard deviation) | 0.103 | 2.96 | 2.25 |

**Table 6.7** An evolved architecture for the MB benchmark

| Layer type | Configuration |
|---|---|
| *An evolved architecture for the MB benchmark* | |
| Conv | Filter size: 2, stride size: 1, feature maps: 26 |
| Conv | Filter size: 6, stride size: 3, feature maps: 82 |
| Conv | Filter size: 8, stride size: 4, feature maps: 114 |
| Conv | Filter size: 7, stride size: 4, feature maps: 107 |
| Full | Neurons: 1686 |
| Full | Neurons: 10 |
| *An evolved architecture for the MDRBI benchmark* | |
| Conv | Filter size: 2, stride size: 1, feature maps: 32 |
| Conv | Filter size: 6, stride size: 3, feature maps: 90 |
| Conv | Filter size: 7, stride size: 4, feature maps: 101 |
| Conv | Filter size: 7, stride size: 4, feature maps: 97 |
| Pool | Kernel size: 4, stride size: 4, type: average |
| Conv | Filter size: 5, stride size: 3, feature maps: 68 |
| Full | Neurons: 1577 |
| Full | Neurons: 10 |
| *An evolved architecture for the CS benchmark* | |
| Conv | Filter size: 1, stride size: 1, feature maps: 11 |
| Conv | Filter size:7, stride size: 4, feature maps: 108 |
| Conv | Filter size: 1, stride size: 1, feature maps: 8 |
| Conv | Filter size: 6, stride size: 3, feature maps: 92 |
| Full | Neurons: 906 |
| Full | Neurons: 2 |

### 6.5.1.3 Visualization

In order to achieve a better understanding of IPPSO, we visualize two parts of the evolutionary process—the accuracy distribution of the PSO vectors, where the architectures of CNNs are encoded, and the PSO trajectory of the evolving process.

In terms of the accuracy distribution, first of all, we obtained the PSO vectors and their corresponding accuracies from 10 runs of the experiments; in addition, the first two principal components from principal component analysis (PCA) are extracted for the usage of visualization. A 3-D triangulated surface with the data containing the first two components and the corresponding accuracy is plotted shown in Fig. 6.6a. It is observed that there are a lot of steep hills on the surface whose summits are at the similar level, which means that there are quite a number of local optima, but most of them are very close to each other, which means that those local optima are acceptable as a good solution of the task.

Regarding the trajectory, the best result of the particles of each generation and the global best in each generation from one run of the experiments are obtained and plotted in blue color and red color, respectively, in Fig. 6.6b. It can be seen that after only a few generations, the global best is found, after which the particles are still flying in the search space, but none of them can obtain a better accuracy, which means the optimum has been reached by PSO after only a few steps. Even though the surface of the optimization task shown in Fig. 6.6a is extremely complicated, the PSO method with only 30 particles can climb up to the optimum very quickly, which proves the effectiveness and efficiency of PSO on optimization tasks.

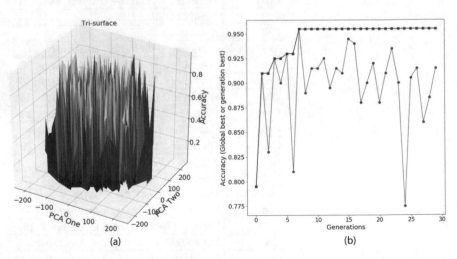

**Fig. 6.6** (a) The surface of CNN accuracies after training 10 epochs with IPPSO encoding. (b) PSO trajectory

## 6.5.2   Results and Analysis of Multi-Objective PSO Method

### 6.5.2.1   Pareto Optimality Analysis

Figures 6.7 and 6.8 show the experimental results of *EXP-20-20* and *EXP-50-10*, respectively, each of which is composed of four sub-figures. From the left to the right, the first sub-figure contains all of the solutions evaluated through the evolutionary process, where the *x*-axis represents the negative value of the FLOPs and *y*-axis shows the accuracy. The non-dominant solutions based on $\epsilon$-Pareto dominance [19] are in orange color and the blue points indicate the others. The second sub-figure illustrates the evolutionary progress of the accuracy of non-dominant solutions based on $\epsilon$-Pareto dominance by each generation with the generation as *x*-axis and the classification accuracy as *y*-axis. The third sub-figure shows the changes of FLOPs of non-dominant solutions based on $\epsilon$-Pareto dominance during the evolutionary process, where the negative value of FLOPs is drawn toward the vertical axis and the generation is plotted toward the horizontal axis. The fourth sub-figure is generated by combining the second and third sub-figures into a 3D figure with *x*-axis, *y*-axis, and *z*-axis represents the generation, the negative FLOPs value, and the classification accuracy, respectively. The level of transparency reflects the depth in the 3D figure, i.e. the negative value of FLOPs

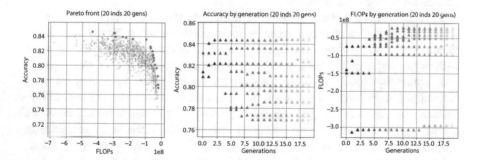

**Fig. 6.7**  Twenty individuals and 20 generations

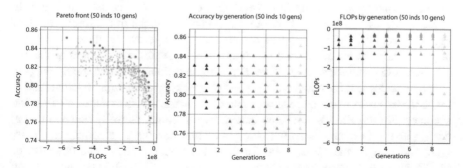

**Fig. 6.8**  Fifty individuals and 10 generations

carried by the point with less transparency is smaller than that represented by the more transparent point.

It can be observed that the negative value of FLOPs is plotted in the figure instead of the positive value, which is because by using the negative value of FLOPs, it converts the optimization of this objective to a maximization problem in order to make it consistent to the other objective of maximizing the classification accuracy. After the conversion, the two objectives have the same optimization direction, which is easier to be understood and analyzed. By looking into the first sub-figure of Figs. 6.7 and 6.8, the non-dominant solutions achieved by both the experiments have formed a clear curve, which defines the Pareto front. When further investigating the Pareto front, it can be found that the two objectives contradict each other at some stage, i.e. the classification accuracy cannot be improved without increasing the FLOPs reflecting the complexity of CNNs, which means the problem of optimizing the two objectives of the classification accuracy and the FLOPs is an obvious multi-objective optimization problem. By comparing the Pareto fronts of the two experiments, especially the points with the lowest FLOPs and the highest accuracy, it can be learnt that *EXP-50-10* provides more diverse non-dominant solutions, which also means the coverage of the non-dominant solutions of *EXP-50-10* is larger than that of *EXP-20-20*, even though the maximum generation of *EXP-50-10* is only half of the generation of *EXP-20-20*, so a larger population size in MOCNN tends to produce more diverse non-dominant solutions, which therefore, provides more options for industrial users to choose.

In regard with the convergence analysis, the second and third sub-figures can be utilized to analyze the convergence of the classification accuracy and FLOPs, respectively, and the fourth sub-figure presents an overview of the convergence of both of the objectives. Firstly, *EXP-20-20* can be considered to be converged for both of the objectives. The classification accuracy changes a lot during the first 7 generations of evolution, and starts to fluctuate a bit until the end of the evolutionary process. As after the 12th generation, only very few non-dominant solutions shift a little bit, so *EXP-20-20* can be deemed converged in terms of the classification accuracy. As shown in the third sub-figure of Fig. 6.7, the number of non-dominant solutions grows fast and the value of FLOPs quickly spreads to both directions before the 8th generation, but it is stabilizing until the 14th generation, after which the FLOPs hardly shift. Therefore, the FLOPs of *EXP-20-20* is converged as well. The convergence progress of both objectives can be noticed in the fourth figure of Fig. 6.7. Secondly, *EXP-50-10* was not able to converge due to the small number of generations. From the second sub-figure, there are obvious changes at the 1st, 3rd, and 10th generations, and between those generations, the shifts rarely happen, which indicates that the convergence speed of the experiment with 50 individuals is much slower and it needs more generations to converge in terms of the classification accuracy. For the FLOPs, the same pattern can be found as well, which is that at the 1st and 10th generations, the changes of non-dominant solutions are clearly seen, and rare changes take place for the other generations, so the objective of FLOPs also needs more time to converge. Therefore, *EXP-50-10* has not reached the convergence, which can also be observed in the 3D sub-figure of Fig. 6.8. To summarize, the experiment with 20 individuals converges faster than that with 50 individuals, but the experiment with 50 individuals tend to provide more non-dominant solutions, which gains more coverage of the potential solutions.

### 6.5.2.2   MOCNN vs DenseNet-121

As described in Sect. 6.3.2.2, DenseNet-121 was set as the maximum complexity of the optimized CNNs, so DenseNet-121 is set as a benchmark, which is used as a comparison to the optimized non-dominant solution that has the best accuracy. As the classification accuracy of DenseNet-121 on CIFAR-10 was not reported in their paper, DenseNet-121 needs to be evaluated and compared with the optimized MOCNN. The same training process and the common-used data augmentation specified in [12] are adopted to train both DenseNet-121 and the optimized MOCNN. The classification accuracy of DenseNet-121 is 94.77% and the classification accuracy for the optimized MOCNN is 95.51%, which shows that the optimized MOCNN outperforms DenseNet-121 on CIFAR-10 dataset in terms of both the classification accuracy and the computational cost. The classification accuracies of DenseNet($k = 12$) of 40 layers (DenseNet40) and DenseNet($k = 12$) of 100 layers (DenseNet100_12) are reported in [12], which are 94.76% and 95.90%, respectively. The optimized MOCNN performs better than (DenseNet40), while a bit worse than (DenseNet100_12). However, (DenseNet100_12) is beyond the search space because it is more complex than DenseNet-121. Therefore, the optimized MOCNN has achieved a promising result among the DenseNets within the search space, and it may possibly outperform (DenseNet100_12) if the search space is extended to include (DenseNet100_12).

### 6.5.2.3   Computational Cost

As described in Sect. 6.3.2.4, the CNNs represented by individuals are fully trained by Adam optimization, which consumes quite a large amount of computation. At the beginning, the experiment *EXP-20-20* was tried on one GPU card, which took almost 3 weeks to finish the experiment, so a server-client infrastructure [38] was adopted to leverage as many as GPU cards across multiple machines to reduce the wall-clock running time. The experiment *EXP-20-20* ran for about 3 days to finish the evolutionary process on 8 GPU cards, and the result of the experiment *EXP-50-10* was achieved by running the program on 10 GPU cards for 3 days as well.

## 6.6   Conclusions

The first goal of this chapter was to develop a new PSO approach with variable length to automatically evolve the architectures of CNNs for image classification problems. This goal has been successfully achieved by proposing a new encoding scheme of using a network interface containing an IP address and its corresponding subnet to carry the configurations of a CNN layers, the design of four subnets including a disabled subnet in order to simulate a pseudo-variable-length PSO, and an efficient fitness evaluation method by using a surrogate dataset. This approach was examined and compared with 12 peer competitors including the most state-of-the-art algorithms on three benchmark datasets commonly used in deep learning and the experimental results show that IPPSO can achieve a very competitive accuracy by outperforming all others on the MDRBI benchmark dataset, being the second best on the MNIST benchmark dataset and ranking above the middle line on the CS benchmark dataset.

The second goal of this chapter was to propose a multi-objective EC method called MOCNN to search for the non-dominant solutions at the Pareto front by optimizing the two objectives of both the classification accuracy and the FLOPs reflecting the computational cost. MOCNN was designed and developed by designing a new encoding strategy to encode CNNs, choosing the two objectives that are critical to measuring the performance of CNNs, and applying a multi-objective particle swarm optimization algorithm called OMOPSO. As non-dominant solutions generated by MOCNN can be provided to the industrial users for them to choose one that suits their usage best, the overall goal of streamlining the usage of the state-of-the-art CNNs for image classification has been achieved.

This chapter only provides two example pieces of work for automated design of CNN architectures using PSO, and the results have shown that evolutionary computation techniques can play a major role in such *AutoML* tasks. Clearly, this only represents an early stage of such work, but the potential of EC techniques has been demonstrated. In the future, we will continue such investigations to make EC a general tool/approach to AutoML.

# References

1. Abadi, M., Agarwal, A., Barham, P., Brevdo, E., Chen, Z., Citro, C., Corrado, G.S., Davis, A., Dean, J., Devin, M., et al.: TensorFlow: large-scale machine learning on heterogeneous distributed systems. Preprint. arXiv:1603.04467 (2016)
2. Assunção, F., Lourenço, N., Machado, P., Ribeiro, B.: Evolving the topology of large scale deep neural networks. In: European Conference on Genetic Programming, pp. 19–34. Springer, Berlin (2018)
3. Bottou, L.: Large-scale machine learning with stochastic gradient descent. In: Proceedings of COMPSTAT'2010, pp. 177–186. Springer, Berlin (2010)
4. Bruna, J., Mallat, S.: Invariant scattering convolution networks. IEEE Trans. Pattern Anal. Mach. Intell. **35**(8), 1872–1886 (2013)
5. Chan, T.H., Jia, K., Gao, S., Lu, J., Zeng, Z., Ma, Y.: PCANet: a simple deep learning baseline for image classification? IEEE Trans. Image Process. **24**(12), 5017–5032 (2015)
6. Deb, K., Pratap, A., Agarwal, S., Meyarivan, T.: A fast and elitist multiobjective genetic algorithm: NSGA-II. IEEE Trans. Evol. Comput. **6**(2), 182–197 (2002)
7. Fuller, V., Li, T., Yu, J., Varadhan, K.: Classless inter-domain routing (CIDR): an address assignment and aggregation strategy (1993)

8. Glorot, X., Bengio, Y.: Understanding the difficulty of training deep feedforward neural networks. In: Proceedings of the Thirteenth International Conference on Artificial Intelligence and Statistics, pp. 249–256 (2010)
9. Glorot, X., Bordes, A., Bengio, Y.: Deep sparse rectifier neural networks. In: Proceedings of the Fourteenth International Conference on Artificial Intelligence and Statistics, pp. 315–323 (2011)
10. He, K., Zhang, X., Ren, S., Sun, J.: Deep residual learning for image recognition. CoRR abs/1512.03385 (2015). http://arxiv.org/abs/1512.03385
11. Hinton, G.E.: A practical guide to training restricted Boltzmann machines. In: Neural Networks: Tricks of the Trade, pp. 599–619. Springer, Berlin (2012)
12. Huang, G., Liu, Z., Weinberger, K.Q.: Densely connected convolutional networks. CoRR abs/1608.06993 (2016). http://arxiv.org/abs/1608.06993
13. Ioffe, S., Szegedy, C.: Batch normalization: accelerating deep network training by reducing internal covariate shift. Preprint. arXiv:1502.03167 (2015)
14. Jones, M.T.: Deep Learning Architectures and the Rise of Artificial Intelligence. IBM DeveloperWorks, Armonk (2017)
15. Kingma, D.P., Ba, J.: Adam: a method for stochastic optimization. Preprint. arXiv:1412.6980 (2014)
16. Krizhevsky, A., Hinton, G.: Learning multiple layers of features from tiny images. Technical Report. Citeseer (2009)
17. Krizhevsky, A., Sutskever, I., Hinton, G.E.: ImageNet classification with deep convolutional neural networks. In: Advances in Neural Information Processing Systems, pp. 1097–1105 (2012)
18. Larochelle, H., Erhan, D., Courville, A., Bergstra, J., Bengio, Y.: An empirical evaluation of deep architectures on problems with many factors of variation. In: Proceedings of the 24th International Conference on Machine Learning, pp. 473–480. ACM, New York (2007)
19. Laumanns, M., Thiele, L., Deb, K., Zitzler, E.: Combining convergence and diversity in evolutionary multiobjective optimization. Evol. Comput. 10(3), 263–282 (2002)
20. LeCun, Y., Boser, B., Denker, J.S., Henderson, D., Howard, R.E., Hubbard, W., Jackel, L.D.: Backpropagation applied to handwritten zip code recognition. Neural Comput. 1(4), 541–551 (1989)
21. LeCun, Y., Bottou, L., Bengio, Y., Haffner, P., et al.: Gradient-based learning applied to document recognition. Proc. IEEE 86(11), 2278–2324 (1998)
22. Li, X.: A non-dominated sorting particle swarm optimizer for multiobjective optimization. In: Genetic and Evolutionary Computation Conference, pp. 37–48. Springer, Berlin (2003)
23. Postel, J.: DoD standard internet protocol (1980)
24. Real, E., Moore, S., Selle, A., Saxena, S., Suematsu, Y.L., Tan, J., Le, Q.V., Kurakin, A.: Large-scale evolution of image classifiers. In: Proceedings of the 34th International Conference on Machine Learning, vol. 70, pp. 2902–2911. JMLR. org (2017)
25. Rifai, S., Vincent, P., Muller, X., Glorot, X., Bengio, Y.: Contractive auto-encoders: explicit invariance during feature extraction. In: Proceedings of the 28th International Conference on International Conference on Machine Learning, pp. 833–840. Omnipress, Madison (2011)
26. Sierra, M.R., Coello, C.A.C.: Improving PSO-based multi-objective optimization using crowding, mutation and $\varepsilon$-dominance. In: International Conference on Evolutionary Multi-Criterion Optimization, pp. 505–519. Springer, Berlin (2005)
27. Simonyan, K., Zisserman, A.: Very deep convolutional networks for large-scale image recognition. CoRR abs/1409.1556 (2014). http://arxiv.org/abs/1409.1556
28. Sohn, K., Lee, H.: Learning invariant representations with local transformations. Preprint. arXiv:1206.6418 (2012)
29. Sohn, K., Zhou, G., Lee, C., Lee, H.: Learning and selecting features jointly with point-wise gated Boltzmann machines. In: International Conference on Machine Learning, pp. 217–225 (2013)
30. Stanley, K.O., Miikkulainen, R.: Evolving neural networks through augmenting topologies. Evol. Comput. 10(2), 99–127 (2002)

31. Suganuma, M., Shirakawa, S., Nagao, T.: A genetic programming approach to designing convolutional neural network architectures. In: Proceedings of the Genetic and Evolutionary Computation Conference, pp. 497–504. ACM, New York (2017)

32. Sun, Y., Xue, B., Zhang, M., Yen, G.G.: Evolving deep convolutional neural networks for image classification. IEEE Trans. Evol. Comput. **24**(2), 394–407 (2019). https://doi.org/10.1109/TEVC.2019.2916183

33. Sun, Y., Yen, G.G., Yi, Z.: Evolving unsupervised deep neural networks for learning meaningful representations. IEEE Trans. Evol. Comput. **23**(1), 89–103 (2019). https://doi.org/10.1109/TEVC.2018.2808689

34. Szegedy, C., Liu, W., Jia, Y., Sermanet, P., Reed, S., Anguelov, D., Erhan, D., Vanhoucke, V., Rabinovich, A.: Going deeper with convolutions. In: Proceedings of the IEEE Conference on Computer Vision and Pattern Recognition, pp. 1–9 (2015)

35. Van den Bergh, F., Engelbrecht, A.P.: A study of particle swarm optimization particle trajectories. Inf. Sci. **176**(8), 937–971 (2006)

36. Wang, B., Sun, Y., Xue, B., Zhang, M.: Evolving deep convolutional neural networks by variable-length particle swarm optimization for image classification. In: 2018 IEEE Congress on Evolutionary Computation (CEC), pp. 1–8. IEEE, Piscataway (2018)

37. Wang, B., Sun, Y., Xue, B., Zhang, M.: A hybrid differential evolution approach to designing deep convolutional neural networks for image classification. In: Australasian Joint Conference on Artificial Intelligence, pp. 237–250. Springer, Berlin (2018)

38. Wang, B., Sun, Y., Xue, B., Zhang, M.: Evolving deep neural networks by multi-objective particle swarm optimization for image classification. In: Proceedings of the Genetic and Evolutionary Computation Conference, GECCO '19, pp. 490–498. ACM, New York (2019). https://doi.org/10.1145/3321707.3321735

39. Wang, B., Sun, Y., Xue, B., Zhang, M.: A hybrid GA-PSO method for evolving architecture and short connections of deep convolutional neural networks. In: Nayak, A.C., Sharma, A. (eds.) PRICAI 2019: Trends in Artificial Intelligence, pp. 650–663. Springer International Publishing, Cham (2019)

# Chapter 7
# Designing Convolutional Neural Network Architectures Using Cartesian Genetic Programming

**Masanori Suganuma, Shinichi Shirakawa, and Tomoharu Nagao**

**Abstract** Convolutional neural networks (CNNs), among the deep learning models, are making remarkable progress in a variety of computer vision tasks, such as image recognition, restoration, and generation. The network architecture in CNNs should be manually designed in advance. Researchers and practitioners have developed various neural network structures to improve performance. Despite the fact that the network architecture considerably affects the performance, the selection and design of architectures are tedious and require trial-and-error because the best architecture depends on the target task and amount of data. Evolutionary algorithms have been successfully applied to automate the design process of CNN architectures. This chapter aims to explain how evolutionary algorithms can support the automatic design of CNN architectures. We introduce a method based on Cartesian genetic programming (CGP) for the design of CNN architectures. CGP is a form of genetic programming and searches the network-structured program. We represent the CNN architecture via a combination of pre-defined modules and search for the high-performing architecture based on CGP. The method attempts to find better architectures by repeating the architecture generation, training, and evaluation. The effectiveness of the CGP-based CNN architecture search is demonstrated through two types of computer vision tasks: image classification and image restoration. The experimental result for image classification shows that the method can find a well-performing CNN architecture. For the experiment on image restoration tasks, we show that the method can find a simple yet high-performing architecture of a convolutional autoencoder that is a type of CNN.

M. Suganuma
Tohoku University, RIKEN Center for AIP, Sendai, Miyagi, Japan
e-mail: suganuma@vision.is.tohoku.ac.jp

S. Shirakawa (✉) · T. Nagao
Yokohama National University, Yokohama, Kanagawa, Japan
e-mail: shirakawa-shinichi-bg@ynu.ac.jp; nagao@ynu.ac.jp

© Springer Nature Singapore Pte Ltd. 2020
H. Iba, N. Noman (eds.), *Deep Neural Evolution*, Natural Computing Series,
https://doi.org/10.1007/978-981-15-3685-4_7

## 7.1   Introduction

Many types of CNN architecture have been developed by researchers during the last few years aiming at achieving good scores on computer vision tasks. Despite the success of CNNs, a question remains given recent developments: *what CNN architectures are good and how can we design such architectures?* One possible direction to address this question is neural architecture search (NAS) [5], in which CNN architectures are automatically designed by an algorithm such as evolutionary computation and reinforcement learning to maximize performance on targeted tasks. NAS can automate the design process of neural networks and aids in reducing the trial-and-error of developers.

This chapter is based on the works of [34–36] and explains a genetic programming-based approach to automatically design CNN architectures. In the next section, we briefly review NAS methods by categorizing them into three approaches: evolutionary computation, reinforcement learning, and gradient-descent-based approaches. Then, we describe the Cartesian genetic programming (CGP)-based NAS method for a CNN, which is categorized as an evolutionary-computation-based approach. In Sect. 7.3, the CGP-based architecture search method for image classification, termed CGP-CNN, is explained. In Sect. 7.4, the CGP-based architecture search method is extended to the convolutional autoencoder (CAE), a type of CNN, for image restoration.

## 7.2   Progress of Neural Architecture Search

Automatic design of neural network structures is an active topic initially presented several decades ago, e.g., [30, 33, 45]. These methods optimize the connection weights and/or network structure of low-level neurons using an evolutionary algorithm, and are also known as evolutionary neural networks. These traditional structure optimization methods target relatively small neural networks whereas recent deep neural networks, including CNNs, have greater than one million parameters though the architectures are still designed by human experts. Aiming at the automatic design of deep neural network architectures, various architecture search methods have been developed since 2017. Nowadays, the automatic design method of deep neural network architectures is termed a neural architecture search (NAS) [5].

To address large-scale architectures, neural network architectures are designed using a certain search method but the network weights are optimized by a stochastic gradient descent method through back-propagation. Evolutionary algorithms are often used to search the architectures. Real et al. [28] optimized large-scale neural networks using an evolutionary algorithm and achieved better performance than that of modern CNNs in image classification tasks. In this method, they represent the CNN architecture as a graph structure and optimize it via the evolutionary algorithm.

The connection weights of the reproduced architecture are optimized by stochastic gradient descent as typical neural network training; the accuracy for the architecture evaluation dataset is assigned as the fitness. Miikkulainen et al. [20] proposed a method termed CoDeepNEAT that is an extended version of NeuroEvolution of Augmenting Topologies (NEAT). This method designs the network architectures using blueprints and modules. The blueprint chromosome is a graph in which each node has a pointer to a particular module species. Each module chromosome is a graph that represents a small DNN. Specifically, each node in the blueprint is replaced with a module selected from a particular species to which that node points. During the evaluation phase, the modules and blueprints are combined to generate assembled networks and the networks are evaluated. Xie and Yuille [42] designed CNN architectures using the genetic algorithm with a binary string representation. They proposed a method for encoding a network structure in which the connectivity of each layer is defined by a binary string representation. The type of each layer, number of channels, and size of a receptive field are not evolved in this method. The method explained in this chapter is also an evolutionary-algorithm-based NAS. Different from the aforementioned methods, it optimizes the architecture based on genetic programming and adopts well-designed modules as the node function.

Another approach is to use reinforcement learning to search the neural architectures. In [49], a recurrent neural network (RNN) was used to generate neural network architectures. The RNN was trained with policy-gradient-based reinforcement learning to maximize the expected accuracy on a learning task. Baker et al. [2] proposed a meta-modeling approach based on reinforcement learning to produce CNN architectures. A Q-learning agent explores and exploits a space of model architectures with an $\epsilon$-greedy strategy and experience replay.

As these methods need neural network training to evaluate the candidate architectures, they often require a considerable computational cost. For instance, the work of [49] used 800 graphics processing units (GPUs). To reduce the computational cost of NAS is an active topic. A promising approach is jointly optimizing the architecture parameter and connection weights. This approach, termed one-shot NAS (aka weight sharing), finds better architecture during single training. In one-shot NAS, the non-differentiable objective function consisting of discrete architecture parameters is transformed into a differentiable objective by continuous [17, 43] or stochastic relaxation [1, 27, 31]; both the architecture parameters and connection weights are optimized by gradient-based optimizers.

## 7.3   Designing CNN Architecture for Image Classification

In this section, we introduce the architecture search method based on CGP for image classification. We term the method CGP-CNN. In CGP-CNN, we directly encode the CNN architectures based on CGP and use highly functional modules as node functions. The CNN architecture defined by CGP is trained by a stochastic gradient descent using a model training dataset and assigns the fitness value based on the

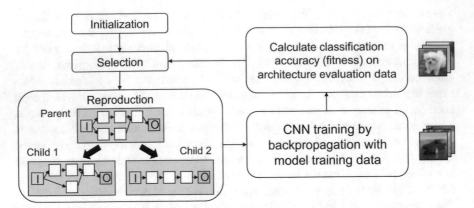

**Fig. 7.1** Overview of CGP-CNN. The method represents the CNN architectures based on CGP. The CNN architecture is trained on a learning task and assigned a fitness based on the accuracies of the trained model for the architecture evaluation dataset. The evolutionary algorithm searches for better architectures

accuracies of another training dataset (i.e. the architecture evaluation dataset). Then, the architecture is optimized to maximize the accuracy of the architecture evaluation dataset using the evolutionary algorithm. Figure 7.1 shows an overview of CGP-CNN. In the following, we describe the network representation and the evolutionary algorithm used in CGP-CNN.

### 7.3.1 Representation of CNN Architectures

For CNN architecture representation, we use the CGP encoding scheme that represents an architecture of CNNs as directed acyclic graphs with a two-dimensional grid. CGP was proposed as a general form of genetic programming in [22]. The graph corresponding to a phenotype is encoded to a string termed a genotype and optimized using the evolutionary algorithm.

Let us assume that the grid has $N_r$ rows by $N_c$ columns; then, the number of intermediate nodes is $N_r \times N_c$ and the number of inputs and outputs depends on the task. The genotype consists of a string of integers of a fixed length and each gene determines the function type of the node and the connection between nodes. The $c$-th column's node is only allowed to be connected from the $(c - 1)$ to $(c - l)$-th column's nodes, in which $l$ is termed a level-back parameter. Figure 7.2 shows an example of the genotype, phenotype, and corresponding CNN architecture. As seen in Fig. 7.2, the CGP encoding scheme has a possibility that not all of the nodes are connected to the output nodes (e.g., node No. 5 in Fig. 7.2). We term these nodes *inactive nodes*. Whereas the genotype in CGP is a fixed-length representation, the number of nodes in the phenotypic network varies because of the inactive nodes.

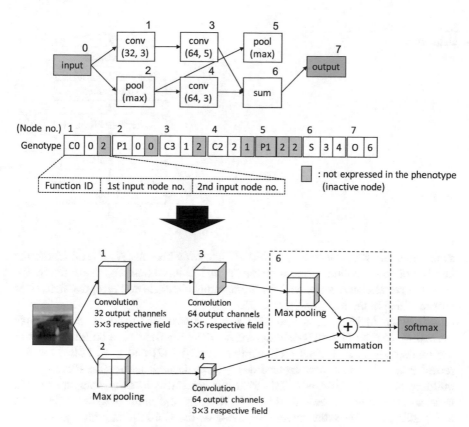

**Fig. 7.2** Examples of a genotype and phenotype. The genotype (left) defines the CNN architecture (right). Node No. 5 on the left is inactive and does not appear in the path from the inputs to the outputs. The summation node applies max pooling to downsample the first input to the same size as the second input

This is a desirable feature because the number of layers can be determined using the evolutionary algorithm.

Referring to modern CNN architectures, we select the highly functional modules as the node function. The frequently used processes in the CNN are convolution and pooling; the convolution processing uses local connectivity and spatially shares the learnable weights and the pooling is nonlinear downsampling. We prepare the six types of node functions, termed ConvBlock, ResBlock, max pooling, average pooling, concatenation, and summation. These nodes operate on the three-dimensional (3-D) tensor (also known as the feature map) defined by the dimensions of the row, column, and channel.

The ConvBlock consists of a convolutional layer with a stride of one followed by the batch normalization [10] and the rectified linear unit (ReLU) [23]. To maintain the size of the input, we pad the input with zero values around the border before the convolutional operation. Therefore, the ConvBlock takes the $M \times N \times C$ tensor

**Fig. 7.3** The ResBlock
architecture

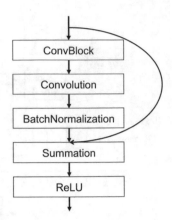

as an input and produces the $M \times N \times C'$ tensor, where $M$, $N$, $C$, and $C'$ are the number of rows, columns, input channels, and output channels, respectively. We prepare several ConvBlocks with different output channels and receptive field sizes (kernel sizes) in the function set of CGP.

As shown in Fig. 7.3, the ResBlock is composed of the ConvBlock, batch normalization, ReLU, and tensor summation. The ResBlock is a building block of the modern successful CNN architectures, e.g., [8, 47] and [13]. Following this recent trend of human architecture design, we decided to use ResBlock as the building block in CGP-CNN. The ResBlock performs identity mapping via the shortcut connection as described in [8]. The row and column sizes of the input are preserved in the same manner as those of the ConvBlock after convolution. As shown in Fig. 7.3, the output feature maps of the ResBlock are calculated via the ReLU activation and the summation with the input. The ResBlock takes the $M \times N \times C$ tensor as an input and produces the $M \times N \times C'$ tensor. We prepare several ResBlocks with different output channels and receptive field sizes (kernel sizes) in the function set of CGP.

The max and average poolings perform the maximum and average operations, respectively, over the local neighbors of the feature maps. We use the pooling with a $2 \times 2$ receptive field size and a stride of two. The pooling layer takes the $M \times N \times C$ tensor and produces the $M' \times N' \times C$ tensor, where $M' = \lfloor M/2 \rfloor$ and $N' = \lfloor N/2 \rfloor$.

The concatenation function takes two feature maps and concatenates them in the channel dimension. When concatenating the feature maps with different numbers of rows and columns, we downsample the larger feature map by max pooling to make them the same sizes as the inputs. Let us assume that we have two inputs of size $M_1 \times N_1 \times C_1$ and $M_2 \times N_2 \times C_2$, then the size of the output feature maps is $\min(M_1, M_2) \times \min(N_1, N_2) \times (C_1 + C_2)$.

The summation performs element-wise summation of two feature maps, channel-by-channel. Similar to the concatenation, when summing the two feature maps with different numbers of rows and columns, we downsample the larger feature map by max pooling. In addition, if the inputs have different numbers of channels, we

**Table 7.1** Node functions and abbreviated symbols used in the experiments

| Node type | Symbol | Variation |
|---|---|---|
| ConvBlock | CB $(C', k)$ | $C' \in \{32, 64, 128\}$ |
| | | $k \in \{3 \times 3, 5 \times 5\}$ |
| ResBlock | RB $(C', k)$ | $C' \in \{32, 64, 128\}$ |
| | | $k \in \{3 \times 3, 5 \times 5\}$ |
| Max pooling | MP | – |
| Average pooling | AP | – |
| Concatenation | Concat | – |
| Summation | Sum | – |

$C'$: Number of output channels
$k$: Receptive field size (kernel size)

expand the channels of the feature maps with a smaller channel size by filling with zero. Let us assume that we have two inputs of size $M_1 \times N_1 \times C_1$ and $M_2 \times N_2 \times C_2$, then the sizes of the output feature maps are $\min(M_1, M_2) \times \min(N_1, N_2) \times \max(C_1, C_2)$. In Fig. 7.2, the summation node applies the max pooling to downsample the first input to the same size as the second input. By using the summation and concatenation operations, our method can express the shortcut connection or branch layers, such as those used in GoogLeNet [37] and residual network (ResNet) [8].

The output node represents the softmax function to produce a distribution over the target classes. The outputs fully connect to all elements of the input. The node functions used in the experiments are listed in Table 7.1.

## 7.3.2 Evolutionary Algorithm

Following the standard CGP, we use a point mutation as the genetic operator. The function and the connection of each node randomly change to valid values according to the mutation rate. The fitness evaluation of the CNN architecture involves CNN training and requires approximately 0.5 to 1 h in our setting. Therefore, we need to efficiently evaluate some candidate solutions in parallel at each generation. To efficiently use the computational resource, we repeatedly apply the mutation operator while an active node does not change and obtain the candidate solutions to be evaluated. We term this mutation forced mutation. Moreover, to maintain a neutral drift, which is effective for CGP evolution [21, 22], we modify a parent by neutral mutation if the fitness of the offspring do not improve. The neutral mutation operates only on the genes of inactive nodes without modification of the phenotype. We use the modified $(1 + \lambda)$ evolution strategy (with $\lambda = 2$ in the experiment) using the aforementioned artifice. The procedure of our evolutionary algorithm is listed in Algorithm 1.

The $(1 + \lambda)$ evolution strategy, the default evolutionary algorithm in CGP, is an algorithm with fewer strategy parameters: the mutation rate and offspring size. We

---

**Algorithm 1** Evolutionary algorithm for CGP-CNN and CGP-CAE

---

1: **Input:** $G$ (number of generations), $r$ (mutation probability), $\lambda$ (children size), $S$ (Training set), $V$ (architecture evaluation set).

2: **Initialization:** (i) Generate a *parent*, (ii) train the model on the $S$, and (iii) assign the fitness $F_p$ using the set $V$.

3: **for** $g = 1$ **to** $G$ **do**

4:    **for** $i = 1$ **to** $\lambda$ **do**

5:       $children_i \leftarrow$ Mutation(*parent*, $r$) # forced mutation

6:       $model_i \leftarrow$ Train($children_i$, $S$)

7:       $fitness_i \leftarrow$ Evaluate($model_i$, $V$)

8:    **end for**

9:    $best \leftarrow \text{argmax}_{i=1,2,\ldots,\lambda} \{fitness_i\}$

10:   **if** $fitness_{best} \geq F_p$ **then**

11:      $parent \leftarrow children_{best}$

12:      $F_p \leftarrow fitness_{best}$

13:   **else**

14:      $parent \leftarrow$ Modify(*parent*, $r$) # neutral mutation

15:   **end if**

16: **end for**

17: **Output:** *parent* (the best architecture found by the evolutionary search).

---

do not need to expend considerable effort to tune such strategy parameters. Thus, we use the $(1 + \lambda)$ evolution strategy in CGP-CNN.

### 7.3.3 Experiment on Image Classification Tasks

#### 7.3.3.1 Experimental Setting

We apply CGP-CNN to the CIFAR-10 and CIFAR-100 datasets consisting of 60,000 color images ($32 \times 32$ pixels) in 10 and 100 classes, respectively. Each dataset is split into a training set of 50,000 images and a test set of 10,000 images. We randomly sample 45,000 examples from the training set to train the CNN and the remaining 5000 examples are used for architecture evaluation (i.e. fitness evaluation of CGP).

To assign the fitness value to the candidate CNN architecture, we train the CNN by stochastic gradient descent (SGD) with a mini-batch size of 128. The softmax cross-entropy loss is used as the loss function. We initialize the weights using the method described in [7] and use the Adam optimizer [11] with an initial learning rate $\alpha = 0.01$ and momentum $\beta_1 = 0.9$ and $\beta_2 = 0.999$. We train each CNN for 50 epochs and use the maximum accuracy of the last 10 epochs as the fitness value. We reduce the learning rate by a factor of 10 at the 30th epoch.

We preprocess the data with pixel-mean subtraction. To prevent overfitting, we use a weight decay with the coefficient $1.0 \times 10^{-4}$. We also use data augmentation based on [8]: padding 4 pixels on each size and randomly cropping a $32 \times 32$ patch from the padded image or its horizontally flipped image.

**Table 7.2** Parameter setting
for the CGP-CNN on image
classification tasks

| Parameters | Values |
|---|---|
| Mutation rate | 0.05 |
| # Offspring ($\lambda$) | 2 |
| # Rows ($N_r$) | 5 |
| # Columns ($N_c$) | 30 |
| Minimum number of active nodes | 10 |
| Maximum number of active nodes | 50 |
| Levels-back ($l$) | 10 |

The parameter setting for CGP is shown in Table 7.2. We use a relatively large number of columns to generate deep architectures. The number of active nodes in the individual of CGP is restricted. Therefore, we apply the mutation operator until the CNN architecture that satisfies the restriction of the number of active nodes is generated. The offspring size $\lambda$ is two, the same number of GPUs in our experimental machines. We test two node function sets termed ConvSet and ResSet for CGP-CNN. The ConvSet contains ConvBlock, max pooling, average pooling, summation, and concatenation in Table 7.1 and the ResSet contains ResBlock, max pooling, average pooling, summation, and concatenation. The difference between these two function sets is whether the set contains ConvBlock or ResBlock. The number of generations is 500 for ConvSet and 300 for ResSet.

The best CNN architecture from the CGP process is retrained using all 50,000 images in the training set. Then, we compute the test accuracy. We optimize the weights of the obtained architecture for 500 epochs using a different training procedure; we use SGD with a momentum of 0.9, a mini-batch size of 128, and a weight decay of $5.0 \times 10^{-4}$. Following the learning rate schedule in [8], we start with a learning rate of 0.01 and set it to 0.1 at the 5th epoch. We reduce it by a factor of 10 at the 250th and 370th epochs. We report the test accuracy at the 500th epoch as the final performance.

We implement CGP-CNN using the Chainer framework [40] (version 1.16.0) and run it on a machine with two NVIDIA GeForce GTX 1080 or two GTX 1080 Ti GPUs. We use a GTX 1080 and 1080 Ti for the experiments on the CIFAR-10 and 100 datasets, respectively. Because of the memory limitation, the candidate CNNs occasionally take up the GPU memory, and the network training process fails because of an out-of-memory error. In this case, we assign a zero fitness to the candidate architecture.

### 7.3.3.2   Experimental Result

We run CGP-CNN 10 times on each dataset and report the classification errors. We compare the classification performance to the hand-designed CNNs and automatically designed CNNs using the architecture search methods on the CIFAR-10 and 100 datasets. A summary of the classification performances is provided in

**Table 7.3** Comparison of the error rates (%), number of learnable weight parameters, and search costs on the CIFAR-10 dataset

| Model | # Params | Test error | GPU days |
|---|---|---|---|
| Maxout [6] | – | 9.38 | – |
| Network in network [15] | – | 8.81 | – |
| VGG [32] | 15.2 M | 7.94 | – |
| ResNet [8] | 1.7 M | 6.61 | – |
| FractalNet [14] | 38.6 M | 5.22 | – |
| Wide ResNet [47] | 36.5 M | 4.00 | – |
| CoDeepNEAT [20] | – | 7.30 | – |
| Genetic CNN [42] | – | 7.10 | 17 |
| MetaQNN [2] | 3.7 M | 6.92 | 80–100 |
| Large-scale evolution [28] | 5.4 M | 5.40 | 2750 |
| Neural architecture search [49] | 37.4 M | **3.65** | 16,800–22,400 |
| CGP-CNN (ConvSet) | 1.50 M | 5.92 (6.48 ± 0.48) | 31 |
| CGP-CNN (ResSet) | 2.01 M | 5.01 (6.10 ± 0.89) | 30 |

The classification error is reported in the format of "best (mean ± std)." In CGP-CNN, the number of learnable weight parameters of the best architecture is reported. The values of other models are referenced from the literature. The bold value indicates the best test error among the compared models

Tables 7.3 and 7.4. The models, Maxout, Network in Network, VGG, ResNet, FractalNet, and Wide ResNet, are the hand-designed CNN architectures whereas MetaQNN, Neural Architecture Search, Large-Scale Evolution, Genetic CNN, and CoDeepNEAT are the models obtained using the architecture search methods. The values of other models, except for VGG and ResNet on CIFAR-100, are referenced from the literature. We implement the VGG net and ResNet for CIFAR-100 because they were not applied to the dataset in [32] and [8]. The architecture of VGG is identical to that of configuration D in [32]. In Tables 7.3 and 7.4, the number of learnable weight parameters in the models is also listed. In CGP-CNN, the number of learnable weight parameters of the best architecture is reported.

On the CIFAR-10 dataset, the CGP-CNNs outperform most of the hand-designed models and show a good balance between the classification errors and the number of parameters. CGP-CNN (ResSet) shows better performance compared to that of CGP-CNN (ConvSet). Compared to other architecture search methods, CGP-CNN (ConvSet and ResSet) outperforms MetaQNN [2], Genetic CNN [42], and CoDeep-NEAT [20]. The best architecture of CGP-CNN (ResSet) outperforms Large-Scale Evolution [28]. The Neural Architecture Search [49] achieved the best error rate, but this method used 800 GPUs and required considerable computational costs to search for the best architecture. Table 7.3 also lists the number of GPU days (the computational time multiplied by the number of GPUs used during the experiments) for the architecture search. As seen, CGP-CNN can find a good architecture at

**Table 7.4** Comparison of the error rates (%) and number of learnable weight parameters on the CIFAR-100 dataset

| Model | # Params | Test error |
|---|---|---|
| Maxout [6] | – | 38.57 |
| Network in network [15] | – | 35.68 |
| VGG [32] | 15.2 M | 33.45 |
| ResNet [8] | 1.7 M | 32.40 |
| FractalNet [14] | 38.6 M | 23.30 |
| Wide ResNet [47] | 36.5 M | **19.25** |
| CoDeepNEAT [20] | – | – |
| Neural architecture search [49] | 37.4 M | – |
| Genetic CNN [42] | – | 29.03 |
| MetaQNN [2] | 3.7 M | 27.14 |
| Large-scale evolution [28] | 40.4 M | 23.0 |
| CGP-CNN (ConvSet) | 2.01 M | 26.7 (28.1 ± 0.83) |
| CGP-CNN (ResSet) | 4.60 M | 25.1 (26.8 ± 1.21) |

The classification errors are reported in the format of "best (mean ± std)." In CGP-CNN, the number of learnable weight parameters of the best architecture is reported. The values of other models except for VGG and ResNet are referenced from the literature. The bold value indicates the best test error among the compared models

a reasonable computational cost. We assume that CGP-CNN, particularly with ResSet, could reduce the search space and find better architectures in an early iteration by using the highly functional modules. The CIFAR-100 dataset is a very challenging task because there are many classes. CGP-CNN finds the competitive network architectures within a reasonable computational time. Even though the obtained architecture is not at the same level as the state-of-the-art architectures, it shows a good balance between the classification errors and number of parameters.

The error rates of the architecture search methods (not only CGP-CNN) do not reach those of Wide ResNet, a human-designed architecture. However, these human-designed architectures are developed with the expenditure of tremendous human effort. An advantage of architecture search methods is that they can automatically find a good architecture for a new dataset. Another advantage of CGP-CNN is that the number of weight parameters in the discovered architectures is less than that in the human-designed architectures, which is beneficial when we want to implement CNN on a mobile device. Note that we did not introduce any criteria for the architecture complexity in the fitness function. It might be possible to find more compact architectures by introducing the penalty term into the fitness function, which is an important research direction, such as in [4, 29, 39].

Figure 7.4 shows the examples of the CNN architectures obtained by CGP-CNN (ConvSet and ResSet). Figure 7.4 shows the complex architectures that are difficult to manually design. Specifically, CGP-CNN (ConvSet) uses the summation and concatenation nodes leading to a wide network and allowing for the formation of skip connections. Therefore, the CGP-CNN (ConvSet) architecture is wider than

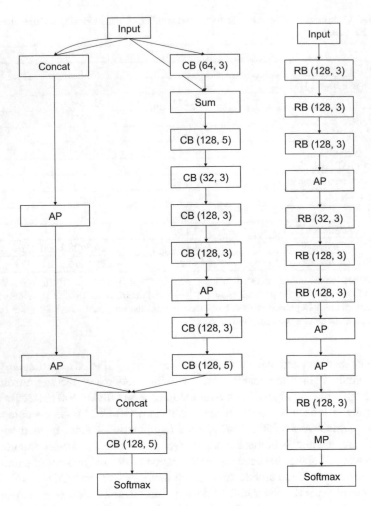

**Fig. 7.4** CNN architectures obtained by CGP-CNN with ConvSet (left) and ResSet (right) on the CIFAR-10 dataset

that of CGP-CNN (ResSet). Additionally, we also observe that CGP-CNN (ResSet) has a similar structure to that of ResNet [8]. ResNet consists of a series of two types of modules: a module with several convolutions and shortcut connections without downsampling and a downsampling convolution with a stride of 2. Although CGP-CNN cannot downsample in the ConvBlock and ResBlock, we see that CGP-CNN (ResSet) uses a pooling layer as an alternative to the downsampling convolution. We can say that CGP-CNN can find an architecture similar to that designed by human experts.

## 7.4 Designing CNN Architectures for Image Restoration

In this section, we apply the CGP-based architecture search method to an image restoration task of recovering a clean image from its degraded version. We term this method CGP-CAE. Recently, learning-based approaches based on CNNs have been applied to image restoration tasks and have significantly improved the state-of-the-art performance. Researchers have approached this problem mainly from three directions: designing new network architectures, loss functions, and training strategies. In this section, we focus on designing a new network architecture for image restoration and report that simple convolutional autoencoders (CAEs) designed by evolutionary algorithms can outperform existing image restoration methods which are designed manually.

### 7.4.1 Search Space of Network Architectures

In this work, we consider CAEs that are built only on convolutional layers with downsampling and skip connections. In addition, we use *symmetric* CAEs such that their first half (encoder part) is symmetric to the second half (decoder part). The final layer is attached to top of the decoder part to obtain images of fixed channels (i.e. single-channel grayscale or three-channel color images), for which either one or three filters of $3 \times 3$ size are used. Therefore, specifying the encoder part of a CAE solely determines its entire architecture. The encoder part can have an arbitrary number of convolutional layers up to a specified maximum, which is selected by the evolutionary algorithm. Each convolutional layer can have an arbitrary number and size of filters, and is followed by ReLU [23]. In addition, each layer can have an optional skip connection [8, 18] that connects the layer to its mirrored counterpart in the decoder part. Specifically, the output feature maps (obtained after ReLU) of the layer are passed to and are added element-wise to the output feature maps (obtained before ReLU) of the counterpart layer. We can use additional downsampling after each convolutional layer depending on the task. Whether to use downsampling is determined in advance and thus it is not selected by the architectural search, as explained later.

### 7.4.2 Representation of CAE Architectures

Following [34], we represent architectures of CAEs via a directed acyclic graph which is defined on a two-dimensional grid. This graph is optimized by the evolutionary algorithm, in which the graph is termed a phenotype and is encoded by a data structure termed a genotype.

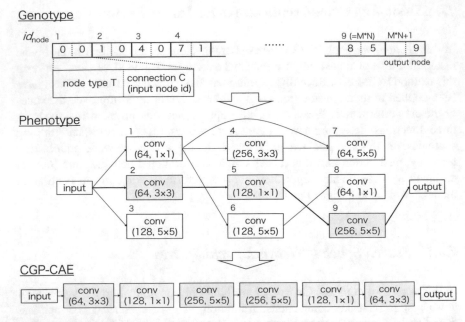

**Fig. 7.5** An example of a genotype and a phenotype of CGP-CAE. A phenotype is a graph representation of a network architecture and a genotype encodes a phenotype. They encode only the encoder part of a CAE and its decoder part is automatically created such that it is symmetrical to the encoder part. In this example, the phenotype is defined on a grid of three rows and three columns

Figure 7.5 shows an example of a genotype and a phenotype of CGP-CAE. Each node of the graph represents a convolutional layer followed by a ReLU in a CAE. An edge connecting two nodes represents the connectivity of the two corresponding layers. The graph has two additional special nodes termed input and output nodes. The former represents the input layer of the CAE and the latter represents the output of the encoder part, or equivalently the input of the decoder part of the CAE. As the input of each node is connected to at most one node, there is a single unique path starting from the input node and ending at the output node. This unique path identifies the architecture of the CAE, as shown in the middle row of Fig. 7.5. Note that the nodes depicted in the neighboring two columns are not necessarily connected. Thus, the CAE can have a different number of layers depending on how the nodes are connected. Because the maximum number of layers (of the encoder part) of the CAE is $N_{\max}$, the total number of layers is $2N_{\max} + 1$ including the output layer. To control how the number of layers will be chosen, we introduce a hyper-parameter termed level-back $l$, such that nodes given in the $c$-th column are allowed to be connected from nodes given in the columns ranging from $c - l$ to $c - 1$. If we use a smaller $l$, then the resulting CAEs will tend to be deeper.

A genotype encodes a phenotype and is manipulated by the evolutionary algorithm. The genotype encoding a phenotype with $N_r$ rows and $N_c$ columns has

$N_r N_c + 1$ genes, each of which represents attributes of a node with two integers (i.e. type and connection). The type specifies the number $F$ and size $k$ of the filters of the node, and whether the layer has skip connections or not, by an integer encoding their combination. The connection specifies the node that is connected to the input of this node. The last $(N_r N_c + 1)$-st gene represents the output node that stores only the connection determining the node connected to the output node. An example of a genotype is shown in the top row of Fig. 7.5, where $F \in \{64, 128, 256\}$ and $k \in \{1 \times 1, 3 \times 3, 5 \times 5\}$.

We use the same evolutionary algorithm as used in the previous section to perform a search in the architecture space (see Algorithm 1).

### 7.4.3 Experiment on Image Restoration Tasks

We conducted experiments to test the effectiveness of CGP-CAE. We chose two tasks: image inpainting and denoising.

#### 7.4.3.1 Experimental Settings

**Inpainting**
We followed the procedures suggested in [46] for experimental design. We used three benchmark datasets: the CelebFaces Attributes Dataset (CelebA) [16], the Stanford Cars Dataset (Cars) [12], and the Street View House Numbers (SVHN) [24]. The CelebA contains 202,599 images, from which we randomly selected 100,000, 1000, and 2000 images for training, architecture evaluation, and testing, respectively. All images were cropped to properly contain the entire face and resized to $64 \times 64$ pixels. For Cars and SVHN, we used the provided training and testing split. The images of Cars were cropped according to the provided bounding boxes and resized to $64 \times 64$ pixels. The images of SVHN were resized to $64 \times 64$ pixels.

We generated images with missing regions of the following three types: a central square block mask (*Center*), random pixel masks such that 80% of all the pixels were randomly masked (*Pixel*), and half-image masks such that a randomly chosen vertical or horizontal half of the image was masked (*Half*). For the latter two, a mask was randomly generated for each training mini-batch and each test image.

Considering the nature of this task, we consider CAEs endowed with down-sampling. To be specific, the same counts of downsampling and upsampling with stride $= 2$ were employed such that the entire network had a symmetric hourglass shape. For simplicity, we used a skip connection and downsampling in an exclusive manner; in other words, every layer (in the encoder part) employed either a skip connection or downsampling.

**Denoising**

We followed the experimental procedures described in [18, 38]. We used grayscale 300 and 200 images belonging to the BSD500 dataset [19] to generate training and test images, respectively. For each image, we randomly extracted $64 \times 64$ patches, to each of which Gaussian noise with different $\sigma = 30, 50$, and 70 are added. As utilized in the previous studies, we trained a single model for all different noise levels.

For this task, we used CAE models without downsampling following the previous studies [18, 38]. We zero-padded the input feature maps computed in each convolution layer not to change the size of the input and output feature space of the layer.

**Configurations of the Architectural Search**

For the evolutionary algorithm, we chose the mutation probability as $r = 0.1$, number of children as $\lambda = 4$, and number of generations as $G = 250$. For the phenotype, we used the graph with $N_r = 3$, $N_c = 20$, and level-back $l = 5$. For the number $F$ and size $k$ of the filters at each layer, we chose them from {64, 128, 256} and {$1 \times 1, 3 \times 3, 5 \times 5$}, respectively. During an evolution process, we trained each CAE for $I = 20,000$ iterations with a mini-batch of size $b = 16$. We set the learning rate of the ADAM optimizer to be 0.001. For the training loss, we used the mean squared error (MSE) between the restored images and their ground truths:

$$L(\theta_D) = \frac{1}{|S|} \sum_{i=1}^{|S|} ||D(y_i; \theta_D) - x_i||_2^2, \tag{7.1}$$

where the CAE and its weight parameters are $D$ and $\theta_D$, respectively; $S$ is the training set, $x_i$ is a ground truth image, and $y_i$ is a corrupted image. For the fitness function of the evolutionary algorithm, we use the peak signal-to-noise ratio (PSNR) of which the higher value indicates the better image restoration.

Following completion of the evolution process, we fine-tuned the best CAE using the training set of images for additional 500,000 iterations, in which the learning rate is reduced by a factor of 10 at the 200,000 and 400,000 iterations. We then calculated its performance using the test set of images. We implemented CGP-CAE using PyTorch [25] and performed the experiments using four P100 GPUs. Execution of the evolutionary algorithm and the fine-tuning of the best model took approximately 3 days for the inpainting tasks and 4 days for the denoising tasks.

### 7.4.3.2  Results of the Inpainting Tasks

We use two standard evaluation measures, the PSNR and structural similarity index (SSIM) [41], to evaluate the restored images. Higher values of these measures indicate better image restoration.

**Table 7.5** Inpainting results

| Dataset | Type | PSNR | | | | | SSIM | | | | |
|---|---|---|---|---|---|---|---|---|---|---|---|
| | | Rand | BASE | CE | SII | CGP-CAE | Rand | BASE | CE | SII | CGP-CAE |
| CelebA | Center | 15.3 | 27.1 | 28.5 | 19.4 | **29.9** | 0.740 | 0.883 | 0.912 | 0.907 | **0.934** |
| | Pixel | 25.5 | 27.5 | 22.9 | 22.8 | **27.8** | 0.766 | 0.836 | 0.730 | 0.710 | **0.887** |
| | Half | 12.7 | 11.8 | 19.9 | 13.7 | **21.1** | 0.549 | 0.604 | 0.747 | 0.582 | **0.771** |
| Cars | Center | 17.1 | 19.5 | 19.6 | 13.5 | **20.9** | 0.704 | 0.767 | 0.767 | 0.721 | **0.846** |
| | Pixel | 17.0 | 19.2 | 15.6 | 18.9 | **19.5** | 0.533 | 0.679 | 0.408 | 0.412 | **0.738** |
| | Half | 13.0 | 11.6 | 14.8 | 11.1 | **16.2** | 0.511 | 0.541 | 0.576 | 0.525 | **0.610** |
| SVHN | Center | 23.5 | 29.9 | 16.4 | 19.0 | **33.3** | 0.819 | 0.895 | 0.791 | 0.825 | **0.953** |
| | Pixel | 29.0 | 40.1 | 30.5 | 33.0 | **40.4** | 0.687 | 0.899 | 0.888 | 0.786 | **0.969** |
| | Half | 11.3 | 12.9 | 21.6 | 14.6 | **24.8** | 0.574 | 0.617 | 0.756 | 0.702 | **0.848** |

Comparison of two baseline architectures (RAND and BASE), Context Autoencoder (CE) [26], Semantic Image Inpainting (SII) [46], and CAEs designed by CGP-CAE using three datasets and three masking patterns. The bold values indicate the best performance among the compared architectures

As previously mentioned, we follow the experimental procedure employed in [46]. In the paper, the authors reported the performances of their proposed method, Semantic Image Inpainting (SII), and Context Autoencoder (CE) [26]. However, we found that CE can provide considerably better results than those reported in [46] in terms of PSNR. Thus, we report here PSNR and SSIM values for CE that we obtained by running the code provided by the authors.[1] To calculate SSIM values of SII, which were not reported in [46], we run the authors' code[2] for SII.

To further validate the effectiveness of the evolutionary search, we evaluate two baseline architectures; an architecture generated by a random search (RAND) and an architecture with same depth as the best-performing architecture found by CGP-CAE but having a constant number (64) of fixed size (3 × 3) filters in each layer with a skip connection (BASE). In the random search, we generate 10 architectures at random in the same search space as ours and report their average PSNR and SSIM values. All other experimental setups are the same.

Table 7.5 shows the PSNR and SSIM values obtained using five methods on three datasets and three masking patterns. We run the evolutionary algorithm three times and report the average accuracy values of the three optimized CAEs. As shown, CGP-CAE outperforms the other four methods for each of the dataset-mask combinations. Notably, CE and SII use mask patterns for inference. To be specific, their networks estimate only pixel values of the missing regions specified by the provided masks, and then they are merged with the unmasked regions of clean pixels. Thus, the pixel intensities of the unmasked regions are identical to their ground truths. On the other hand, CGP-CAE does not use masks yet outputs

---

[1] https://github.com/pathak22/context-encoder.

[2] https://github.com/moodoki/semantic_image_inpainting.

**Fig. 7.6** Examples of inpainting results obtained by CGP-CAE (CAEs designed by the evolutionary algorithm)

**Table 7.6** Denoising results on BSD200

| Noise $\sigma$ | PSNR | | | | | SSIM | | | | |
|---|---|---|---|---|---|---|---|---|---|---|
| | Rand | BASE | RED | MemNet | CGP-CAE | Rand | BASE | RED | MemNet | CGP-CAE |
| 30 | 27.25 | 27.00 | 27.95 | 28.04 | **28.23** | 0.7491 | 0.7414 | 0.8019 | **0.8053** | 0.8047 |
| 50 | 25.11 | 24.88 | 25.75 | 25.86 | **26.17** | 0.6468 | 0.6229 | 0.7167 | 0.7202 | **0.7255** |
| 70 | 23.50 | 23.22 | 24.37 | 24.53 | **24.83** | 0.5658 | 0.5349 | 0.6551 | 0.6608 | **0.6636** |

Comparison of results of two baseline architectures (RAND and BASE), RED [18], MemNet [38], and CGP-CAE. The bold values indicate the best performance among the compared architectures

complete images such that the missing regions are hopefully correctly inpainted. We then calculate the PSNR of the output image against the ground truth without identifying missing regions. This difference should help CE and SII to achieve high PSNR and SSIM values, but nevertheless CGP-CAE performs better.

Sample inpainted images obtained by CGP-CAE along with the masked inputs and the ground truths are shown in Fig. 7.6. It is observed that overall CGP-CAE stably performs; the output images do not have large errors for all types of masks. It performs particularly well for random pixel masks (the middle column of Fig. 7.6); the images are realistic and sharp. It is also observed that CGP-CAE tends to yield less sharp images for those with a filled region of missing pixels. However, CGP-CAE can accurately infer their contents, as shown in the examples of inpainting images of numbers (the rightmost column of Fig. 7.6).

### 7.4.3.3 Results of the Denoising Task

We compare CGP-CAE to two baseline architectures (i.e. RAND and BASE described in Sect. 7.4.3.2) and two state-of-the-art methods RED [18] and MemNet [38]. Table 7.6 shows the PSNR and SSIM values for three versions of the BSD200 test set with different noise levels $\sigma = 30, 50$, and 70, in which the performance values of RED and MemNet are obtained from [38]. CGP-CAE again achieves the best performance for all cases except for a single case (MemNet for $\sigma = 30$). It is worth noting that the networks of RED and MemNet have 30 and 80 layers, respectively, whereas our best CAE has only 15 layers (including the decoder part

Input                    CGP-CAE                 Ground truth

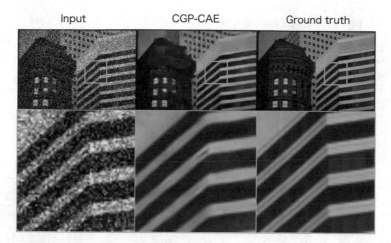

**Fig. 7.7** Examples of images reconstructed by CGP-CAE for the denoising task. The first column shows the input image with noise level $\sigma = 50$

and output layer), showing that our evolutionary method was able to find simpler architectures that can provide more accurate results.

An example of an image recovered by CGP-CAE is shown in Fig. 7.7. As we can see, CGP-CAE correctly removes the noise and produces an image as sharp as the ground truth.

### 7.4.3.4 Analysis of Optimized Architectures

Table 7.7 shows the top five best-performing architectures designed by CGP-CAE for the image inpainting task using center masks on the CelebA dataset and the denoising task, along with their performances measured on their test datasets. One of the best-performing architectures for each task is shown in Fig. 7.8. We can see that although their overall structures do not appear unique, mostly because of the limited search space of CAEs, the number and size of filters are quite different across layers, which is difficult to manually determine. Although it is difficult to provide a general interpretation of why the parameters of each layer are selected, we can make the following observations: (1) regardless of the task, almost all networks have a skip connection in the first layer, implying that the input images contain essential information to yield accurate outputs; (2) $1 \times 1$ convolution seems to be an important ingredient for both tasks; $1 \times 1$ convolution layers dominate the denoising networks, and all the inpainting networks employ two $1 \times 1$ convolution layers; (3) when comparing the inpainting networks to the denoising networks, the following differences are apparent: the largest filters of size $5 \times 5$ tend to be employed by the former more often than the latter (2.8 vs. 0.8 layers on average), and $1 \times 1$ filters tend to be employed by the former less often than the latter (2.0 vs. 3.2 layers on average).

**Table 7.7** Best-performing five architectures of CGP-CAE

| Architecture (Inpainting) | PSNR | SSIM |
|---|---|---|
| $CS(128, 3) - C(64, 3) - CS(128, 5) - C(128, 1) - CS(256, 5) -$ $C(256, 1) - CS(64, 5)$ | 29.91 | 0.9344 |
| $C(256, 3) - CS(64, 1) - C(128, 3) - CS(256, 5) - CS(64, 1) -$ $C(64, 3) - CS(128, 5)$ | 29.91 | 0.9343 |
| $CS(128, 5) - CS(256, 3) - C(64, 1) - CS(128, 3) - CS(64, 5) -$ $CS(64, 1) - C(128, 5) - C(256, 5)$ | 29.89 | 0.9334 |
| $CS(128, 3) - CS(64, 3) - C(64, 5) - CS(256, 3) - C(128, 3) -$ $CS(128, 5) - CS(64, 1) - CS(64, 1)$ | 29.88 | 0.9346 |
| $CS(64, 1) - C(128, 5) - CS(64, 3) - C(64, 1) - CS(256, 5) - C(128, 5)$ | 29.63 | 0.9308 |

| Architecture (Denoising) | PSNR | SSIM |
|---|---|---|
| $CS(64, 3) - C(64, 1) - C(128, 3) - CS(64, 1) - CS(128, 5) -$ $C(128, 3) - C(64, 1)$ | 26.67 | 0.7313 |
| $CS(64, 5) - CS(256, 1) - C(256, 1) - C(64, 3) - CS(128, 1) -$ $C(64, 3) - CS(128, 1) - C(128, 3)$ | 26.28 | 0.7113 |
| $CS(64, 3) - C(64, 1) - C(128, 3) - CS(64, 1) - CS(128, 5) -$ $C(128, 3) - C(64, 1)$ | 26.28 | 0.7107 |
| $CS(128, 3) - CS(64, 1) - C(64, 3) - C(64, 3) - CS(64, 1) - C(64, 3)$ | 26.20 | 0.7047 |
| $CS(64, 5) - CS(128, 1) - CS(256, 3) - CS(128, 1) - CS(128, 1) -$ $C(64, 1) - CS(64, 3)$ | 26.18 | 0.7037 |

$C(F, k)$ indicates that the layer has $F$ filters of size $k \times k$ without a skip connection. $CS$ indicates that the layer has a skip connection. This table shows only the encoder part of CAEs. For denoising, the average PSNR and SSIM values of three noise levels are shown

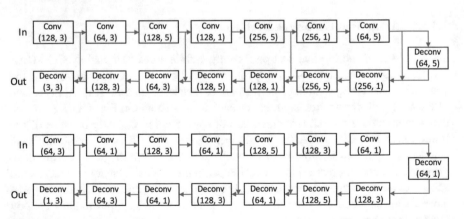

**Fig. 7.8** One of the best-performing architectures given in Table 7.7 for inpainting (upper) and denoising (lower) tasks

## 7.5 Summary

This chapter introduced a neural architecture search for CNNs: a CGP-based approach for designing deep CNN architectures. Specifically, the methods, CGP-CNN for image classification and CGP-CAE for image restoration, were explained. The methods generate CNN architectures based on the CGP encoding scheme with highly functional modules and use the evolutionary algorithm to find good architectures. The effectiveness and potential of CGP-CNN and CGP-CAE were verified through numerical experiments. The experimental results of image classification showed that CGP-CNN can find a well-performing CNN architecture. In the experiment on image restoration tasks, we showed that CGP-CAE can find a simple yet high-performing architecture of a CAE. We believe that evolutionary computation is a promising solution for NAS.

The bottleneck of the architecture search of DNN is the computational cost. Simple yet effective acceleration techniques, termed rich initialization and early termination of network training, can be found in [36]. Another possible acceleration technique is starting with a small data size and increasing the training data for the neural networks as the generation progresses. Moreover, to simplify and compact the CNN architectures, we may introduce regularization techniques to the architecture search process. Alternatively, we may be able to manually simplify the obtained CNN architectures by removing redundant or less effective layers.

Considerable room remains for exploration of search spaces of architectures of classical convolutional networks, which may apply to other tasks such as single image colorization [48], depth estimation [3, 44], and optical flow estimation [9].

## References

1. Akimoto, Y., Shirakawa, S., Yoshinari, N., Uchida, K., Saito, S., Nishida, K.: Adaptive stochastic natural gradient method for one-shot neural architecture search. In: Proceedings of the 36th International Conference on Machine Learning (ICML), vol. 97, pp. 171–180 (2019)
2. Baker, B., Gupta, O., Naik, N., Raskar, R.: Designing neural network architectures using reinforcement learning. In: Proceedings of the 5th International Conference on Learning Representations (ICLR) (2017)
3. Eigen, D., Puhrsch, C., Fergus, R.: Depth map prediction from a single image using a multi-scale deep network. In: Advances in Neural Information Processing Systems 27 (NIPS '14), pp. 2366–2374 (2014)
4. Elsken, T., Metzen, J.H., Hutter, F.: Efficient multi-objective neural architecture search via Lamarckian evolution. In: Proceedings of the 7th International Conference on Learning Representations (ICLR) (2019)
5. Elsken, T., Metzen, J.H., Hutter, F.: Neural architecture search: a survey. Journal of Machine Learning Research 20(55), 1–21 (2019)
6. Goodfellow, I.J., Warde-Farley, D., Mirza, M., Courville, A., Bengio, Y.: Maxout networks. In: Proceedings of the 30th International Conference on Machine Learning (ICML), pp. 1319–1327 (2013)

7. He, K., Zhang, X., Ren, S., Sun, J.: Delving deep into rectifiers: surpassing human-level performance on ImageNet classification. In: Proceedings of the International Conference on Computer Vision (ICCV), pp. 1026–1034 (2015)

8. He, K., Zhang, X., Ren, S., Sun, J.: Deep residual learning for image recognition. In: Proceedings of the IEEE Conference on Computer Vision and Pattern Recognition (CVPR), pp. 770–778 (2016)

9. Ilg, E., Mayer, N., Saikia, T., Keuper, M., Dosovitskiy, A., Brox, T.: Flownet 2.0: Evolution of optical flow estimation with deep networks. In: Proceedings of the IEEE Conference on Computer Vision and Pattern Recognition (CVPR) (2017)

10. Ioffe, S., Szegedy, C.: Batch normalization: accelerating deep network training by reducing internal covariate shift. In: Proceedings of the 32nd International Conference on Machine Learning (ICML), pp. 448–456 (2015)

11. Kingma, D.P., Ba, J.: Adam: a method for stochastic optimization. In: Proceedings of the 3rd International Conference on Learning Representations (ICLR) (2015)

12. Krause, J., Stark, M., Deng, J., Fei-Fei, L.: 3D object representations for fine-grained categorization. In: Proceedings of the International Conference on Computer Vision Workshops (ICCVW), pp. 554–561 (2013)

13. Kupyn, O., Budzan, V., Mykhailych, M., Mishkin, D., Matas, J.: DeblurGAN: blind motion deblurring using conditional adversarial networks. In: Proceedings of the IEEE Conference on Computer Vision and Pattern Recognition (CVPR), pp. 8183–8192 (2018)

14. Larsson, G., Maire, M., Shakhnarovich, G.: FractalNet: ultra-deep neural networks without residuals. In: Proceedings of the 5th International Conference on Learning Representations (ICLR) (2017)

15. Lin, M., Chen, Q., Yan, S.: Network in network. In: Proceedings of the 2nd International Conference on Learning Representations (ICLR) (2014)

16. Liu, Z., Luo, P., Wang, X., Tang, X.: Deep learning face attributes in the wild. In: Proceedings of the International Conference on Computer Vision (ICCV), pp. 3730–3738 (2015)

17. Liu, H., Simonyan, K., Yang, Y.: Darts: differentiable architecture search. In: Proceedings of the International Conference on Learning Representations (ICLR) (2019)

18. Mao, X., Shen, C., Yang, Y.: Image restoration using very deep convolutional encoder-decoder networks with symmetric skip connections. In: Advances in Neural Information Processing Systems (NIPS), pp. 2802–2810 (2016)

19. Martin, D., Fowlkes, C., Tal, D., Malik, J.: A database of human segmented natural images and its application to evaluating segmentation algorithms and measuring ecological statistics. In: Proceedings of the International Conference on Computer Vision (ICCV), pp. 416–423 (2001)

20. Miikkulainen, R., Liang, J.Z., Meyerson, E., Rawal, A., Fink, D., Francon, O., Raju, B., Shahrzad, H., Navruzyan, A., Duffy, N., Hodjat, B.: Evolving deep neural networks. Preprint. arXiv:1703.00548 (2017)

21. Miller, J.F., Smith, S.L.: Redundancy and computational efficiency in Cartesian genetic programming. IEEE Trans. Evol. Comput. **10**(2), 167–174 (2006)

22. Miller, J.F., Thomson, P.: Cartesian genetic programming. In: Proceedings of the European Conference on Genetic Programming (EuroGP), pp. 121–132 (2000)

23. Nair, V., Hinton, G.E.: Rectified linear units improve restricted Boltzmann machines. In: Proceedings of the 27th International Conference on Machine Learning (ICML), pp. 807–814 (2010)

24. Netzer, Y., Wang, T., Coates, A., Bissacco, A., Wu, B., Ng, A.Y.: Reading digits in natural images with unsupervised feature learning. In: Advances in Neural Information Processing Systems (NIPS) Workshop on Deep Learning and Unsupervised Feature Learning (2011)

25. Paszke, A., Chanan, G., Lin, Z., Gross, S., Yang, E., Antiga, L., Devito, Z.: Automatic differentiation in PyTorch. In: Autodiff Workshop in Thirty-first Conference on Neural Information Processing Systems (NIPS) (2017)

26. Pathak, D., Krahenbuhl, P., Donahue, J., Darrell, T., Efros, A.A.: Context encoders: feature learning by inpainting. In: Proceedings of the IEEE Conference on Computer Vision and Pattern Recognition (CVPR), pp. 2536–2544 (2016)

27. Pham, H., Guan, M.Y., Zoph, B., Le, Q.V., Dean, J.: Efficient neural architecture search via parameter sharing. In: Proceedings of the 35th International Conference on Machine Learning (ICML), vol. 80, pp. 4095–4104 (2018)
28. Real, E., Moore, S., Selle, A., Saxena, S., Suematsu, Y.L., Le, Q.V., Kurakin, A.: Large-scale evolution of image classifiers. In: Proceedings of the 34th International Conference on Machine Learning (ICML), pp. 2902–2911 (2017)
29. Saito, S., Shirakawa, S.: Controlling model complexity in probabilistic model-based dynamic optimization of neural network structures. In: Proceedings of the 28th International Conference on Artificial Neural Networks (ICANN), Part II (2019)
30. Schaffer, J.D., Whitley, D., Eshelman, L.J.: Combinations of genetic algorithms and neural networks: a survey of the state of the art. In: Proceedings of International Workshop on Combinations of Genetic Algorithms and Neural Networks (COGANN '92), pp. 1–37 (1992)
31. Shirakawa, S., Iwata, Y., Akimoto, Y.: Dynamic optimization of neural network structures using probabilistic modeling. In: Proceedings of the 32nd AAAI Conference on Artificial Intelligence (AAAI-18), pp. 4074–4082 (2018)
32. Simonyan, K., Zisserman, A.: Very deep convolutional networks for large-scale image recognition. In: Proceedings of the 3rd International Conference on Learning Representations (ICLR) (2015)
33. Stanley, K.O., Miikkulainen, R.: Evolving neural networks through augmenting topologies. Evol. Comput. **10**(2), 99–127 (2002)
34. Suganuma, M., Shirakawa, S., Nagao, T.: A genetic programming approach to designing convolutional neural network architectures. In: Proceedings of the Genetic and Evolutionary Computation Conference (GECCO), pp. 497–504 (2017)
35. Suganuma, M., Ozay, M., Okatani, T.: Exploiting the potential of standard convolutional autoencoders for image restoration by evolutionary search. In: Proceedings of the 35th International Conference on Machine Learning (ICML), vol. 80, pp. 4771–4780 (2018)
36. Suganuma, M., Kobayashi, M., Shirakawa, S., Nagao, T.: Evolution of deep convolutional neural networks using Cartesian genetic programming. Evol. Comput. (2019). https://doi.org/10.1162/evco_a_00253. Early access
37. Szegedy, C., Liu, W., Jia, Y., Sermanet, P., Reed, S., Anguelov, D., Erhan, D., Vanhoucke, V., Rabinovich, A.: Going deeper with convolutions. In: Proceedings of the IEEE Conference on Computer Vision and Pattern Recognition (CVPR), pp. 1–9 (2015)
38. Tai, Y., Yang, J., Liu, X., Xu, C.: MemNet: A persistent memory network for image restoration. In: Proceedings of the IEEE Conference on Computer Vision and Pattern Recognition (CVPR), pp. 4539–4547 (2017)
39. Tan, M., Chen, B., Pang, R., Vasudevan, V., Sandler, M., Howard, A., Le, Q.V.: MnasNet: platform-aware neural architecture search for mobile. In: Proceedings of the IEEE Conference on Computer Vision and Pattern Recognition (CVPR) (2019)
40. Tokui, S., Oono, K., Hido, S., Clayton, J.: Chainer: a next-generation open source framework for deep learning. In: Proceedings of Workshop on Machine Learning Systems (LearningSys) in The Twenty-ninth Annual Conference on Neural Information Processing Systems (NIPS) (2015)
41. Wang, Z., Bovik, A., Sheikh, H., Simoncelli, E.: Image quality assessment: from error visibility to structural similarity. IEEE Trans. Image Process. **13**(4), 600–612 (2004)
42. Xie, L., Yuille, A.: Genetic CNN. In: Proceedings of the International Conference on Computer Vision (ICCV), pp. 1388–1397 (2017)
43. Xie, S., Zheng, H., Liu, C., Lin, L.: SNAS: stochastic neural architecture search. In: Proceedings of the International Conference on Learning Representations (ICLR) (2019)
44. Xu, D., Ricci, E., Ouyang, W., Wang, X., Sebe, N.: Multi-scale continuous CRFs as sequential deep networks for monocular depth estimation. In: Proceedings of the IEEE Conference on Computer Vision and Pattern Recognition (CVPR), pp. 5354–5362 (2017)
45. Yao, X.: Evolving artificial neural networks. Proc. IEEE **87**(9), 1423–1447 (1999)

46. Yeh, R.A., Chen, C., Lim, T.Y., Schwing, A.G., Hasegawa-Johnson, M., Do, M.N.: Semantic image inpainting with deep generative models. In: Proceedings of the IEEE Conference on Computer Vision and Pattern Recognition (CVPR), pp. 6882–6890 (2017)
47. Zagoruyko, S., Komodakis, N.: Wide residual networks. In: Proceedings of the British Machine Vision Conference (BMVC), pp. 87.1–87.12 (2016)
48. Zhang, R., Isola, P., Efros, A.A.: Colorful image colorization. In: European Conference on Computer Vision (ECCV) 2016. Lecture Notes in Computer Science, vol. 9907, pp. 649–666. Springer, Berlin (2016)
49. Zoph, B., Le, Q.V.: Neural architecture search with reinforcement learning. In: Proceedings of the 5th International Conference on Learning Representations (ICLR) (2017)

# Chapter 8
# Fast Evolution of CNN Architecture for Image Classification

Ali Bakhshi, Stephan Chalup, and Nasimul Noman

**Abstract** The performance improvement of Convolutional Neural Network (CNN) in image classification and other applications has become a yearly event. Generally, two factors are contributing to achieving this envious success: stacking of more layers resulting in gigantic networks and use of more sophisticated network architectures, e.g. modules, skip connections, etc. Since these state-of-the-art CNN models are manually designed, finding the most optimized model is not easy. In recent years, evolutionary and other nature-inspired algorithms have become human competitors in designing CNN and other deep networks automatically. However, one challenge for these methods is their very high computational cost. In this chapter, we investigate if we can find an optimized CNN model in the classic CNN architecture and if we can do that automatically at a lower cost. Towards this aim, we present a genetic algorithm for optimizing the number of blocks and layers and some other network hyperparameters in classic CNN architecture. Experimenting with CIFAR10, CIFAR100, and SVHN datasets, it was found that the proposed GA evolved CNN models which are competitive with the other best models available.

## 8.1 Introduction

In recent years many pieces of research have been directed towards designing deep neural networks (DNNs). The performance of DNNs is very depended on its architecture and its hyperparameters' setting. The state-of-the-art DNN models are designed by qualified experts in various areas of machine learning. Moreover, all of these networks are designed for specific problems or data. For example, convolutional neural networks (CNNs) are most widely used in various image related applications in computer vision. Although the state-of-the-art DNNs proposed in the literature can be used for solving similar problems using some

A. Bakhshi · S. Chalup · N. Noman (✉)
The University of Newcastle, Newcastle, NSW, Australia
e-mail: ali.bakhshi@uon.edu.au; stephan.chalup@newcastle.edu.au;
nasimul.noman@newcastle.edu.au

© Springer Nature Singapore Pte Ltd. 2020
H. Iba, N. Noman (eds.), *Deep Neural Evolution*, Natural Computing Series,
https://doi.org/10.1007/978-981-15-3685-4_8

techniques like transfer learning, the same network model is not suitable for a diverse class of problems. For best performance, we need to design a DNN tailored to the problem under consideration. Consequently, many researchers are working towards automatic methods that can identify suitable DNN architecture as well as hyperparameters for a certain problem.

Evolutionary algorithms (EAs) are a class of the generic population-based meta-heuristic optimization algorithms that can be used to identify the suitable network architecture and hyperparameters [1]. There are remarkable efforts in the literature that used the variants of evolutionary algorithms such as the genetic algorithm (GA) and particle swarm optimization (PSO) to solve a variety of optimization problems [2]. Considering the promising success of the artificial neural networks (ANNs) and evolutionary algorithm in solving different machine learning problems, finding efficient ways to combine these two methods has been an active research area for the past two decades. There exists a good survey that classifies different approaches for combining ANNs and GAs into two categories: the supportive combination and collaborative combination [3]. In supportive combination, either GA or ANN is the main problem solver, and the other assists it in accomplishing that, whereas, in collaborative combination, both GA and ANN work in synergy to solve the problem. An example of a supporting combination is using GA for selecting features for a neural network, and an example of a collaborative combination is designing the ANN topology using GA.

With the emergence of DNNs as a powerful machine learning method for solving different problems, there has been a growing interest in designing and training these networks using evolutionary algorithms. Considering the success of gradient base algorithms in training the DNNs, and due to other considerations such as very large search space, there has been limited interest in training the DNNs using evolutionary algorithms. Although there exist examples of outstanding efforts in training deep neural networks for reinforcement learning using GA [4], the majority of researches concentrated on evolving DNN architectures and finding the best combination of hyperparameters for a range of classification and regression tasks [1].

Convolutional Neural Network (CNN) is one of the most successful deep architectures as manifested by its remarkable achievements in many real-world applications. The state-of-the-art CNN architectures such as VGGNet [5], ResNet [6], GoogLeNet [7], designed by experienced researchers, exhibited performance competitive to humans. However, crafting such powerful and well-designed networks requires extensive domain knowledge and expertise in neural network design. These requirements often make it a difficult task for inexperienced researchers and application engineers to design a suitable architecture according to the problem and available data. Hence, in recent years, we have seen several attempts to automatically designing CNN architectures as well as network hyperparameters using evolutionary algorithms.

In this work, we used a conventional GA to evolve optimized CNN architectures and to find the best combination of hyperparameters for the image classification task on multiple datasets. We considered the optimization of the classical CNN architecture (i.e., VGG-like networks) consisting of blocks of convolutional layers.

The proposed GA was used to optimize the number of convolutional blocks, as well as the number of layers in each block. Using a fixed-sized chromosome, we explored the search space of CNN architectures with the variable number of layers. The algorithm also searched for the optimal set of hyperparameters for the network from the selected ranges.

One particular challenge in the evolution of all kinds of DNNs is the high computational cost. The computational burden originates from the fitness evaluation of each individual in the evolutionary algorithm, which requires training of many deep neural networks. Recent research has shown that partial training is sufficient for estimating the quality of CNN architecture [8, 9]. In this work, we adopted this strategy and trained the CNN architectures for a few epochs in the evolution phase. Later, the best evolved CNN model was trained completely for evaluating its performance. The proposed GA was applied to three well-known datasets, and the evolved CNN models were compared with many existing models designed by human and automatically.

The rest of the chapter is organized as follows. A brief overview of CNN is presented in Sect. 8.2. Section 8.3 reviews the related work. The proposed GA is described in Sect. 8.4. Section 8.5 details the experimental setup and the experimental results are presented in Sect. 8.6. Section 8.7 contains a brief discussion on the results and Sect. 8.8 concludes the chapter.

## 8.2 A Brief Overview of CNNs

Convolutional neural networks (CNNs) that were inspired by the organization of the animal cortex [10] are mostly used for two-dimensional data like images. CNNs consist of three major types of network layers, namely: convolutional, pooling, and fully connected. The learning in a convolutional layer depends on three concepts: sparse interaction, equivariant representation, and parameter sharing [11]. Unlike the feed-forward neural network layers that utilize layer-wide matrix multiplication for relating the inputs with the outputs, convolutional layers implement sparse interactions by using filters smaller than the inputs. By sharing the same filter across the input surface, convolutional layers can achieve spatial equivariance as well as reduce the computational volume considerably. Using multiple learnable filters the convolutional layer can learn different features from the input. Pooling layers are usually placed after one or more convolutional layers to reduce the dimensionality of the data. Multiple blocks of convolutional and pooling layers are used to extract hierarchical features from the data. Also, depending on the nature of the problem, the final convolutional or pooling layer is followed by one or more fully connected or recurrent layers.

There are many hyperparameters involved in designing a CNN architecture, such as the number of layers in a convolutional block, the kernel size, stride size, and channel size (number of feature maps) of a convolution layer, the stride size of the pooling layer, type of pooling operation, number of fully connected layers, the number of nodes in a fully connected layer, etc. In human-designed CNNs, these hyperparameters are selected on a trial and error basis with the help of prior knowledge about the functionality of these layers. As discussed in Sect. 8.3, meta-heuristic algorithms can help to automatically select an optimal set of hyperparameters for a CNN in a given task.

## 8.3 Related Works

In recent years, we have seen a increasing interest in the evolutionary computation community in evolving DNN architectures and their hyperparameters. David and Greental [12] proposed a simple GA-assisted method to improve the performance of the deep autoencoder on the MNIST dataset. They stored the sets of weights of an autoencoder in the chromosomes of individuals in their GA population. Then, by calculating the root mean square error (RMSE) of each chromosome for the training sample, they set the fitness score of each individual as the inverse of RMSE. After sorting all chromosomes from the fittest to the least fit, they tuned the weights of the high ranking chromosomes using backpropagation and replaced the low-ranking members with the offspring of high ranking ones. However, they just used the fitness score as a criterion for removing the low ranking members, and selection is implemented uniformly and applied to the outstanding chromosomes with equal likelihood. The authors showed that compared to the traditional backpropagation, the GA-assisted method gives better reconstruction error and network sparsity.

Suganuma et al. [8] used Cartesian Genetic Programming (CGP) to construct the CNN structure and network connectivities. To reduce the search space, high-level functional modules, such as convolutional block and tensor concatenation, were used as the node functions of CGP. Following the training of the network using the training data, they utilized the validation accuracy as the fitness score. By evaluating the performance of the evolved CNN models on the CIFAR10 dataset, they achieved the error rates of 6.34 and 6.05% for the CGP-CNN (ConvSet) and the CGP-CNN (ResSet), respectively. Loshchilov and Hutter [13] introduced the Covariance Matrix Adaptation Evolution Strategy (CMA-ES) as an optimization method for selecting the hyperparameters of the deep neural networks (DNNs). In their experiments, the performance of CMA-ES and other state-of-the-art algorithms were evaluated for tuning 19 hyperparameters of a DNN on the MNIST dataset. They pointed out that the CMA-ES algorithm shows competitive performance, especially in parallel evaluations.

In another study, Sun et al. [14] used a GA to design the CNN architectures automatically for image classification. They used skip layers, composed of two convolutional layers and one skip connection borrowed from ResNet [6], to increase the depth of the network. Moreover, they used the same filter size and stride for all convolutional layers, and the number of feature maps was selected by their method. The fully connected layers were omitted in their model, but the pooling layers were used. They evaluated the performance of their model on several popular benchmarks, such as CIFAR10 and CIFAR100. The other work conducted by Sun et al. [15] utilized the ResNet and DenseNet blocks [16] for automatically evolving the CNN architectures. In their approach, a combination of three different units, ResNet block units, DenseNet block units, and pooling layer units, have been used to generate the CNN architecture. In their encoding strategy, to increase the depth of the network as well as the speed of heuristic search by changing the depth of the network, each ResNet or DenseNet unit composed of multiple ResNet and DenseNet blocks. They showed the superiority of their model by comparing their model generated results with 18 state-of-the-art algorithms on CIFAR10 and CIFAR100 datasets. Ali et al. [17] proposed a GA model to evolve a CNN architecture and other network hyperparameters. They used a generic GA to find the best combination of network hyperparameters, such as the number of layers, learning rate, and weight decay factor. Using some design rules and constraints for genotype to phenotype mapping, they evolved a CNN architecture on CIFAR10 dataset. They compared the performance of the best CNN architecture evolved by their method with 13 other models in terms of classification accuracy and GPU days. Sun et al. [9] introduced a GA model for evolving CNN architecture as well as connection weight initialization values in image classification tasks. They used an efficient variable-length gene encoding method, representing various building blocks, to find the optimal depth of the CNN. Furthermore, to avoid trapping in local minima, a major problem in gradient-based training, they introduced a new representation scheme for initializing the connection weights of DNN. They showed the effectiveness of their proposed method by comparing their results with 22 existing algorithms involving state-of-the-art models on nine popular image classification tasks.

## 8.4    The Proposed Genetic Algorithm for Designing CNNs

In this work, using a GA, we evolved CNN architectures with the best combination of hyperparameters for the image classification task. Our GA operates in the search space of VGG-like architectures, i.e., we assumed that the CNN architecture consists of a sequence of convolutional blocks, each followed by a pooling layer, and a fully connected layer at the end. The GA is used to optimize the number of convolutional blocks, the number of convolutional layers in each block as well as some other hyperparameters of the CNN architecture. The assumption about the organization of the CNN architecture confines the GA to discover only the classical CNN models, i.e., it does not allow to design CNN architectures with more sophisticated modules like residual blocks [6] or inception modules [7]. However, by exploring the limited search space, the proposed GA was able to design CNNs, which exhibited competitive performance with the other state-of-the-art CNN models.

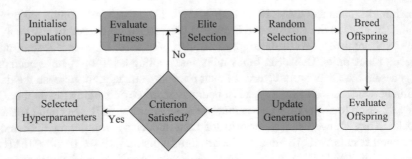

**Fig. 8.1** The flowchart of the genetic algorithm for evolving CNN model

The proposed algorithm for optimizing CNN architecture works in the generic GA framework with standard GA operations (Fig. 8.1). The algorithm begins with an initial population created by the random selection of genes for each individual. The chromosome of each individual uniquely determines the architecture of a CNN as well as some of its hyperparameters. Then, each CNN model is trained and validated using the training dataset, and the average classification accuracy of the network in the validation phase is used as the individual's fitness score. Next, the population of individuals is sorted in descending order of their fitness scores. Then, by applying a chain of genetic operations such as elite selection, random selection, and breeding of the new members, the next generation of the population is created. This process is repeated until the satisfaction of the termination criterion. The pseudocode of the genetic algorithm for CNN architecture optimization is shown in Algorithm 1, and the details of each module are presented in the following subsections.

---

**Algorithm 1:** Proposed GA framework for evolving CNN models

---

**Input**: Population size ($N_P$), Maximum number of generation ($G_{max}$), the range of values
   for the selected hyperparameters ($L_H$), the RGB images of training dataset
**Output**: The best CNN architecture with its hyperparameters
1  Initialize the population using a random combination of hyperparameters [Algorithm 2]
2  Train the CNN model designated by each individual in the population, and calculate the
   corresponding fitness score [Algorithm 3]
3  Store the population of individuals and their fitness scores in a list called $P$
4  $N_G \leftarrow 0$
5  **while** $N_G < G_{max}$ **do**
6  |   Create a new generation $P_{new}$ consisting of elite individuals, random individuals and
   |   offspring created from $P$ [Algorithm 4]
7  |   Evaluate individuals in $P_{new}$
8  |   Set $P \leftarrow P_{new}$ and $N_G \leftarrow N_G + 1$
9  |   **end**
10 Return the best CNN architecture in $P$ along with its hyperparameters

---

| LR | WD | M | DR | B1 | B2 | B3 | B4 | B5 | F1 | F2 | F3 | F4 | F5 |
|----|----|----|----|----|----|----|----|----|----|----|----|----|----|
| 0.01 | 0.001 | 0.8 | 0.25 | 2 | 0 | 3 | 2 | 4 | 256 | 128 | 32 | 512 | 256 |

**Fig. 8.2** The chromosome of an individual showing different hyperparameters of the CNN model

## 8.4.1 Population Initialization

As mentioned earlier, the proposed GA is used to seek the optimal CNN model in the search space of classical CNN architectures. We limited the maximum number of convolutional layers in a CNN to 20 divided into a maximum of 5 blocks $B1, B2, B3, B4, B5$. Each of the blocks can have any number of layers between 0 and 4. The number of feature maps for each block is also optimized by GA, which can be chosen from $\{32, 64, 128, 256, 512\}$. The other hyperparameters, optimized by our GA, are learning rate (LR), weight decay (WD), momentum (M), and dropout rate (DR). The structure of the chromosome is shown in Fig. 8.2.

From the chromosome structure, it becomes clear that the proposed GA works with a fixed size chromosome. However, by allowing a block size to be zero, the GA can actually search for variable-length CNN models having variable number of blocks with any number of layers between 0 and 20. The example in Fig. 8.2 shows a CNN architecture with 11 layers where the first block consists of 2 convolutional layers with a feature map size of 256. The second block does not exist, the third, fourth, and fifth block have 3, 2, and 4 layers and their corresponding feature map sizes are 32, 512, and 256, respectively.

In genotype to phenotype mapping, a couple of additional layers are added in each CNN model. Each convolutional block is followed by a max-pooling layer with kernel size 2 and stride size 2, an average-pooling layer with a kernel size 2 and a stride size 1 is added after the final max-pooling layer, and a linear fully connected layer is placed at the end of the network. Moreover, each convolutional layer is followed by a batch normalization layer [18] and a Rectified Linear Unit (ReLU) layer [19], and a dropout layer is added at the rear of each convolutional block.

Algorithm 2 summarizes the process of population initialization for the GA. Each gene in a chromosome can take a range of values, and the proposed GA searches for the optimal combination of these values through the evolution process. The range of possible values for each gene (shown in Table 8.1) is selected according to the previous experiences in different classification problems using CNNs. Following the random selection of the hyperparameters, the CNN architecture is created without any constraints on the number or the order of convolutional layers or feature maps. Often, human-designed CNN architectures are created following some rules, e.g., an increasing number of feature maps are used in successive convolutional blocks. However, in the proposed GA model, no such restriction was imposed, and the architecture of a CNN is completely guided by the chromosome.

---

**Algorithm 2:** For generating initial population

---

**Input**: The population size $N_P$
**Output**: The initialized population $P_{init}$
**Data**: The ranges of values for different hyperparameter are stored in a list called $L_H$

1  $P_{init} \leftarrow \emptyset$
2  **while** $|P_{init}| < N_P$ **do**
3       Select the learning rate ($lr$) randomly from the $L_H$ [LR]
4       Select the weight decay factor ($wd$) randomly from the $L_H$ [WD]
5       Select the momentum ($m$) randomly from the $L_H$ [M]
6       Select the dropout ($d$) randomly from the $L_H$ [DR]
7       Select the number of convolutional layer in each block $\{B1, B2, B3, B4, B5\}$ randomly from the $L_H$ [NL]
8       Select the number of feature maps corresponding to each block $\{F1, F2, F3, F4, F5\}$ randomly from the $L_H$ [NF]
9       Create an individual ($Ind$) with the selected hyperparameters
10      $P_{init} \leftarrow P_{init} \cup Ind$

11 Return $P_{init}$

---

**Table 8.1** The range of possible values for different hyperparameters to be searched by the GA

| Hyperparameter | Values |
| --- | --- |
| Learning rate (LR) | 0.1, 0.01, 0.001, 0.0001 |
| Weight decay (WD) | 0.1, 0.01, 0.001, 0.0001, 0.00001 |
| Momentum (M) | 0.5, 0.55, 0.6, 0.65, 0.7, 0.75, 0.8, 0.85, 0.9 |
| Dropout rate (DR) | 0.25, 0.5, 0.75 |
| Block size (NL) in B1-B5 | 0, 1, 2, 3, 4 |
| Feature map size (NF) in F1-F5 | 32, 64, 128, 256, 512 |

Finally, it should be noted that the chromosome structure and the algorithm are flexible enough to make it more general by considering the optimization of additional hyperparameters, e.g., activation function, individual kernel size for each convolutional block, etc. Nevertheless, increasing the search space size will necessitate more extensive searching. We did some preliminary study with some other variants of the chromosome, but later fixed those hyperparameter values (e.g., fixed the kernel size of convolutional layers to 3) to reduce the computational burden.

## 8.4.2 Fitness Evaluation

In order to assess the quality of a CNN model constructed from the chromosome of an individual, we need to train it and evaluate its classification performance. Training and evaluating a deep neural network is the computationally most expensive part of any deep neuroevolution algorithm. Recent studies have suggested that it is possible to roughly assess the architectural quality of a CNN model based on its evaluation after partial training [8, 9]. Henceforth, during the evolution process, we evaluated the performance of the CNN networks after partially training them for only $N_{epoch} = 10$ epoch, which significantly accelerated the genetic algorithm.

---

**Algorithm 3:** For fitness evaluation of an individual

**Input**: The individual ($Ind$), training data ($D_{train}$), validation data ($D_{valid}$), the number of
    epoch in training phase ($N_{epoch}$)
**Output**: The fitness score of the individual
1  Create the CNN model ($net$) from the hyperparameters of $Ind$ augmented with pooling,
    fully connected, batch-normalization, ReLU and dropout layers (details in Sect. 8.4.1)
2  $Acc \leftarrow \emptyset$
3  $step \leftarrow 0$
4  $Acc_{avg} \leftarrow 0$
5  **while** $step < N_{epoch}$ **do**
6      Train the model $net$ using the $D_{train}$
7      Calculate the classification accuracy ($acc$) using the $D_{valid}$
8      $Acc \leftarrow Acc \cup acc$
9      $step \leftarrow step + 1$
10   **end**
11 $Acc_{avg} \leftarrow$ Average of accuracies in $Acc$
12 Return $Acc_{avg}$

---

We used the average validation accuracy of the constructed CNN model as the fitness score of the corresponding individual. 90% of the training data is used during the training phase, and the rest 10% is utilized for validation. The constructed CNN model is trained by the stochastic gradient descent (SGD) algorithm [20], for a constant number of epochs ($N_{epoch} = 10$), and the average classification accuracy of the validation phase is used as the fitness score of the individual. In all experiments, during the training phase, the cross-entropy loss is used as the loss function, and the learning rate is reduced by a factor of 10 in every 10 epochs during complete training for the model after the evolutionary phase. The details of the fitness evaluation of an individual are summarized in Algorithm 3.

## 8.4.3  Creating New Generation

In the proposed GA, the next generation of individuals is created from the current generation using elite selection, random selection, and offspring generation. First, the individuals in the current generation are sorted based on their fitness scores. Top $e\%$ of the individuals, known as elites, are selected from the current population and added to the next generation. To maintain the population diversity and to prevent premature convergence, some random individuals are also added [21, 22]. Specifically, from the rest of the current population, individuals are randomly selected with a probability of $p_r$ and added to the next generation. Finally, the selected elite and random individuals form the parent pool to breed offspring.

---

**Algorithm 4:** For creating a new generation of individuals

---

   **Input**: The current population of individuals with their fitness scores ($P$), the percentage of
           population preserved as the elite ($e$), the probability of preserving an individual from
           the non-elite section of the current population ($p_r$), the probability of mutation ($p_m$),
           and the population size ($N_p$)

   **Output**: The new population ($P_{new}$)

**1**  $P_{new} \leftarrow \emptyset$

**2**  Sort the individuals in $P$ in descending order of their fitness scores

**3**  Add top $e\%$ individuals from $P$ to the new population $P_{new}$

**4**  Select the individuals from the bottom $(1 - e)\%$ of $P$ with probability $p_r$ and add them to
    $P_{new}$

**5**  $P_{parents} \leftarrow P_{new}$

**6**  **while** $|P_{new}| < N_p$ **do**

**7**        $Par_1 \leftarrow$ An individual randomly selected from $P_{parents}$

**8**        $Par_2 \leftarrow$ An individual randomly selected from $P_{parents}$

**9**        **if** $Par_1 \neq Par_2$ **then**

**10**            Create two children from the selected parents using uniform crossover operation
              and save them in *Children*

**11**            **for** *each Child in Children* **do**

**12**                r $\leftarrow$ Randomly generate a number from the range (0,1)

**13**                **if** $p_m > r$ **then**

**14**                    Randomly replace a gene in *Child* with the randomly selected value

**15**                **end**

**16**            **end**

**17**            $P_{new} \leftarrow P_{new} \cup Children$

**18**        **end**

**19**     **end**

**20**  Return $P_{new}$

---

The process of generating the offspring starts with the random selection of two dissimilar individuals from the parent pool. The selected parents participate in uniform crossover operation to create two offspring. Each child may inherit various combinations of genes from the parents because its genes are randomly selected from the parents. Then, the child undergoes the mutation operation with a predefined mutation probability of $p_m$. If the mutation condition is met, one randomly selected gene in the offspring chromosome is randomly modified from a set of predefined values (shown in Table 8.1). Newly created offspring are added to the next generation. The process of creating new children is repeated until the number of individuals in the new generation reaches the population size. Finally, the new generation, consisting of parent pool and children pool, replaces the current generation. The process of generation alternation is repeated $G_{max}$ times so that the GA can search for the best network architecture and hyperparameters. The process of creating a new generation from the current generation is shown in Algorithm 4.

## 8.5   Experimental Setup

### 8.5.1   Datasets

In this work, we used three popular datasets CIFAR10, CIFAR100, and SVHN as the benchmark for image classification problems. These datasets have been used in many pieces of research for evaluation of the state-of-the-art deep neural network models which makes our evolved models comparable with those models. The CIFAR10 dataset includes 60,000 color RGB images belonging to 10 classes and is mostly used for image classification tasks. These images are of dimension $32 \times 32$ and are divided into training and testing parts. The training set contains 50,000 images, and the rest of 10,000 images are used as the testing set. There is an equal number of training and testing samples for each class.

The CIFAR100 dataset is similar to the CIFAR10, except with 100 classes that are categorized into 20 superclasses each of which contains five classes. There exist only 500 training images and 100 testing images, per class, making classification more challenging in this dataset.

The SVHN (Street View House Numbers) dataset that can be considered in essence similar to the MNIST dataset but with more labeled data contains 73257 and 26032 digits for training and testing, respectively. Compared to the MNIST dataset, the SVHN dataset originates from a more difficult real-world problem. This dataset contains the original, colored, and variable resolution images from the house numbers in Google Street View images. However, all digits of the house numbers have been resized to a fixed resolution $32 \times 32$ and originate from 10 different classes [23].

### 8.5.2   Experimental Environment

In this work, we used the Pytorch framework (Version 1.2.0) of the python programming language (Version 3.7) in all experiments. Besides, we ran the codes using both high-performance computing (HPC) services and the DGX station machine at the University of Newcastle. All codes ran with two GPUs at all stages including the evolution of various CNN architectures, complete training of the selected models with a larger epoch, and testing the best-performing models.

### 8.5.3   Parameter Selection

As stated before, our proposed framework is very flexible for increasing the search space. In other words, many hyperparameters can be evolved through the GA framework, but because of the limitation imposed by computational resources, we just evolved some of the selected hyperparameters. Also, according to our earlier experimental results, a specific value is always selected by the GA for some of these hyperparameters, namely activation function, optimizer. Hence, in all experiments, the ReLU activation function and SGD optimizer have been used. Besides, the kernel size of the convolutional layers was set to 3 and the stride size of the convolutional layer and the max-pooling layer was set to 2.

The GA parameters were set as follows: maximum number of generation $G_{max} = 40$, population size $N_P = 30$, the percentage of population retained as the elite $e = 40\%$, the probability of retaining an individual from the non-elite part of the population $p_r = 0.1$, and the probability of the mutation $p_m = 0.2$. These parameters were set based on our experience with evolutionary algorithms and using some primary studies. Moreover, to decrease the computational burden and speed up the evolution process, during evolution, we trained the networks with a smaller number of the epochs $N_{epoch} = 10$. After the evolutionary phase, the best CNN model is trained completely before it is evaluated with the test dataset. Precisely, the best model evolved by the GA is trained for a higher number of epoch $N_{epoch} = 350$ using the full training set.

## 8.6 Experimental Results

In our experiments, we applied the proposed GA for evolving the CNN architectures for each dataset. Considering the stochastic nature of the algorithm, we repeated each experiment 5 times and the best CNN model evolved in each experiment is later trained completely and tested on the corresponding dataset. Table 8.2 shows the performance of the evolved models in different datasets in terms of their average accuracy, standard deviation, best accuracy, and the worst accuracy. The CNN models designed by the GA was able to achieve a good performance in all three datasets. Considering the average, best, and worst accuracies as well as the standard deviations shown in Table 8.2, it is evident that the evolutionary algorithm was pretty reliable in finding CNN models of similar quality over multiple experimental runs.

The average convergence graph of GA from a single representative run is shown in Figs. 8.3 and 8.4 for all three datasets. Note that the fitness score in these graphs is the average validation accuracy of the CNN models in each generation which were

**Table 8.2** The average accuracy, standard deviation, best and worst accuracy of the best CNN models evolved by multiple GA runs in each dataset

| Dataset | Average accuracy | STD | Best accuracy | Worst accuracy |
|---------|------------------|-------|---------------|----------------|
| CIFAR10 | 94.75 | 0.650 | 95.82 | 94.01 |
| CIFAR100 | 75.90 | 0.521 | 76.79 | 75.34 |
| SVHN | 95.11 | 0.48 | 95.57 | 94.52 |

Each network was trained for 350 epochs

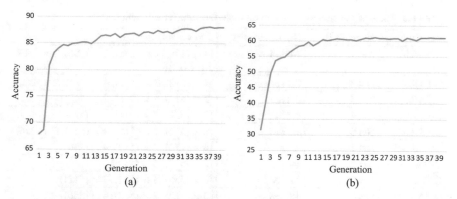

**Fig. 8.3** The average convergence graph of the GA population in one representative run on (**a**) CIFAR10 dataset and (**b**) CIFAR100 dataset

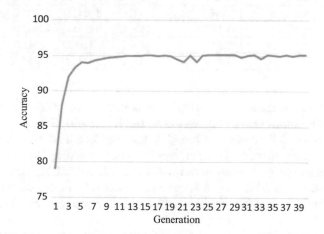

**Fig. 8.4** The average convergence graph of the GA population in one representative run on SVHN dataset

trained only for 10 epochs. As shown in these graphs, the proposed algorithm was successful to improve the overall fitness of the population on all datasets. The fitness of the population improved quickly in the first 10 generations and then slowed down gradually. However, this behavior is expected because of the high selection pressure from the elitism strategy and the participation of the elite individuals in offspring generation.

Table 8.3 shows the structures of the best networks evolved in different GA runs for each dataset. Each row of Table 8.3 shows the CNN architecture in terms of the number of blocks, number of convolutional layers in each block, and the feature map size of each block. The convolutional blocks are separated by comma and in each block we show the number of layers and the feature map size (in parenthesis) for each layer in that block. For example in the first row of Table 8.3, $4 \times (128)$ represents that the first convolutional block of the CNN consists

**Table 8.3** The best structures evolved in different GA runs for different datasets

| Dataset | Network ID | No. parameters | Evolved architecture |
|---|---|---|---|
| CIFAR10 | Net 1 | 14.3 M | [4 × (128), 4 × (256), 2 × (256), 3 × (512), 2 × (512)] |
| CIFAR10 | Net 2 | 25.4 M | [4 × (512), 3 × (512), 4 × (512), 2 × (256)] |
| CIFAR10 | Net 3 | 5.8 M | [2 × (512), 4 × (256), 2 × (128)] |
| CIFAR10 | Net 4 | 6.7 M | [2 × (256), 3 × (512), 2 × (32)] |
| CIFAR10 | Net 5 | 20.7 M | [3 × (512), 2 × (512), 4 × (512), 2 × (256)] |
| CIFAR100 | Net 1 | 1.7 M | [4 × (256), 2 × (512), 4 × (256)] |
| CIFAR100 | Net 2 | 14.2 M | [3 × (512), 4 × (512)] |
| CIFAR100 | Net 3 | 11.2 M | [2 × (256), 4 × (512), 4 × (512)] |
| CIFAR100 | Net 4 | 18.3 M | [3 × (512), 3 × (512), 4 × (256)] |
| CIFAR100 | Net 5 | 14.8 M | [2 × (256), 2 × (512), 3 × (512)] |
| SVHN | Net 1 | 19 M | [3 × (512), 4 × (512), 4 × (256)] |
| SVHN | Net 2 | 7.8 M | [3 × (32), 3 × (512), 4 × (512), 2 × (512)] |
| SVHN | Net 3 | 17.1 M | [3 × (32), 3 × (512), 4 × (512), 4 × (512)] |
| SVHN | Net 4 | 23.8 M | [3 × (256), 3 × (256), 2 × (256), 4 × (512)] |
| SVHN | Net 5 | 11.9 M | [3 × (32), 3 × (512), 4 × (512), 2 × (512)] |

of 4 convolutional layer each having a feature map size of 128. The complete CNN model is constructed by adding ReLU, batch normalization, dropout, average pooling layer and fully connected layers as described in Sect. 8.4.1. This table also shows the number of trainable parameters for different evolved models. The other hyperparameters of each evolved network for CIFAR10, CIFAR100, and SVHN dataset are shown in Tables 8.4, 8.5, and 8.6, respectively. Figure 8.5 visualizes the architecture of the best CNN models evolved by GA over five repeated runs in three datasets.

It can be seen from Table 8.3 that different evolutionary runs evolved quite different CNN architectures in terms of the number of blocks, number of layers, and number of trainable parameters even for the same dataset. However, with these different architecture the models achieved very similar accuracy in the respective datasets as shown in Tables 8.4, 8.5, and 8.6. One interesting observation in evolved architectures by the GA is the feature map sizes in different blocks. In human-designed architectures, usually the feature map size increases in later convolutional blocks as we have seen in case of different VGG models. In some of the evolved models we notice the same characteristics, e.g. Net 1 for CIFAR10, however, in general, this order was not maintained in the evolved models. Some architecture has it in decreasing order (Net 3 for CIFAR10) and some has it in no specific order (Net 4 for CIFAR10). From this observation we can infer that use of increased feature map size in later convolutional block is not absolutely necessary for good CNN architecture design.

In order to further assess the merit of the evolved models, we trained each evolved model with the other datasets and tested its accuracy. Specifically, Table 8.4 shows the best evolved models for the CIFAR10 datasets and then applied in CIFAR10,

**Table 8.4** Hyperparameters of the top five CNN models evolved by different GA runs for CIFAR10 dataset

| Hyperparameters | | | | | | | CIFAR10 | CIFAR100 | SVHN |
|---|---|---|---|---|---|---|---|---|---|
| Net name | LR | WD | M | DR | No. blocks | No. layers | accuracy | accuracy | accuracy |
| Net 1 | 0.01 | 0.01 | 0.65 | 0.5 | 5 | 15 | 95.82 | 79.48 | 96.93 |
| Net 2 | 0.01 | 0.001 | 0.8 | 0.5 | 4 | 13 | 95.09 | 76.95 | 96.60 |
| Net 3 | 0.1 | 0.001 | 0.7 | 0.25 | 3 | 8 | 94.64 | 74.41 | 95.64 |
| Net 4 | 0.01 | 0.01 | 0.7 | 0.5 | 3 | 7 | 94.21 | 71.53 | 92.80 |
| Net 5 | 0.01 | 0.0001 | 0.8 | 0.5 | 4 | 11 | 94.01 | 76.07 | 96.27 |

**Table 8.5** Hyperparameters of the top five CNN models evolved by different GA runs for CIFAR100 dataset

| Hyperparameters | | | | | | | CIFAR10 | CIFAR100 | SVHN |
|---|---|---|---|---|---|---|---|---|---|
| Net name | LR | WD | M | DR | No. blocks | No. layers | accuracy | accuracy | accuracy |
| Net 1 | 0.01 | 0.001 | 0.8 | 0.5 | 3 | 10 | 93.28 | 76.79 | 96.03 |
| Net 2 | 0.1 | 0.0001 | 0.75 | 0.5 | 2 | 7 | 94.14 | 76.04 | 95.87 |
| Net 3 | 0.01 | 0.0001 | 0.8 | 0.5 | 3 | 10 | 94.66 | 75.59 | 96.54 |
| Net 4 | 0.01 | 0.001 | 0.8 | 0.25 | 3 | 10 | 94.56 | 75.52 | 96.35 |
| Net 5 | 0.01 | 0.0001 | 0.9 | 0.5 | 3 | 7 | 94.34 | 75.34 | 96.03 |

**Table 8.6** Hyperparameters of the top five CNN models evolved by different GA runs for SVHN dataset

| Hyperparameters | | | | | | | CIFAR10 | CIFAR100 | SVHN |
|---|---|---|---|---|---|---|---|---|---|
| Net name | LR | WD | M | DR | No. blocks | No. layers | accuracy | accuracy | accuracy |
| Net 1 | 0.01 | 0.0001 | 0.9 | 0.25 | 3 | 11 | 95.10 | 73.95 | 95.57 |
| Net 2 | 0.01 | 0.0001 | 0.7 | 0.5 | 4 | 12 | 93.21 | 73.53 | 95.51 |
| Net 3 | 0.01 | 0.0001 | 0.7 | 0.5 | 4 | 14 | 92.58 | 72.89 | 95.43 |
| Net 4 | 0.01 | 0.001 | 0.7 | 0.5 | 4 | 12 | 94.19 | 73.18 | 94.53 |
| Net 5 | 0.01 | 0.0001 | 0.8 | 0.5 | 4 | 12 | 94.05 | 73.10 | 94.52 |

CIFAR100 and SVHN datasets. Similarly, Tables 8.5 and 8.6 show the performance of the evolved models for the CIFAR100 and SVHN datasets, respectively, in all three datasets.

It was expected that the model evolved for a particular dataset will exhibit the best performance in that dataset. However, we notice that in general the best performance was exhibited by the models evolved for CIFAR10 dataset (Table 8.4). We hypothesize that for more complex datasets like CIFAR100 and SVHN, the training of 10 epochs is not sufficient to assess the quality of the network architecture, therefore, the evolved model did not perform the best in the respective datasets. One particular point to note in Tables 8.4, 8.5, and 8.6 is that evolutionary runs selected network hyperparameters pretty consistently. For example almost in every model, dropout rate was chosen as 0.5 and learning rate was chosen as 0.01.

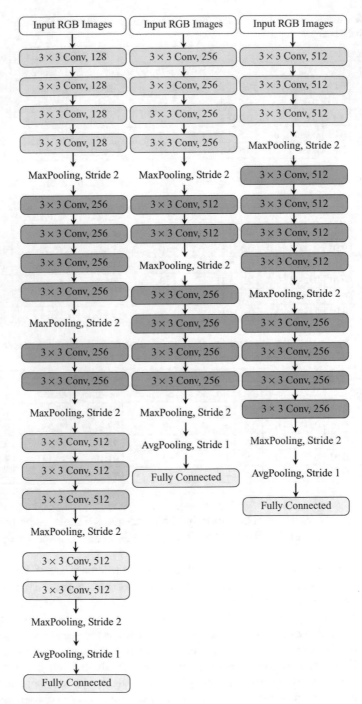

**Fig. 8.5** Architectures of top CNN models evolved by GA (left to right for: CIFAR10, CIFAR100, and SVHN datasets)

**Table 8.7** The comparisons between the GA-evolved CNN model and the state-of-the-art CNN algorithms in terms of the classification accuracy (%)

| Algorithm name | Accuracy CIFAR10 | Accuracy CIFAR100 | GPU days | Parameter setting |
|---|---|---|---|---|
| VGG16 | 93.05 | 74.94 | – | Manually |
| VGG19 | 92.59 | 74.04 | – | Manually |
| ResNet101 | 94.08 | 75.39 | – | Manually |
| DenseNet | 94.52 | 76.61 | – | Manually |
| Maxout[a] | 90.70 | 61.40 | – | Manually |
| Genetic CNN[a] | 92.90 | 70.97 | 17 | Semi-auto |
| Hierarchical evol.[a] | **96.37** | – | 300 | Semi-auto |
| Block-QNN-S[a] | 95.62 | 79.35 | 90 | Semi-auto |
| Large-scale evol.[a] | 94.60 | 77 | 2750 | Automatic |
| CGP-CNN[a] | 94.02 | – | 27 | Automatic |
| NAS[a] | 93.99 | – | 22,400 | Automatic |
| Meta-QNN[a] | 93.08 | 72.86 | 100 | Automatic |
| CNN-GA[a] | 95.22 | 77.97 | 35 | Automatic |
| Fast-CNN [17] | 94.70 | 75.63 | 14 | Automatic |
| This work (CIFAR10) | 95.82 | **79.48** | 6 | Automatic |

[a]The values of this algorithm reported in [14]

The purpose of partial training (with a smaller number of epochs) of different models during evolutionary phase was to reduce the computation burden. Some other work [8, 9] and our previous experiments [17] showed that such partial training can be sufficient for assessing network architecture in image classification. In this work we aimed to investigate it further by training the network for only 10 epochs. No doubt it accelerates the evolution process greatly—the average time for evolving the networks was only 6 GPU days. However, based on the performance of the networks evolved for CIFAR100 and SVHN datasets, we infer that such minimum training will be useful for simpler dataset but will not be sufficient for assessing the network's quality in complex datasets.

Finally, to assess the competitiveness of the GA-evolved models with the other state-of-the-art CNN models, we compared their performance in Table 8.7 (best performances are shown in bold). We contrasted the performance of three categories of networks, manually designed, designed semi-automatically, and designed completely automatically [14]. Among the manually designed networks are VGG16 [5], VGG19 [5], ResNet101 [6], DenseNet [16], and Maxout [24]. Among the semi-automatically designed networks are Genetic CNN, Hierarchical Evolution, Block-QNN-S and from the fully automatically designed networks are Large-scale Evolution, CGP-CNN, NAS, Meta-QNN, CNN-GA, and Fast-CNN [17]. We compared the performance of the best model evolved by the proposed GA for CIFAR10 dataset on both CIFAR10 and CIFAR100 datasets (last row). Besides comparing these models in terms of their accuracy in CIFAR10 and CIFAR100 datasets, for the automatically designed networks, we compared them in terms of

GPU days required to design those networks. The GPU days are a rough estimation for determining the speed of the algorithm, but it is not applicable to the manually designed models. It should be noted that some of the results have been reproduced by us, while others (indicated in the table footnote) were just copied from [14].

From Table 8.7 it can be seen that the CNN model evolved by our proposed GA was obviously better than VGG models as well as other human designed CNN models in CIFAR10 dataset. The CNN model was also better than all other models designed automatically. Its performance was second best among the compared models in CIFAR10 dataset. The best performance was exhibited by the model designed by hierarchical evolution which is a semi-automatically designed network. In terms of performance on CIFA100 dataset, the evolved model in this work exhibited the best accuracy compared to all other CNN models whether designed manually, semi-automatically, or automatically. Finally, when compared in terms of required computational power, the proposed GA was really fast, requiring only 6 GPU days, in finding the optimal CNN architecture compared to other semi-automatic and automatic methods. Although the Hierarchical Evolutionary model in the semi-automatic category shows better classification accuracy on CIFAR10, its GPU days is 50 times bigger than that required by the proposed method in this work.

## 8.7   Discussion

This chapter basically investigates if a genetic algorithm can help to find an optimized VGG-like CNN model for image classification. Using a fixed length chromosome, the proposed GA explored the search space of CNN architectures consisting of variable number of layers divided into a variable number of blocks. Experiments with three widely used datasets show that the proposed GA is able to design CNN models optimizing both its structure and hyperparameters. The evolved models were better than the classic VGG models and several other human-designed CNN models. Despite having a VGG-like architecture, the GA designed models were also very competitive with other state-of-the-art CNN models designed by semi-automatic and automatic methods. Additionally, the GA designed CNN models sometimes had structural characteristics different from those designed by humans. We also evaluated the performance of the CNN model, evolved by GA for one dataset, on other datasets. The high-quality performance of the models on other datasets indicates the superiority of the architectures evolved by GA. Based on these results, we conclude that the proposed GA is capable of optimizing classic CNN models for higher performance.

For the answer to our second question, if we can reduce the high cost of evolving CNN architecture by using partial training of the models, we evaluated the performance of networks after a few epochs of training. From our experiments with the CIFAR10 dataset, we found that partial training of the models with only a few epochs was good for finding very good architectures. However, the CNN models evolved for CIFAR10 dataset exhibited an overall better performance in CIFAR100 and SVHN datasets than those evolved for these two datasets. Based on these results, we hypothesize that perhaps partial training was not effective in evaluating network's performance in complex datasets. Nevertheless, a more detailed study by varying the training epochs on multiple datasets is required for a more general and accurate conclusion.

In our experiments, some of the structural parameters and hyperparameters of CNN were kept fixed. Although the presented framework is ready to be extended for the evolution of those parameters, the expansion of the search space will necessitate more computational power to find the optimal CNN model. On the other hand, at that expense, it might be possible to find a more efficient CNN model. Besides, the size of the evolved networks is big compared to many other architectures because no measure was taken to restrict the network size. The current framework can also be extended to incorporate that criterion either in a single or multi-objective setup. Optimizing a larger set of parameters may also help in finding smaller network models.

## 8.8   Conclusion and Future Work

In this chapter, we showed how a simple genetic algorithm (GA) can be used to automatically discover the optimized CNN model. Exploring the search space of the classic CNN models, the proposed GA optimized the number of convolutional blocks, number of convolutional layers in each blocks, the size of feature maps for each block and other training related hyperparameters such as dropout rate, learning rate, weight decay, and momentum. To reduce the computational burden in model training, which is a common challenge for all deep neuroevolution algorithms, we trained the models partially during evolution. The proposed GA, when evaluated in three popular datasets, CIFAR10, CIFAR100 and SVHN, designed very high quality CNN models over multiple repeated experiments. Performance of the evolved CNN model is compared with 14 state-of-the-art models, chosen from different categories, in terms of classification accuracy and GPU days. The best CNN model evolved on the CIFAR10 dataset was found very competitive with other human designed and automatically designed CNN models in terms of the classification accuracy and better in terms of GPU days to evolve them.

The proposed GA framework can be used for searching more structure and training related hyperparameters of CNN, e.g. kernel size, stride size, activation function, optimizer choice, etc., with very minimum changes. Besides, the framework can be extended for finding a smaller network model in terms of parameter numbers. Additionally, the performance of proposed GA can be tested in other applications of CNN as well as for optimizing other types of deep neural networks.

# References

1. Darwish, A., Hassanien, A.E., Das, S.: A survey of swarm and evolutionary computing approaches for deep learning. Artif. Intell. Rev., 1–46 (2019)
2. Iba, H., Noman, N.: New Frontier in Evolutionary Algorithms: Theory and applications. Imperial College Press, London (2011)
3. Schaffer, J.D., Whitley, D., Eshelman, L.J.: Combinations of genetic algorithms and neural networks: a survey of the state of the art. In: International Workshop on Combinations of Genetic Algorithms and Neural Networks, 1992. OGANN-92, pp. 1–37. IEEE, Piscataway (1992)
4. Such, F.P., Madhavan, V., Conti, E., Lehman, J., Stanley, K.O., Clune, J.: Deep Neuroevolution: genetic algorithms are a competitive alternative for training deep neural networks for reinforcement learning. Preprint. arXiv:1712.06567 (2017)
5. Simonyan, K., Zisserman, A.: Very deep convolutional networks for large-scale image recognition. Preprint. arXiv:1409.1556 (2014)
6. He, K., Zhang, X., Ren, S., Sun, J.: Deep residual learning for image recognition. In: Proceedings of the IEEE Conference on Computer Vision and Pattern Recognition, pp. 770–778 (2016)
7. Szegedy, C., Liu, W., Jia, Y., Sermanet, P., Reed, S., Anguelov, D., Erhan, D., Vanhoucke, V., Rabinovich, A.: Going deeper with convolutions. In: The IEEE Conference on Computer Vision and Pattern Recognition (CVPR) (2015)
8. Suganuma, M., Shirakawa, S., Nagao, T.: A genetic programming approach to designing convolutional neural network architectures. In: Proceedings of the Genetic and Evolutionary Computation Conference, pp. 497–504. ACM, New York (2017)
9. Sun, Y., Xue, B., Zhang, M., Yen, G.G.: Evolving deep convolutional neural networks for image classification. IEEE Trans. Evol. Comput. 24(2), 394–407 (2020)
10. Matsugu, M., Mori, K., Mitari, Y., Kaneda, Y.: Subject independent facial expression recognition with robust face detection using a convolutional neural network. Neural Netw. 16(5–6), 555–559 (2003)
11. Goodfellow, I., Bengio, Y., Courville, A., Bengio, Y.: Deep Learning, vol. 1. MIT Press, Cambridge (2016)
12. David, O.E., Greental, I.: Genetic algorithms for evolving deep neural networks. In: Proceedings of the Companion Publication of the 2014 Annual Conference on Genetic and Evolutionary Computation, pp. 1451–1452. ACM, New York (2014)
13. Loshchilov, I., Hutter, F.: CMA-ES for hyperparameter optimization of deep neural networks. Preprint. arXiv:1604.07269 (2016)
14. Sun, Y., Xue, B., Zhang, M., Yen, G.G.: Automatically designing CNN architectures using genetic algorithm for image classification. Preprint. arXiv:1808.03818 (2018)
15. Sun, Y., Xue, B., Zhang, M.: Automatically evolving CNN architectures based on blocks. Preprint. arXiv:1810.11875 (2018)
16. Huang, G., Liu, Z., Van Der Maaten, L., Weinberger, K.Q.: Densely connected convolutional networks. In: 2017 IEEE Conference on Computer Vision and Pattern Recognition (CVPR), pp. 2261–2269. IEEE, Piscataway (2017)
17. Bakhshi, A., Noman, N., Chen, Z., Zamani, M., Chalup, S.: Fast automatic optimisation of CNN architectures for image classification using genetic algorithm. In: 2019 IEEE Congress on Evolutionary Computation (CEC), pp. 1283–1290. IEEE, Piscataway (2019)
18. Ioffe, S., Szegedy, C.: Batch normalization: accelerating deep network training by reducing internal covariate shift. Preprint. arXiv:1502.03167 (2015)
19. Glorot, X., Bordes, A., Bengio, Y.: Deep sparse rectifier neural networks. In: Proceedings of the Fourteenth International Conference on Artificial Intelligence and Statistics, pp. 315–323 (2011)
20. Bottou, L.: Large-scale machine learning with stochastic gradient descent. In: Proceedings of COMPSTAT'2010, pp. 177–186. Springer, Berlin (2010)

21. Anderson-Cook, C.M.: Practical genetic algorithms. J. Am. Stat. Assoc. **100**(471), 1099–1099 (2005)
22. Malik, S., Wadhwa, S.: Preventing premature convergence in genetic algorithm using DGCA and elitist technique. Int. J. Adv. Res. Comput. Sci. Softw. Eng. **4**(6) (2014)
23. Netzer, Y., Wang, T., Coates, A., Bissacco, A., Wu, B., Ng, A.Y.: Reading digits in natural images with unsupervised feature learning. In: NIPS Workshop on Deep Learning and Unsupervised Feature Learning (2011)
24. Goodfellow, I.J., Warde-Farley, D., Mirza, M., Courville, A., Bengio, Y.: Maxout networks. Preprint. arXiv:1302.4389 (2013)

# Part IV
# Deep Neuroevolution

# Chapter 9
# Discovering Gated Recurrent Neural Network Architectures

Aditya Rawal, Jason Liang, and Risto Miikkulainen

**Abstract** Gated recurrent networks such as those composed of Long Short-Term Memory (LSTM) nodes have recently been used to improve state of the art in many sequential processing tasks such as speech recognition and machine translation. However, the basic structure of the LSTM node is essentially the same as when it was first conceived 25 years ago. Recently, evolutionary and reinforcement-learning mechanisms have been employed to create new variations of this structure. This chapter proposes a new method, evolution of a tree-based encoding of the gated memory nodes, and shows that it makes it possible to explore new variations more effectively than other methods. The method discovers nodes with multiple recurrent paths and multiple memory cells, which lead to significant improvement in the standard language modeling benchmark task. The chapter also shows how the search process can be speeded up by training an LSTM network to estimate performance of candidate structures, and by encouraging exploration of novel solutions. Thus, evolutionary design of complex neural network structures promises to improve performance of deep learning architectures beyond human ability to do so.

## 9.1 Introduction

In many areas of engineering design, the systems have become so complex that humans can no longer optimize them, and instead, automated methods are needed. This has been true in VLSI design for a long time, but it has also become compelling in software engineering: The idea in "programming by optimization" is that humans should design only the framework and the details should be left for automated methods such as optimization [1]. Recently similar limitations

A. Rawal (✉)
Uber AI Labs, San Francisco, CA, USA
e-mail: aditya.rawal@uber.com

J. Liang · R. Miikkulainen (✉)
Cognizant Technologies, Teaneck, NJ, USA
e-mail: Jason.Liang@cognizant.com; risto@cs.utexas.edu; risto@cognizant.com

© Springer Nature Singapore Pte Ltd. 2020
H. Iba, N. Noman (eds.), *Deep Neural Evolution*, Natural Computing Series,
https://doi.org/10.1007/978-981-15-3685-4_9

233

have started to emerge in deep learning. The neural network architectures have grown so complex that humans can no longer optimize them; hyperparameters and even entire architectures are now optimized automatically through gradient descent [2], Bayesian parameter optimization [3], reinforcement learning [4, 5], and evolutionary computation [6–8]. Improvements from such automated methods are significant: the structure of the network matters.

This work shows that the same approach can be used to improve architectures that have been used essentially unchanged for decades. The case in point is the Long Short-Term Memory (LSTM) network [9]. It was originally proposed in 1997; with the vastly increased computational power, it has recently been shown a powerful approach for sequential tasks such as speech recognition, language understanding, language generation, and machine translation, in some cases improving performance 40% over traditional methods [10]. The basic LSTM structure has changed very little in this process, and thorough comparisons of variants concluded that there is little to be gained by modifying it further [11, 12].

However, very recent studies on meta-learning methods such as neural architecture search and evolutionary optimization have shown that LSTM performance can be improved by complexifying it further [4, 6]. This chapter develops a new method along these lines, recognizing that a large search space where significantly more complex node structures can be constructed could be beneficial. The method is based on a tree encoding of the node structure so that it can be efficiently searched using genetic programming. Indeed, the approach discovers significantly more complex structures than before, and they indeed perform significantly better: Performance in the standard language modeling benchmark, where the goal is to predict the next word in a large language corpus, is improved by 6 perplexity points over the standard LSTM [13], and 0.9 perplexity points over reinforcement-learning based neural architecture search [4].

These improvements are obtained by constructing a homogeneous layered network architecture from a single gated recurrent node design. A second innovation in this work shows that further improvement can be obtained by constructing such networks from multiple different designs. As a first step, allocation of different kinds of LSTM nodes into slots in the network is shown to improve performance by another 0.5 perplexity points. This result suggests that further improvements are possible with more extensive network-level search.

A third contribution of this work is to show that evolution of neural network architectures in general can be speeded up significantly by using an LSTM network to predict the performance of candidate neural networks. After training the candidate for a few epochs, such a Meta-LSTM network predicts what performance a fully trained network would have. That prediction can then be used as fitness for the candidate, speeding up evolution fourfold in these experiments. A fourth contribution is to encourage exploration by using an archive of already-explored areas. The effect is similar to that of novelty search, but does not require a separate novelty objective, simplifying the search.

Interestingly, when the recurrent node evolved for language modeling was applied to another task, music modeling, it did not perform well. However, it was possible to evolve another solution for that task that did. As a fifth contribution, the results in this work demonstrate that it is not simply the added complexity in the nodes that matter, but that it is the right kind, i.e. complexity customized for each task.

Thus, evolutionary optimization of complex deep learning architectures is a promising approach that can yield significant improvements beyond human ability to do so.

## 9.2 Background and Related Work

In recent years, LSTM-based recurrent networks have been used to achieve strong results in the supervised sequence learning problems such as in speech recognition [10] and machine translation [10]. Further techniques have been developed to improve performance of these models through ensembling [13], shared embeddings [14], and dropouts [15].

In contrast, previous studies have shown that modifying the LSTM design itself did not provide any significant performance gains [12, 16, 17]. However, a recent paper from Zoph and Le [4] showed that policy gradients can be used to train an LSTM network to find better LSTM designs. The network is rewarded based on the performance of the designs it generates. While this approach can be used to create new designs that perform well, its exploration ability is limited (as described in more detail in Sect. 9.3.3). The setup detailed in Zoph and Le [4] is used for comparison in this work. In a subsequent paper [18], the same policy gradient approach is used to discover new recurrent highway networks to achieve even better results.

Neuroevolution methods like NEAT [19] are an alternative to policy gradient approaches, and have also been shown to be successful in the architecture search problem [6, 7]. For instance, Cartesian genetic programming was recently used to achieve state of the art results in CIFAR-10 [20]. Along similar lines, a tree-based variant of genetic programming is used in this work to evolve recurrent nodes. These trees can grow in structure and can be pruned as well, thus providing a flexible representation.

Novelty search is a particularly useful technique to increase exploration in evolutionary optimization [21]. Novelty is often cast as a secondary objective to be optimized. It allows searching in areas that do not yield immediate benefit in terms of fitness, but make it possible to discover stepping stones that can be combined to form better solutions later. This work proposes an alternative approach: keeping an archive of areas already visited and exploited, achieving similar goals without additional objectives to optimize.

Most architecture search methods reduce compute time by evaluating individuals only after partial training [7, 20]. This chapter proposes a meta-LSTM framework to predict final network performance based on partial training results.

These techniques are described in detail in the next section.

## 9.3   Methods

Evolving recurrent neural networks is an interesting problem because it requires searching the architecture of both the node and the network. As shown by recent research [4, 14], the recurrent node in itself can be considered a deep network. In this chapter, Genetic Programming (GP) is used to evolve such node architectures. In the first experiment, the overall network architecture is fixed, i.e. constructed by repeating a single evolved node to form a layer (Fig. 9.1b). In the second, it is evolved by combining several different types of nodes into a layer (Fig. 9.1c). In the future more complex coevolution approaches are also possible.

Evaluating the evolved node and network is costly. Training the network for 40 epochs takes 2 h on a 1080 NVIDIA GPU. A sequence to sequence model called meta-LSTM is developed to speed up evaluation. Following sections describe these methods in detail.

### 9.3.1   Genetic Programming for Recurrent Nodes

As shown in Fig. 9.1a, a recurrent node can be represented as a tree structure, and GP can, therefore, be used to evolve it. However, standard GP may not be sufficiently powerful to do it. In particular, it does not maintain sufficient diversity in the population. Similar to the GP-NEAT approach by Trujillo et al. [22], it can be augmented with ideas from NEAT speciation.

A recurrent node usually has two types of outputs. The first, denoted by symbol $h$ in Fig. 9.1a, is the main recurrent output. The second, often denoted by $c$, is the native memory cell output. The $h$ value is weighted and fed to three locations: (1) to the higher layer of the network at the same time step, (2) to other nodes in

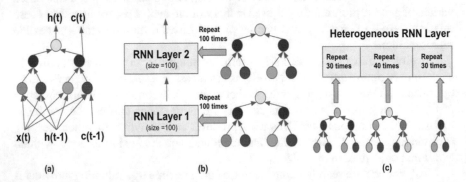

**Fig. 9.1** (a) Tree-based representation of the recurrent node. Tree outputs $h(t)$ and $c(t)$ are fed as inputs in the next time step. (b) In standard recurrent network, the tree node is repeated several times to create each layer in a multi-layered network. Different node colors depict various element activations. (c) The heterogeneous layer consists of different types of recurrent nodes

the network at the next time step, and (3) to the node itself at the next time step. Before propagation, $h$ are combined with weighted activations from the previous layer, such as input word embeddings in language modeling, to generate eight node inputs (termed as base eight by Zoph and Le [4]). In comparison, the standard LSTM node has four inputs (see Fig. 9.6a). The native memory cell output is fed back, without weighting, only to the node itself at the next time step. The connections within a recurrent cell are not trainable by backpropagation and they all carry a fixed weight of 1.0.

Thus, even without an explicit recurrent loop, the recurrent node can be represented as a tree. There are two type of elements in the tree: (1) linear activations with arity two (add, multiply), and (2) non-linear activations with arity one (tanh, sigmoid, ReLU).

There are three kind of mutation operations in the experiments: (1) Mutation to randomly replace an element with an element of the same type, (2) Mutation to randomly insert a new branch at a random position in the tree. The subtree at the chosen position is used as child node of the newly created subtree. (3) Mutation to shrink the tree by choosing a branch randomly and replacing it with one of the branch's arguments (also randomly chosen). These mutations are also depicted in Fig. 9.2b.

One limitation of standard tree is that it can have only a single output: the root. This problem can be overcome by using a modified representation of a tree that consists of Modi outputs [23]. In this approach, with some probability $p_m$ (termed Modi rate), non-root nodes can be connected to any of the possible outputs. A higher

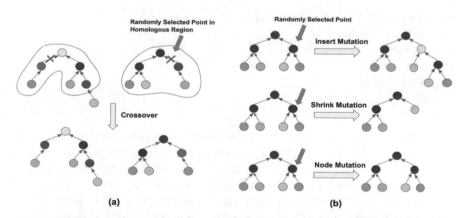

**Fig. 9.2** (a) **Homologous crossover:** the two trees on the top look different but in-fact they are almost mirror images of each other. These two trees will therefore belong in the same species. The line drawn around the trees marks the homologous regions between the two. A crossover point is randomly selected and one point crossover is performed. The bottom two networks are the resultant offsprings. (b) **Mutations:** Three kinds of mutation operators are shown from top to bottom. For each mutation operator, a node in the existing tree is randomly selected. Insert mutation can add new subtree at the selected node. Shrink mutation can replace the selected node with one of its branches. Node mutation can replace the selected node with another node of the same type

Modi rate would lead to many subtree nodes connected to different outputs. A node is assigned Modi (i.e., connected to memory cell outputs $c$ or $d$) only if its subtree has a path from native memory cell inputs.

This representation allows searching for a wide range of recurrent node structures with GP.

### 9.3.2 Speciation and Crossover

One-point crossover is the most common type of crossover in GP. However, since it does not take into account the tree structure, it can often be destructive. An alternative approach, called homologous crossover [24], is designed to avoid this problem by crossing over the common regions in the tree. Similar tree structures in the population can be grouped into species, as is often done in NEAT [22]. Speciation achieves two objectives: (1) it makes homologous crossover effective, since individuals within species are similar, and (2) it helps keep the population diverse, since selection is carried out separately in each species. A tree distance metric proposed by Tujillo et al. [22] is used to determine how similar the trees are

$$\delta(T_i, T_j) = \beta \frac{N_{i,j} - 2n_{S_{i,j}}}{N_{i,j} - 2} + (1 - \beta) \frac{D_{i,j} - 2d_{S_{i,j}}}{D_{i,j} - 2}, \qquad (9.1)$$

where

$n_{T_x}$ = number of nodes in GP tree $T_x$,
$d_{T_x}$ = depth of GP tree $T_x$,
$S_{i,j}$ = shared tree between $T_i$ and $T_j$,
$N_{i,j} = n_{T_i} + n_{T_j}$,
$D_{i,j} = d_{T_i} + d_{T_j}$,
$\beta \in [0, 1]$,
$\delta \in [0, 1]$.

On the right-hand side of Eq. (9.1), the first term measures the difference with respect to size, while the second term measures the difference in depth. Thus, setting $\beta = 0.5$ gives an equal importance to size and depth. Two trees will have a distance of zero if their structure is the same (irrespective of the actual element types).

In most GP implementations, there is a concept of the left and the right branch. A key extension in this work is that the tree distance is computed by comparing trees after all possible tree rotations, i.e. swaps of the left and the right branch. Without such a comprehensive tree analysis, two trees that are mirror images of each other might end up into different species. This approach reduces the search space by not searching for redundant trees. It also ensures that crossover can be truly homologous Fig. 9.2a.

The structural mutations in GP, i.e. insert and shrink, can lead to recycling of the same structure across multiple generations. In order to avoid such repetitions,

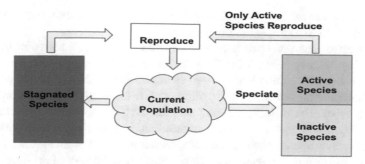

**Fig. 9.3 Hall of Shame:** An archive of stagnant species called Hall of Shame (shown in red) is built during evolution. This archive is looked up during reproduction, to make sure that newly formed offsprings do not belong to any of the stagnant species. At a time, only 10 species are actively evaluated (shown in green). This constraint ensures that active species get enough spawns to ensure a comprehensive search in its vicinity before it is added to the Hall of Shame. Offsprings that belong to new species are pushed into an inactive species list (shown in yellow) and are only moved to the active list whenever an active species moves to Hall of Shame

an archive called Hall of Shame is maintained during evolution (Fig. 9.3). This archive consists of individuals representative of stagnated species, i.e. regions in the architecture space that have already been discovered by evolution but are no longer actively searched. During reproduction, new offsprings are repeatedly mutated until they result in an individual that does not belong to Hall of Shame. Mutations that lead to Hall of Shame are not discarded, but instead used as stepping stones to generate better individuals. Such memory based evolution is similar to novelty search. However, unlike novelty search [21], there is no additional fitness objective, simply an archive.

## 9.3.3   Search Space: Node

GP evolution of recurrent nodes starts with a simple fully connected tree. During the course of evolution, the tree size increases due to insert mutations and decreases due to shrink mutations. The maximum possible height of the tree is fixed at 15. However, there is no restriction on the maximum width of the tree.

The search space for the nodes is more varied and several orders of magnitude larger than in previous approaches. More specifically, the main differences from the state-of-the-art Neural Architecture Search (NAS) [4] are: (1) NAS searches for trees of fixed height 10 layers deep; GP searches for trees with height varying between six (the size of fully connected simple tree) and 15 (a constraint added to GP). (2) Unlike in NAS, different leaf elements can occur at varying depths in GP. (3) NAS adds several constraints to the tree structure. For example, a linear element in the tree is always followed by a non-linear element. GP prevents only consecutive non-linearities (they would cause loss of information since the connections within a cell are not weighted). (4) In NAS, inputs to the tree are used only once; in GP, the inputs can be used multiple times within a node.

Most gated recurrent node architectures consist of a single native memory cell (denoted by output $c$ in Fig. 9.1a). This memory cell is the main reason why LSTMs perform better than simple RNNs. One key innovation introduced in this work is to allow multiple native memory cells within a node (for example, see outputs $c$ and $d$ in Fig. 9.6c). The memory cell output is fed back as input in the next time step without any modification, i.e. this recurrent loop is essentially a skip connection. Adding another memory cell in the node, therefore, does not affect the number of trainable parameters: it only adds to the representational power of the node.

Also, since the additional recurrent skip connection introduced as a result of the extra memory cell is local within the recurrent node, no overhead logic is required when combining different recurrent nodes with varying number of memory cells into a single heterogeneous layer (as described in the next section).

### 9.3.4   Search Space: Network

Standard recurrent networks consist of layers formed by repetition of a single type of node. However, the search for better recurrent nodes through evolution often results in solutions with similar task performance but very different structure. Forming a recurrent layer by combining such diverse node solutions is potentially a powerful idea, related to the idea of ensembling, where different models are combined together to solve a task better.

In this chapter, such heterogenous recurrent networks are constructed by combining diverse evolved nodes into a layer (Fig. 9.1c). A candidate population is created that consists of top-performing evolved nodes that are structurally very different from other nodes. The structure difference is calculated using the tree distance formula detailed previously. Each heterogenous layer is constructed by selecting nodes randomly from the candidate population. Each node is repeated 20 times in a layer; thus, if the layer size is, e.g. 100, it can consist of five different node types, each of cardinality 20.

The random search is an initial test of this idea. As described in Sect. 9.5, in the future the idea is to search for such heterogenous recurrent networks using a genetic algorithm as well.

### 9.3.5   Meta-LSTM for Fitness Prediction

In both node and network architecture search, it takes about 2 h to fully train a network until 40 epochs. With sufficient computing power it is possible to do it: for instance, Zoph and Le [4] used 800 GPUs for training multiple such solutions in

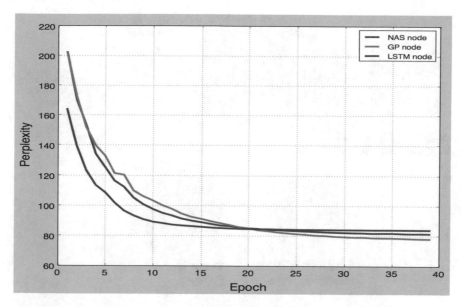

**Fig. 9.4** Learning curve comparison of LSTM node, NAS node, and GP nodes. $Y$-axis is the validation perplexity (lower is better) and $X$-axis is the epoch number. Notice that LSTM node learns quicker than the other two initially but eventually settles at a larger perplexity value. This graph demonstrates that the strategy to determine network fitness using partial training (say based on epoch 10 validation perplexity) is faulty. A fitness predictor model like meta-LSTM can overcome this problem

parallel. However, if training time could be shortened, no matter what resources are available, those resources could be used better.

A common strategy for such situations is early stopping [20], i.e. selecting networks based on partial training. For Example, in case of recurrent networks, the training time would be cut down to one fourth if the best network could be picked based on the 10th epoch validation loss instead of 40th. Figure 9.4 demonstrates that this is not a good strategy, however. Networks that train faster in the initial epochs often end up with a higher final loss.

To overcome costly evaluation and to speed up evolution, a meta-LSTM framework for fitness prediction was developed. Meta-LSTM is a sequence to sequence model [25] that consists of an encoder RNN and a decoder RNN (see Fig. 9.5a). Validation perplexity of the first 10 epochs is provided as sequential input to the encoder, and the decoder is trained to predict the validation loss at epoch 40 (show figure). Training data for these models is generated by fully training sample networks (i.e., until 40 epochs). The loss is the mean absolute error percentage at epoch 40. This error measure is used instead of mean squared error because it is unaffected by the magnitude of perplexity (poor networks can have very large perplexity values that overwhelm MSE). The hyperparameter values of the meta-LSTM were selected based on its performance in the validation dataset. The best

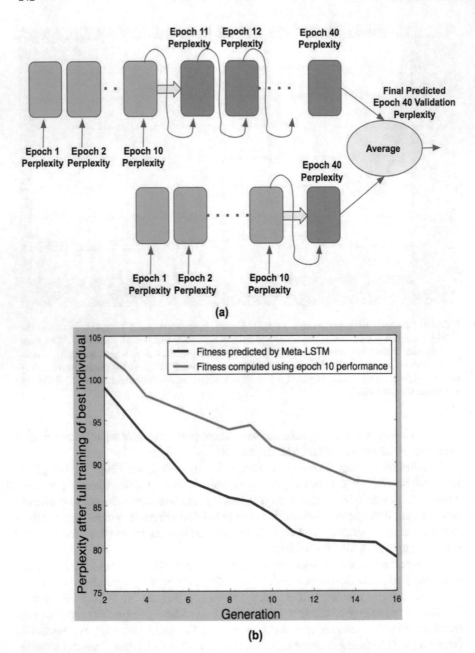

**Fig. 9.5** (a) Meta LSTM model: this is a sequence to sequence (seq2seq) model that takes the validation perplexity of the first 10 epochs as sequential input and predicts the validation perplexity at epoch 40. The green rectangles denote the encoder and the orange rectangles denote the decoder. Two variants of the model are averaged to generate one final prediction. In one variant (top), the decoder length is 30 and in the other variant (bottom), the decoder length is 1.

configuration that achieved an error rate of 3% includes an ensemble of two seq2seq models: one with a decoder length of 30 and the other with a decoder length of 1 (Fig. 9.5a).

Recent approaches to network performance prediction include Bayesian modeling [26] and regression curve fitting [27]. The learning curves for which the above methods are deployed are much simpler as compared to the learning curves of structures discovered by evolution. Note that meta-LSTM is trained separately and only deployed for use during evolution. Thus, networks can be partially trained with a 4× speedup, and assessed with near-equal accuracy as with full training.

## 9.4   Experiments

Neural architectures were constructed for the language modeling and music modeling tasks. In the first experiment, homogeneous networks were constructed from single evolved recurrent nodes, and in the second, heterogeneous networks that consisted of multiple evolved recurrent nodes. Both the experiments were targeted for the language modeling. In the third experiment, homogeneous networks were constructed from single evolved recurrent nodes for the music modeling task.

### 9.4.1   Natural Language Modeling Task

Experiments focused on the task of predicting the next word in the Penn Tree Bank corpus (PTB), a well-known benchmark for language modeling [28]. LSTM architectures in general tend to do well in this task, and improving them is difficult [12, 13, 15]. The dataset consists of 929k training words, 73k validation words, and 82k test words, with a vocabulary of 10k words. During training, successive minibatches of size 20 are used to traverse the training set sequentially.

### 9.4.2   Music Modeling Task

Music consists of a sequence of notes that often exhibit temporal dependence. Predicting future notes based on the previous notes can, therefore, be treated as a

---

**Fig. 9.5** (continued) (**b**) Meta LSTM performance: Two evolution experiments are conducted—one, where epoch 10 validation perplexity of the network is used as the fitness and second, where the value predicted by meta-LSTM is used as the network fitness. After evolution has completed, the best individuals from each generation are picked and fully trained till epoch 40. For both the experiments, this graph plots the epoch 40 performance of the best network in a given generation. The plot shows that as evolution progresses, meta-LSTM framework selects better individuals

sequence prediction problem. Similar to natural language, musical structure can be captured using a music language model (MLM). Just like natural language models form an important component of speech recognition systems, polyphonic music language model is an integral part of automatic music transcription (AMT). AMT is defined as the problem of extracting a symbolic representation from music signals, usually in the form of a time–pitch representation called piano-roll, or in a MIDI-like representation.

MLM predicts the probability distribution of the notes in the next time step. Multiple notes can be turned on at a given time step for playing chords. The input is a piano-roll representation, in the form of an $88 \times T$ matrix $M$, where $T$ is the number of timesteps, and 88 corresponds to the number of keys on a piano, between MIDI notes A0 and C8. $M$ is binary, such that $M[p, t] = 1$ if and only if the pitch $p$ is active at timestep $t$. In particular, held notes and repeated notes are not differentiated. The output is of the same form, except it only has $T - 1$ timesteps (the first timestep cannot be predicted since there is no previous information).

The dataset piano-midi.de is used as the benchmark data. This dataset holds 307 pieces of classical piano music from various composers. It was made by manually editing the velocities and the tempo curve of quantized MIDI files in order to give them a natural interpretation and feeling [29]. MIDI files encode explicit timing, pitch, velocity and instrumental information of the musical score.

### 9.4.3  Network Training Details

During evolution, each network has two layers of 540 units each, and is unrolled for 35 steps. The hidden states are initialized to zero; the final hidden states of the current minibatch are used as the initial hidden states of the subsequent minibatch. The dropout rate is 0.4 for feedforward connections and 0.15 for recurrent connections [15]. The network weights have L2 penalty of 0.0001. The evolved networks are trained for 10 epochs with a learning rate of 1; after six epochs the learning rate is decreased by a factor of 0.9 after each epoch. The norm of the gradients (normalized by minibatch size) is clipped at 10. Training a network for 10 epochs takes about 30 min on an NVIDIA 1080 GPU. The following experiments were conducted on 40 such GPUs.

The meta-LSTM consists of two layers, 40 nodes each. To generate training data for it, 1000 samples from a preliminary node evolution experiment was obtained, representing a sampling of designs that evolution discovers. Each of these sample networks was trained for 40 epochs with the language modeling training set; the perplexity on the language modeling validation set was measured in the first 10 epochs, and at 40 epochs. The meta-LSTM network was then trained to predict the perplexity at 40 epochs, given a sequence of perplexity during the first 10 epochs as input. A validation set of 500 further networks was used to decide when to stop training the meta-LSTM, and its accuracy measured with another 500 networks.

**Table 9.1** Single model perplexity on test set of Penn Tree Bank

| Model | Parameters | Test perplexity |
|---|---|---|
| Gal and Ghahramani [15]—variational LSTM | 66M | 73.4 |
| Zoph and Le [4] | 20M | 71.0 |
| GP node evolution | 20M | 68.2 |
| Zoph and Le [4] | 32M | 68.1 |
| GP node evolution | 32M | 66.5 |
| Zilly et al.[14], shared embeddings | 24M | 66.0 |
| Zoph and Le [4], shared embeddings | 25M | 64.0 |
| GP evolution, shared embeddings | 25M | 63.0 |
| Heterogeneous, shared embeddings | 25M | 62.2 |
| Zoph and Le [4], shared embeddings | 54M | 62.9 |

Node evolved using GP outperforms the node discovered by NAS (Zoph and Le [4]) and Recurrent Highway Network (Zilly et al. [14]) in various configurations

In line with meta-LSTM training, during evolution each candidate is trained for 10 epochs, and tested on the validation set at each epoch. The sequence of such validation perplexity values is fed into the trained meta-LSTM model to obtain its predicted perplexity at epoch 40; this prediction is then used as the fitness for that candidate. The individual with the best fitness after 30 generations is scaled to a larger network consisting of 740 nodes in each layer. This setting matches the 32 Million parameter configuration used by Zoph and Le [4]. A grid search over dropout rates is carried out to fine-tune the model. Its performance after 180 epochs of training is reported as the final result (Table 9.1).

### 9.4.4 Experiment 1: Evolution of Recurrent Nodes

A population of size 100 was evolved for 30 generations with a crossover rate of 0.6, insert and shrink mutation probability of 0.6 and 0.3, respectively, and Modi rate (i.e., the probability that a newly added node is connected to memory cell output) of 0.3. A compatibility threshold of 0.3 was used for speciation; species is marked stagnated and added to the Hall of Shame if the best fitness among its candidates does not improve in four generations. Each node is allowed to have three outputs: one main recurrent output ($h$) and two native memory cell outputs ($c$ and $d$).

The best evolved node is shown Fig. 9.6. The evolved node reuses inputs as well as utilize the extra memory cell pathways. As shown in Table 9.1, the evolved node (called GP Node evolution in the table) achieves a test performance of 68.2 for 20 Million parameter configuration on Penn Tree Bank. This is 2.8 perplexity points better than the test performance of the node discovered by NAS (Zoph and Le [4] in the table) in the same configuration. Evolved node also outperforms NAS in the 32 Million configuration (66.5 v/s. 68.1). Recent work has shown that sharing input and output embedding weight matrices of neural network language models improves

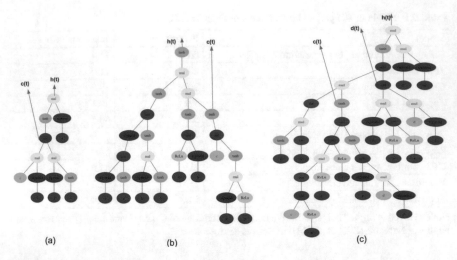

(a)                                       (b)                                                        (c)

**Fig. 9.6** Comparing evolved recurrent node with NASCell and LSTM. The green input elements denote the native memory cell outputs from the previous time step $(c, d)$. The red colored inputs are formed after combining the node output from the previous time step $h(t-1)$ and the new input from the current time step $x(t)$. In all three solutions, the memory cell paths include relatively few non-linearities. The evolved node utilizes the extra memory cell in different parts of the node. GP evolution also reuses inputs unlike the NAS and LSTM solution. Evolved node also discovered LSTM like output gating. (**a**) LSTM. (**b**) NASCell. (**c**) Evolved cell

performance [30]. The experimental results obtained after including this method are marked as shared embeddings in Table 9.1.

It is also important to understand the impact of using meta-LSTM in evolution. For this purpose, an additional evolution experiment was conducted, where each individual was assigned a fitness equal to its 10th epoch validation perplexity. As evolution progressed, in each generation, the best individual was trained fully till epoch 40. Similarly, the best individual from an evolution experiment with meta-LSTM enabled was fully trained. The epoch 40 validation perplexity in these two cases has been plotted in Fig. 9.5b. This figure demonstrates that individuals that are selected based upon meta-LSTM prediction perform better than the ones selected using only partial training.

### 9.4.5   Experiment 2: Heterogeneous Recurrent Networks

Top 10% of the population from 10 runs of Experiment 1 was collected into a pool 100 nodes. Out of these, 20 that were the most diverse, i.e. had the largest tree distance from the others, were selected for constructing heterogeneous layers (as shown in Fig. 9.1c). Nodes were chosen from this pool randomly to form 2000 such networks. Meta-LSTM was again used to speed up evaluation.

After hyperparameter tuning, the best network (for 25 Million parameter configuration) achieved a perplexity of 62.2, i.e. 0.8 better than the homogeneous network constructed from the best evolved node. This network is also 0.7 perplexity point better than the best NAS network double its size (54 Million parameters). Interestingly, best heterogeneous network was also found to be more robust to hyperparameter changes than the homogeneous network. This result suggests that diversity not only improves performance, but also adds flexibility to the internal representations. The heterogeneous network approach, therefore, forms a promising foundation for future work, as discussed next.

## 9.4.6   Experiment 3: Music Modeling

The piano-midi.de dataset is divided into train (60%), test (20%), and validation (20%) sets. The music model consists of a single recurrent layer of width 128. The input and output layers are 88 wide each. The network is trained for 50 epochs with Adam at a learning rate of 0.01. The network is trained by minimizing cross entropy between the output of the network and the ground truth. For evaluation, F1 score is computed on the test data. F1 score is the harmonic mean of precision and recall (higher is better). Since the network is smaller, regularization is not required.

Note, this setup is similar to that of Ycart and Benetos [29]. The goal of this experiment is not to achieve state-of-the-art results but to perform apples-to-apples comparison between LSTM nodes and evolved nodes (discovered for language) in a new domain, i.e. music.

In this transfer experiment, three networks were constructed: the first with LSTM nodes, the second with NAS nodes, and the third with evolved nodes. All the three networks were trained under the same setting as described in the previous section. The F1 score of each of the three models is shown in Table 9.2. LSTM nodes outperform both NAS and evolved nodes. This result is interesting because both NAS and evolved nodes significantly outperformed LSTM nodes in the language modeling task. This result suggests that NAS and evolved nodes are custom solution for a specific domain, and do not necessarily transfer to other domains.

**Table 9.2** F1 scores computed on Piano-Midi dataset

| Model | F1 score |
|---|---|
| LSTM | 0.548 |
| Zoph and Le [4] | 0.48 |
| GP evolution (language) | 0.49 |
| GP evolution (music) | 0.599 |

LSTM outperforms both the evolved node and NAS node for language, but not the node evolved specifically for music, demonstrating that the approach discovers solutions customized for the task

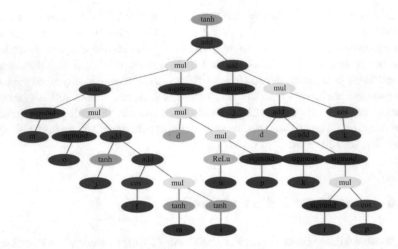

**Fig. 9.7** Evolved node for music. The node evolved to solve the music task is very different from the node for the natural language task. For example, this node only uses a single memory cell (green input $d$ in the figure) unlike the language node that used both $c$ and $d$. This results indicates that 'architecture does matter' and that custom evolved solution perform better than hand-designed ones

However, the framework developed for evolving recurrent nodes for natural language can be transferred to the music domain as well. The setup is the same, i.e. at each generation a population of recurrent nodes represented as trees will be evaluated for their performance in the music domain. The validation performance of the network constructed from the respective tree node will be used as the node fitness. The performance measure of the network in music domain is the F1 score, therefore, it is used as the network fitness value.

The evolution parameters are the same as those used for language modeling. Meta-LSTM is not used for this evolution experiment because the run-time of each network is relatively small (<600 s). The results from evolving custom node for music are shown in Table 9.2. The custom node (GP Evolution (Music)) achieves an improvement of five points in F1 score over LSTM (Fig. 9.7). Thus, evolution was able to discover custom structure for the music modeling domain as well—and it was different from structure in the language domain.

## 9.5 Discussion and Future Work

The experiments in this chapter demonstrate how evolutionary optimization can discover improvements to designs that have been essentially unchanged for 25 years. Because it is a population-based method, it can harness more extensive exploration than other meta-learning techniques such as reinforcement learning, Bayesian parameter optimization, and gradient descent. It is therefore in a position to discover novel, innovative solutions that are difficult to develop by hand or through gradual improvement. Remarkably, the node that performed well in language modeling performed poorly in music modeling, but evolution was able to discover a different node that performed well in music. Apparently, the approach discovers regularities in each task and develops node structures that take advantage of them, thus customizing the nodes separately for each domain. Analyzing what those regularities are and how the structures encode them is an interesting direction of future work.

The GP-NEAT evolutionary search method in this chapter is run in the same search space used by NAS [4], resulting in significant improvements. In a recent paper [18], the NAS search space was extended to include recurrent highway connections as well, improving the results further. An interesting direction of future work is thus to extend the GP-NEAT search space in a similar manner; similar improvements should result.

The current experiments focused on optimizing the structure of the gated recurrent nodes, cloning them into a fixed layered architecture to form the actual network. The simple approach of forming heterogeneous layers by choosing from a set of different nodes was shown to improve the networks further. A compelling next step is thus to evolve the network architecture as well, and further, coevolve it together with the LSTM nodes [6].

## 9.6   Conclusion

Evolutionary optimization of LSTM nodes can be used to discover new variants that perform significantly better than the original 25-year old design. The tree-based encoding and genetic programming approach makes it possible to explore larger design spaces efficiently, resulting in structures that are more complex and more powerful than those discovered by hand or through reinforcement-learning based neural architecture search. Further, these structures are customized to each specific domain. The approach can be further enhanced by optimizing the network level as well, in addition to the node structure, by training an LSTM network to estimate the final performance of candidates instead of having to train them fully, and by encouraging novelty through an archive. Evolutionary neural architecture search is, therefore, a promising approach to extending the abilities of deep learning networks to ever more challenging tasks.

# References

1. Hoos, H.: Programming by optimization. Commun. ACM **55**, 70–80 (2012)
2. Andrychowicz, M., Denil, M., Colmenarejo, S.G., Hoffman, M.W., Pfau, D., Schaul, T., Shillingford, B., de Freitas, N.: Learning to learn by gradient descent by gradient descent. In: Proceedings of the 30th International Conference on Neural Information Processing Systems, NIPS'16, pp. 3988–3996. Curran Associates Inc., Red Hook (2016). Available: http://dl.acm. org/citation.cfm?id=3157382.3157543
3. Malkomes, G., Schaff, C., Garnett, R.: Bayesian optimization for automated model selection. In: Proceedings of the 30th International Conference on Neural Information Processing Systems, NIPS'16, pp. 2900–2908. Curran Associates Inc., Red Hook (2016). Available: http:// dl.acm.org/citation.cfm?id=3157382.3157422
4. Zoph, B., Le, Q.V.: Neural architecture search with reinforcement learning (2016). Available: https://arxiv.org/pdf/1611.01578v1.pdf
5. Baker, B., Gupta, O., Naik, N., Raskar, R.: Designing neural network architectures using reinforcement learning (2016). Available: https://arxiv.org/pdf/1611.02167v2.pdf
6. Miikkulainen, R., Liang, J.Z., Meyerson, E., Rawal, A., Fink, D., Francon, O., Raju, B., Shahrzad, H., Navruzyan, A., Duffy, N., Hodjat, B.: Evolving deep neural networks. CoRR, vol. abs/1703.00548 (2017). Available: http://arxiv.org/abs/1703.00548
7. Real, E., Moore, S., Selle, A., Saxena, S., Suematsu, Y.L., Tan, J., Le, Q.V., Kurakin, A.: Large-scale evolution of image classifiers. In: Proceedings of the 34th International Conference on Machine Learning - Volume 70, ICML'17. JMLR.org, pp. 2902–2911 (2017). Available: http://dl.acm.org/citation.cfm?id=3305890.3305981
8. Fernando, C.: PathNet: evolution channels gradient descent in super neural networks (2017). Available: https://arxiv.org/abs/1701.08734
9. Hochreiter, S., Schmidhuber, J.: Long short-term memory. Neural Comput. **9**(8), 1735–1780 (1997). Available: http://dx.doi.org/10.1162/neco.1997.9.8.1735
10. Bahdanau, D., Cho, K., Bengio, Y.: Neural machine translation by jointly learning to align and translate. Preprint (2014). arXiv:1409.0473
11. Klaus, G., Srivastava, R., Koutník, J., Steunebrink, R., Schmidhuber, J.: LSTM: a search space odyssey. Preprint, vol. arxiv/1503.04069 (2014)
12. Jozefowicz, R., Zaremba, W., Sutskever, I.: An empirical exploration of recurrent network architectures. In: Proceedings of the 32nd International Conference on Machine Learning, pp. 2342–2350 (2015)
13. Zaremba, W., Sutskever, I., Vinyals, O.: Recurrent neural network regularization. Preprint, vol. arxiv/1409.2329 (2014)
14. Zilly, J.G., Srivastava, R.K., Koutník, J., Schmidhuber, J.: Recurrent highway networks. In: Precup, D., Teh, Y.W. (eds.) Proceedings of the 34th International Conference on Machine Learning. Proceedings of Machine Learning Research, 06–11 Aug 2017, vol. 70, pp. 4189–4198. International Convention Centre, Sydney (2017). Available: http://proceedings.mlr.press/ v70/zilly17a.html
15. Gal, Y., Ghahramani, Z.: A theoretically grounded application of dropout in recurrent neural networks. In: Proceedings of the 30th International Conference on Neural Information Processing Systems, NIPS'16, pp. 1027–1035. Curran Associates Inc., Red Hook (2016). Available: http://dl.acm.org/citation.cfm?id=3157096.3157211
16. Bayer, J., Wierstra, D., Togelius, J., Schmidhuber, J.: Evolving memory cell structures for sequence learning. In: Artificial Neural Networks ICANN, pp. 755–764 (2009)
17. Cho, K., van Merrienboer, B., Gülçehre, Ç., Bougares, F., Schwenk, H., Bengio, Y.: Learning phrase representations using RNN encoder-decoder for statistical machine translation. CoRR, vol. abs/1406.1078 (2014). Available: http://arxiv.org/abs/1406.1078

18. Pham, H.,  Guan, M.,  Zoph, B.,  Le, Q.,  Dean, J.: Efficient neural architecture search via parameters sharing. In: Dy, J., Krause, A. (eds.) Proceedings of the 35th International Conference on Machine Learning. Proceedings of Machine Learning Research, PMLR, 10–15 July 2018, vol. 80, pp. 4095–4104. Stockholmsmässan, Stockholm (2018). Available: http://proceedings.mlr.press/v80/pham18a.html
19. Stanley, K.O.,  Miikkulainen, R.: Evolving neural networks through augmenting topologies. Evol. Comput. **10**(2), 99–127 (2002). Available: http://nn.cs.utexas.edu/?stanley:ec02
20. Suganuma, M.,  Shirakawa, S.,  Nagao, T.: A genetic programming approach to designing convolutional neural network architectures. In: Proceedings of the Genetic and Evolutionary Computation Conference, GECCO '17, pp. 497–504. ACM, New York (2017). Available: http://doi.acm.org/10.1145/3071178.3071229
21. Lehman, J.,  Stanley, K.O.: Abandoning objectives: evolution through the search for novelty alone. Evol. Comput. **19**(2), 189–223 (2011)
22. Trujillo, L.,  Muñoz, L.,  López, E.G.,  Silva, S.: neat genetic programming: controlling bloat naturally. Inf. Sci. **333**, 21–43 (2016)
23. Zhang, Y.,  Zhang, M.: A multiple-output program tree structure in genetic programming. Tech. Rep. (2004)
24. Francone, F.D.,  Conrads, M.,  Banzhaf, W.,  Nordin, P.: Homologous crossover in genetic programming. In: Proceedings of the 1st Annual Conference on Genetic and Evolutionary Computation - Volume 2, GECCO'99, pp. 1021–1026. Morgan Kaufmann Publishers Inc., San Francisco (1999). Available: http://dl.acm.org/citation.cfm?id=2934046.2934059
25. Sutskever, I.,  Vinyals, O., Le, Q.V.: Sequence to sequence learning with neural networks. In: Proceedings of the 27th International Conference on Neural Information Processing Systems - Volume 2, NIPS'14, pp. 3104–3112. MIT Press, Cambridge (2014). Available: http://dl.acm.org/citation.cfm?id=2969033.2969173
26. Klein, A.,  Falkner, S.,  Springenberg, J.T.,  Hutter, F.: Learning curve prediction with Bayesian neural networks. In: Conference Paper at ICLR (2017)
27. Baker, B.,  Gupta, O.,  Raskar, R.,  Naik, N.: Practical neural network performance prediction for early stopping. CoRR, vol. abs/1705.10823 (2017). Available: http://arxiv.org/abs/1705.10823
28. Marcus, M.P.,  Marcinkiewicz, M.A.,  Santorini, B.: Building a large annotated corpus of English: the Penn Treebank. Comput. Linguist. **19**(2), 313–330 (1993)
29. Ycart, A.,  Benetos, E.: Polyphonic music sequence transduction with meter-constrained LSTM networks. In: 2018 IEEE International Conference on Acoustics, Speech and Signal Processing (ICASSP), pp. 386–390, April 2018
30. Press, O.,  Wolf, L.: Using the output embedding to improve language models. Preprint, vol. arxiv/1608.05859 (2016)

# Chapter 10
# Investigating Deep Recurrent Connections and Recurrent Memory Cells Using Neuro-Evolution

Travis Desell, AbdElRahman A. ElSaid, and Alexander G. Ororbia

**Abstract** Neural architecture search poses one of the most difficult problems for statistical learning, given the incredibly vast architectural search space. This problem is further compounded for recurrent neural networks (RNNs), where every node in an architecture can be connected to any other node via recurrent connections which pass information from previous passes through the RNN via a weighted connection. Most modern-day RNNs focus on recurrent connections which pass information from the immediately preceding pass by utilizing gated constructs known as memory cells; however, connections farther back in time, or *deep recurrent connections*, are also possible. A novel neuro-evolutionary metaheuristic called EXAMM is utilized to conduct extensive experiments evolving RNNs consisting of a suite of memory cells and simple neurons, with and without deep recurrent connections. These experiments evolved and trained 10.56 million RNNs, with results showing that networks with deep recurrent connections perform significantly better than those without, and in some cases the best evolved RNNs consist of only simple neurons and deep recurrent connections. These results strongly suggest that utilizing complex recurrent connectivity patterns in RNNs deserves further study and also showcases the strong potential for using neuro-evolutionary metaheuristic algorithms as tools for understanding and training effective RNNs.

## 10.1 Introduction

Research in artificial neural networks has exploded in recent years as a result of the many successes achieved using them to solve problems in tasks such as image classification [1, 2], video categorization [3], sentence modeling [4], and speech recognition [5]. This success has largely been driven by highly engineered, carefully hand-crafted networks, e.g., convolutional architectures in computer vision such

---

T. Desell (✉) · A. A. ElSaid · A. G. Ororbia
Rochester Institute of Technology, Rochester, NY, USA
e-mail: tjdvse@rit.edu; aae8800@rit.edu; ago@cs.rit.edu

© Springer Nature Singapore Pte Ltd. 2020
H. Iba, N. Noman (eds.), *Deep Neural Evolution*, Natural Computing Series,
https://doi.org/10.1007/978-981-15-3685-4_10

as AlexNet [1], VGGNet [6], GoogleNet [7], and, more recently, ResNet [8]. In addition, for handling the modeling of sequential data with temporal dependencies, such as in natural language processing, recurrent neural networks (RNNs) have become very useful, especially that those that utilize complex memory cell structures to capture long-term dependencies, e.g., $\Delta$-RNN units [9], gated recurrent units (GRUs) [10], long short-term memory cells (LSTMs) [11], minimal gated units (MGUs) [12], and update gate RNN cells (UGRNNs) [13].

Due to the largely intractable problem of determining an optimal neural network architecture for a given learning task, neuro-evolution [14–29] and neural architecture search algorithms [30–34] have become increasingly popular. Much of this work has focused on feedforward architectures (such as convolutional neural networks), restricting itself to selecting parameters of networks that are hierarchically structured in layers. However, artificial neural networks (ANNs) can be represented as completely unstructured graphs (without layers) with each node representing a neuron, and each edge representing a neuronal connection, which, even argued classically [35], could potentially be more expressive (and exhibit better generalization across data) while at the same time less complex computationally. Unfortunately, when relaxing constraints on the type of topological organization such networks may have, the search space of potential architectures expands dramatically.

RNNs provide particularly unique challenges for neuro-evolution given that they operate over potentially very long temporal sequences of input data. This necessitates designing more complicated internal memory cell structures or employing recurrent connections that transmit information from varying time delays in the past. As a result, the architecture search space becomes even larger now that neurons not only connect to one another via complex feedforward patterns but also through complicated recurrent pathways that could potentially span an indeterminate amount of time. Most modern RNNs avoid complex recurrent connectivity structures by using gated memory cells, resting on the assumption that while the recurrent connections within these memory cells only go back explicitly to the immediately preceding time step, their internal gates might provide a means to accurately latch on to long-term information and ensure good predictive ability. Nonetheless, there exists a body of literature which suggests that recurrent connections which skip more than one time step, which we will call *deep recurrent connections*, can play a vital role in RNN design when aiming to better capture long-term dependencies in sequential data [27, 36–42].

This work investigates the capabilities of both recurrent memory cells and deep recurrent connections through the use of a novel neuro-evolution algorithm called EXAMM (Evolutionary eXploration of Augmenting Memory Models) [43]. Instead of simply testing a few hand-crafted RNNs that employ various memory cell types and that either use or do not use deep recurrent connections, we take a neuro-evolutionary approach to the experimental process—we allow EXAMM to search for connectivity patterns and make memory cell choices, allowing each particular design choice to evolve architectures that best suit it, ultimately providing a more rigorous analysis. A variety of experiments were performed evolving RNNs consisting of LSTM, GRU, MGU, UGRNN, $\Delta$-RNN memory cells as well as simple neurons, allowing EXAMM to design deep recurrent connectivity patterns of varying complexity. The RNNs were evolved to perform time series data prediction on four real world benchmark problems. In total 10.56 million RNNs were trained to collect the results we report in this chapter.

The results of our neuro-evolutionary architecture search demonstrate that RNNs with deep recurrent connections perform significantly better than those without, and in some cases, the best performing RNNs consisted of only simple neurons and deep recurrent connections—in other words, no memory cells were deemed necessary to attain strong predictive performance. These results strongly suggest that utilizing deep recurrent connections in RNNs for time series data prediction not only deserves further study, but also showcases neuro-evolution as a potentially powerful tool for studying, understanding, and training effective RNN models. Further, with respect to memory cell types, we find that EXAMM uncovers that the more recently proposed and far simpler $\Delta$-RNN unit performs better and more reliably than the other memory cells when modeling sequences.

## 10.2 Related Work

### 10.2.1 Elman, Jordan and Arbitrary Recurrent Connections

Elman and Jordan RNNs are traditional RNN architectures that have been studied extensively. Both typically only have a single hidden layer. Elman RNNs [44] have hidden layer nodes that are (recurrently) self-connected as well as fully connected to all other hidden nodes in the same layer. Jordan RNNs [45] have recurrent connections from the output node(s) to all the hidden layer node(s). In both RNN types, each hidden node can be viewed as a simple neuron, meaning that no internal gating or latching structure is employed during neural computation.

In EXAMM, simple neurons are represented as point-wise neurons with potentially both recurrent and feedforward inputs. $I$ is the set of all nodes with a feedforward connection to simple neuron $j$, while $R$ is the set of all nodes with a recurrent connection to simple neuron $j$. At time step $t$, the input signal to a neuron $j$ is a weighted summation of all feedforward inputs, where $w_{ij}$ is the feedforward weight connecting node $i$ to node $j$, plus a weighted summation of recurrent inputs, where $v_{rjk}$ is the recurrent weight from node $r$ at time step $t - k$ to the node $j$, where $k$ is the time span of the recurrent connection. Thus the state function $s$ for computing a simple neuron is[1]

$$s_j(t) = \phi_s \left( \sum_{i \in I} w_{ij} \cdot s_i(t) + \sum_{r \in R, k} v_{rjk} \cdot s_r(t - k) \right) \tag{10.1}$$

where $s_i(t)$ marks the state of the neuronal unit $s_i$ at time step $t$. The overall state is a linear combination of the projected input and an affine transformation of the vector summary of the past. The post-activation function, $\phi_s(\cdot)$, can be any differentiable element-wise function; however, in this chapter it was limited it to be the hyperbolic tangent, $\phi(v) = tanh(v) = (e^{(2v)} - 1)/(e^{(2v)} + 1)$.

Fundamentally, the EXAMM metaheuristic has the capability of evolving Elman-style connections when its *add recurrent edge* mutation operator adds a recurrent edge from a simple neuron back to itself (see Sect. 10.3). Jordan-style connections are added when EXAMM decides to add a recurrent edge from an output to any simple neuron (we generalize this to include connections back to input neurons as well as to other output neurons). EXAMM supports a generalization of Elman and Jordan-style connectivity patterns (which traditionally only span a single time step) by incorporating temporal skipping—deep recurrent connections over varying time delays are possible. Figure 10.1 depicts an RNN that is composed of a variety of connections (all of which could be evolved by EXAMM). In essence, it is possible for EXAMM to automatically craft RNNs that have arbitrary feedforward and recurrent connectivity topologies, allowing the networks to make use of the increased computational power afforded by deep recurrent connections in both time and structure.

### 10.2.2 Recurrent Memory Cells

One of the more popularly used recurrent neural cell types is the long short-term memory (LSTM) cell developed by Hochreiter and Schmidhuber in 1997 [11]. The original motivation behind the LSTM was to implement the "constant error carousal" in order to mitigate the problem of vanishing gradients. This means that long-term memory can be explicitly represented with a separate cell state $c_t$. Using the notation presented earlier (Sect. 10.2.1), the LSTM state function (without

---

[1]The bias is omitted for clarity and simplicity of presentation.

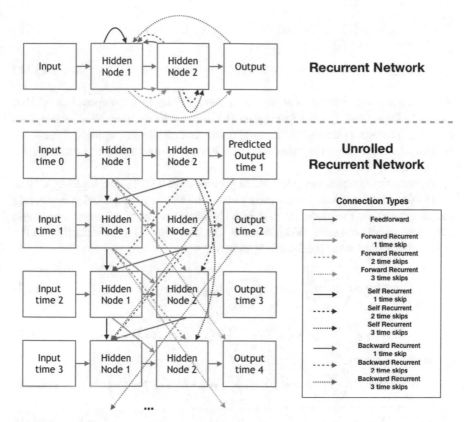

**Fig. 10.1** RNNs can have a wide variety of connections. Most RNNs consist of feedforward, self-recurrent, and backward recurrent connections with a single time step; however, it is also possible to also have forward recurrent connections and recurrent connections which span/skip multiple time steps

extensions, such as "peephole" connections) is implemented by the following equations:

$$f_j = \sigma\left(\sum_{i \in I} w_{ij}^f \cdot s_i(t) + \sum_{r \in R,k} v_{rjk}^f \cdot s_r(t-k)\right) \tag{10.2}$$

$$i_j = \sigma\left(\sum_{i \in I} w_{ij}^i \cdot s_i(t) + \sum_{r \in R,k} v_{rjk}^i \cdot s_r(t-k)\right) \tag{10.3}$$

$$\widetilde{c}_j = tanh\left(\sum_{i \in I} w_{ij}^c \cdot s_i(t) + \sum_{r \in R,k} v_{ijk}^c \cdot s_r(t-k)\right) \tag{10.4}$$

$$o_j = \sigma\left(\sum_{i \in I} w_{ij}^o \cdot s_i(t) + \sum_{r \in R,k} v_{ijk}^o \cdot s_r(t-k)\right) \tag{10.5}$$

$$c_j(t) = f_j(t) \cdot c_j(t-1) + i_j \cdot \widetilde{c}_j, \quad s_j(t) = o_j \cdot \phi_s(c_j(t)) \tag{10.6}$$

where we separate the computation of a neural state as a composition of four internal calculations, i.e., the forget gate ($\mathbf{f}_t$), input gate ($\mathbf{i}_t$), the cell-state proposal ($\widetilde{\mathbf{c}}_t$), and the output gate ($\mathbf{o}_t$). This cellular structure, while conceptually appealing, is computationally complex with each cell having 11 trainable parameters (and 8 weights and 3 biases).

In more recent times, the gated recurrent unit (GRU; [10]), developed by Cho et al. in 2014, can be viewed as an early attempt to simplify the LSTM cell. Among the changes made, the model fuses the LSTM input and forget gates into one single gate, merging the cell state and hidden state back together. The state function for the GRU is computed using the following equations:

$$z_j = \sigma\left(\sum_{i \in I} w_{ij}^z \cdot s_i(t) + \sum_{r \in R,k} v_{ijk}^z \cdot s_r(t-k)\right) \tag{10.7}$$

$$r_j = \sigma\left(\sum_{i \in I} w_{ij}^r \cdot s_i(t) + \sum_{r \in R,k} v_{ijk}^r \cdot s_r(t-k)\right) \tag{10.8}$$

$$\widetilde{s}_j(t) = \phi_s\left(\sum_{i \in I} w_{ij}^s \cdot s_i(t) + \sum_{r \in R,k} v_{ijk}^s \cdot (r_j \cdot s_r(t-k))\right) \tag{10.9}$$

$$s_j(t) = z_j \cdot \widetilde{s}_j + (1 - z_j) \cdot s_j(t-1) \tag{10.10}$$

noting that $\phi_s(v) = tanh(v)$. Note that the GRU requires fewer internal operations than the LSTM cell, and, as a result, requires only adapting 9 trainable parameters.

The MGU model, proposed by Zhou et al. in 2016, is very similar in structure to the GRU, reducing the number of required parameters by merging the reset and update gates into one forget gate [12]. The state computation then proceeds as follows:

$$f_j = \sigma\left(\sum_{i \in I} w_{ij}^f \cdot s_i(t) + \sum_{r \in R,k} v_{ijk}^f \cdot s_r(t-k)\right) \tag{10.11}$$

$$\widetilde{s}_j(t) = \phi_s\left(\sum_{i \in I} w_{ij}^s \cdot s_i(t) + \sum_{r \in R,k} v_{ijk}^s \cdot (f_j \cdot s_r(t-k))\right) \tag{10.12}$$

$$s_j(t) = z_j \cdot \widetilde{s}_j + (1 - z_j) \cdot s_j(t-1). \tag{10.13}$$

The MGU is one of the simpler memory cells, with only 6 trainable parameters.

Update gate RNN cells (UGRNNs) were introduced in 2016 by Collins et al. [13]. UGRNN updates are defined in the following manner:

$$c_j = \phi_s \left( \sum_{i \in I} w_{ij}^c \cdot s_i(t) + \sum_{r \in R, k} v_{ijk}^c \cdot s_r(t - k) \right) \qquad (10.14)$$

$$g_j = \sigma \left( \sum_{i \in I} w_{ij}^g \cdot s_i(t) + \sum_{r \in R, k} v_{ijk}^g \cdot s_r(t - k) \right) \qquad (10.15)$$

$$s_j(t) = g_j \cdot s_j(t - 1) + (1 - g_j) \cdot c_j. \qquad (10.16)$$

UGRNNs require the same number of trainable parameters as MGUs (6); however, it also requires one less multiplication, making it a bit faster computationally. The UGRNN is a simple cell, essentially working very much like an Elman-RNN but with a single update gate. This extra gate decides whether a hidden state is carried over from the previous time step or if the state should be updated.

Most recently, the $\Delta$-RNN unit was derived from recently developed framework known as the differential state framework (DSF) [9], from which one can furthermore derive all other more complex gated RNN cells, including the LSTM, GRU, and MGU. The $\Delta$-RNN is one of the simplest RNN models proposed, featuring only a few extra bias vector parameters to control its own internal gating, making it the closest to an Elman-RNN in terms of computational efficiency while also offering competitive generalization in problem tasks ranging from language modeling [9, 46] to image decoding [47]. With $\{\alpha, \beta_1, \beta_2, b_j\}$ as learnable coefficient scalars, the $\Delta$-RNN state is defined as:

$$e_j^w = \sum_{i \in I} w_{ij} \cdot s_i(t), \quad e_j^v = \sum_{r \in R, k} v_{rjk} \cdot s_r(t - k) \qquad (10.17)$$

$$d_j^1 = \alpha \cdot e_j^v \cdot e_j^w, \quad d_j^2 = \beta_1 \cdot e_j^v + \beta_2 \cdot e_j^w \qquad (10.18)$$

$$\widetilde{s}_j(t) = \phi_s(d_j^1 + d_j^2), \quad r_j = \sigma(e_j^w + b_j) \qquad (10.19)$$

$$s_j(t) = \Phi_s((1 - r_j) \cdot \widetilde{s}_j(t) + r_j \cdot s_j(t - 1) \qquad (10.20)$$

$\Delta$-RNN cells only require 6 trainable parameters, and though in matrix-vector they can be shown to be more efficient than most modern-day gated cell types, from a single neuronal cell point-of-view that are equal in terms of scalar parameter complexity as the UGRNN and MGU. From a functional perspective, the $\Delta$-RNN is more nonlinear than both the UGRNN and MGU given that it can require up to three activation functions, i.e., an inner activation ($tanh$) for the state proposal function, an outer activation ($tanh$) for the state mixing function, and an activation ($\sigma$) for the sigmoidal latch.

### 10.2.3  Temporal Skip Connections

As memory cells have proven to be quite useful in capturing some longer term
dependencies in sequential data, ANN research has largely chosen to focus on them
almost exclusively. However, there exists a great deal of classical work that has
shown the utility of *deep recurrent* or *temporal skip connections*, i.e., recurrent
connections that explicitly go back farther in time than the previous time step.
Some of these prior efforts even suggest that fairly deep recurrent connections yield
the best performance [40, 41], and even showing these kinds of connections can
even yield RNN predictive models that even outperform those that use memory cell
architectures instead [41]. Results from this work provide evidence that, at least in
the case of time series data prediction, deep recurrent connections could be a major
overlooked factor that might significantly improve the forecasting ability of RNNs.

Lin et al. investigated locally recurrent and globally recurrent RNNs as well as
NARX neural networks with increasing *embedded memory orders* (EMO), which
involved recurrent connections to nodes or layers further back in time [36, 37]. For
example, a network with an EMO of one would have its units connected to time step
$t - 1$, units with EMO of two would be connected to time $t - 1$ and $t - 2$, units
with EMO of three would be connected to $t - 1$, $t - 2$, and $t - 3$. The globally
recurrent networks had their hidden layers fully connected back in time according
to their EMO, and, due to this complexity, were only tested up to an EMO of 3.
The locally recurrent RNNs and NARX required fewer recurrent synapses and thus
were tested up to an EMO of 6. These networks were tested on a latching problem,
where a sequence's class only depends on the first few time steps (or sub-sequence
of patterns) of which are then followed by uniform noise (vectors). In addition,
these models were tested on a grammatical inference problem where the sequence
class was only dependent on an input symbol that occurred at a predetermined
time $t$. Increasing the EMO resulted in significant performance improvements for
both test problems. Additional work went on to show that the order of a NARX
network plays a critical role in determining well an RNN generalizes, allowing to
learn complex temporal functions as the EMO approaches the order of the target,
unknown recursive system/data generating process [38, 39].

El Hihi and Bengio also investigated RNNs with recurrent connections that
spanned multiple time steps in order to ascertain if their inclusion improved
an RNN's ability to extract long-term sequential data dependencies [48]. Their
hierarchical RNNs consisted of neurons with connections that looped back at
different time-scales (up to six time steps in the past). Unlike the work of Lin
et al., where nodes and/or layers were connected recurrently to all previous time
steps (up to their embedded memory order), in this work, each neuron's loop-back
connection was to a specific step in time—in other words, a deep recurrent synaptic
connection as focused on in this chapter. Similarly, this work uncovered significant
improvements in network performance when deeper recurrent connections were
employed. According to their experimental results, RNN models that only used

recurrence that explicitly spanned one step in time were not be able to generalize nearly as well those that spanned $K$ steps.

Kalini and Sagiroglu extended the embedded memory NARX RNNs developed by Lin et al. (described above) by combining them with Elman recurrent synapses (which have a fully connected set of recurrent connections between hidden layers at subsequent time steps) [49]. This work utilized EMOs similar to the order of the linear and nonlinear systems that they were learning to recognize, i.e., up to an EMO of 3. By adding embedded memory to the Elman networks, they found identification performance increased when attempting to recognize these systems, improving model training as well as memory storage capacity especially when compared to four other ANNs (which were variations of Elman and Jordan RNNs).

Diaconescu utilized embedded memory NARX RNNs for modeling chaotic time series generated by the chaotic Mackney-Glass, Fractal Weierstrass, and BET-index benchmark time series processes [40]. This work also showed that the EMO (referred to as a time window in this work) of the NARX model was crucial—best results were found in EMO range of 12–30. Furthermore, sparser networks with fewer neurons tended to perform better for these time series prediction problems. Lastly, they noted challenges in training these RNNs due to the wide variance of the backpropagation (backprop) based estimator used to compute gradients (random restarts were also used) as well as due to vanishing and exploding gradients [50].

Chen and Chaudhari later propose a segmented-memory recurrent neural network (SMRNN) [41], which utilizes a two layer recurrent structure which first passes input symbols to a symbol layer, and then connects the symbol layers to a segmentation layer. Both the symbol and segmentation layers have recurrent connections to themselves. The first $1..d$ symbols are passed to the first segment (where $d$ is the segment width), after which the next $d + 1..2d$ symbols are passed to the next segment, and so on and so forth. The state of the first hidden layer (the symbol layer) is updated after each symbol, while the state of the second hidden layer (the segmentation) layer is only updated after every $d$ symbols—in effect, this creates time-scales of different speeds. The SMRNN was compared to both Elman RNNs and LSTM RNNs on the latching problem (as described above), a two-sequence problem which classified input sequences into two classes, and a protein secondary structure (PSS) problem, and was shown to adapt/learn faster while achieving higher accuracy. This work showed that intervals $10 <= d <= 50$ provided the best results on this data, given that a lower $d$ required more computation each iteration, i.e., the segmentation was used too frequently, slowing convergence. In contrast, at higher values of $d$ the model approximated to a conventional RNN, i.e., the segmentation layer was not used at all. In this work, the segment interval $d$ operates quite similarly to a deep recurrent connection where information is efficiently passed from past states forward along the unrolled network. SMRNNs were originally trained in the original study using a variation of real-time recurrent learning (RTRL) [51] called extended RTRL (eRTRL) algorithm. The eRTRL procedure was later replaced with a variation of backprop through time (BPTT) known as extended BPTT (eBPTT) by Gläge et al. , yielding further improvements to model generalization and information latching over longer periods of time [52].

ElSaid et al. later utilized a combination of embedded memory order with LSTM cells to craft RNNs that predicted engine vibration from time series data gathered from aircraft flight data recorders [42]. This dataset was particularly challenging in that data included engine vibration events which occurred in sharp spikes as compared to prior data, and 26 correlated data sequences were used as inputs. This work investigated a number of architectures and found that a two-level architecture with an EMO/time window of 10 was able to provide good predictions of engine vibration up to 20 s in the future. This architecture was quite similar to the SMRNNs proposed by Chen and Chaudhari, except that each neuron was an LSTM cell and the model utilized multiple input (time) series. This work was also interesting in that, unlike most other time series research studies, it investigated an RNN's ability to forecast for horizons greater than 1 (a single time step into the future).

### 10.2.4   Evolving Recurrent Neural Networks

While neuro-evolution (applying evolutionary processes to the automatic development of ANNs) has been well applied to feedforward and even convolutional architectures for tasks involving static inputs [14–22, 34], far less effort has been put into exploring the evolution of recurrent memory structures that operate with complex, temporal data sequences. The forms and structures of optimal models that could be uncovered by neuro-evolution are largely unknown, a primary motivation behind the work that composes this chapter.

Several methods for evolving ANN topologies, along with synaptic weights value, have been searched and deployed, with NeuroEvolution of Augmenting Topologies (NEAT) [21] being perhaps one of the most well-known algorithms. The EXAMM metaheuristic differs quite a bit from NEAT in that it includes more advanced node-level mutations (see Sect. 10.3.1.2), utilizes Lamarckian weight initialization (see Sect. 10.3.2), and integrates BPTT to evolve candidate model weights locally (as opposed to using a genetic strategy to adjust the weights). Notably, a hallmark of EXAMM is its focus on large-scale concurrency in mind, i.e., an asynchronous steady state approach is utilized that allows it to naturally scale to potentially *millions of compute nodes* [53].

Other related work by Rawal and Miikkulainen has investigated an information maximization objective [23] strategy for evolving RNNs. This strategy essentially utilizes NEAT with LSTM cells instead of regular neurons. While powerful, EXAMM offers a stronger potential for a much more in-depth study of RNN model composition given that it already selects both simple neurons as well a larger and easily-extensible library of different cell structures beyond exclusively LSTM units. Rawal and Miikkulainen have also utilized a tree-based encoding [24] to evolve recurrent cellular structures within fixed architectures built from layers of the evolved cell types. Combining this evolution of cell structure along with the overall RNN structure adjustment of EXAMM we view as interesting future work.

Ant colony optimization (ACO) has also been investigated as a way to select which connections should be utilized in RNNs and LSTM RNNs by Desell and ElSaid [25–27]. In particular, this ACO approach was shown to reduce the number of trainable connections in half while providing a significant improvement in the prediction of engine vibration [26]. However, this approach only operated within a fixed RNN architecture and could not evolve an overall RNN structure. Very recently, the ant swarm neuro-evolution metaheuristic was proposed [54], combining a novel nature-inspired optimization procedure that formalized ant colony role-specialization with key elements of neuro-evolution. Notably, this work provided empirical data that corroborated the classical results (described earlier) that demonstrated the value of utilizing deep recurrent connectivity patterns to model long sequences of patterns.

## 10.3 Evolutionary eXploration of Augmenting Memory Models

The EXAMM algorithm presented in this work expands on two earlier algorithms: Evolutionary eXploration of Augmenting LSTM Topologies (EXALT) [25] which can evolve RNNs with either simple neurons or LSTM cells and Evolutionary eXploration of Augmenting Convolutional Topologies (EXACT) which evolves convolutional neural networks for image classification [28, 55]. It further refines EXALT's mutation operations to reduce the need for various hyperparameters by using statistical information extracted from parental RNN genomes. Additionally, EXALT only uses a single steady state population, while EXAMM, on the other hand, expands on this to use multiple island-based populations, which have been shown by Alba and Tomassini to greatly improve performance of distributed evolutionary algorithms, providing potentially even a superlinear speedup [56].

Figure 10.2 provides a high-level view of EXAMM's asynchronous operation. A master process maintains the populations for each island and generates new RNN candidate models from the islands in a round robin manner. Candidate models are locally trained upon request by workers. When a worker completes training an RNN, the RNN is inserted into the island it was generated from if and only if its fitness (mean squared error on the validation data) is better than the worst fitness score in the island. The insertion of this RNN is then followed by removal of the worst RNN in the island. This asynchrony is particularly important as the generated RNNs will have different architectures, each taking a different amount of time to train. It allows the workers to complete the training of the generated RNNs at whatever speed they can, yielding an algorithm that is naturally load balanced. Unlike synchronous parallel evolutionary strategies, EXAMM easily scales up to the number of available processors, allowing population sizes that are independent of processor availability.

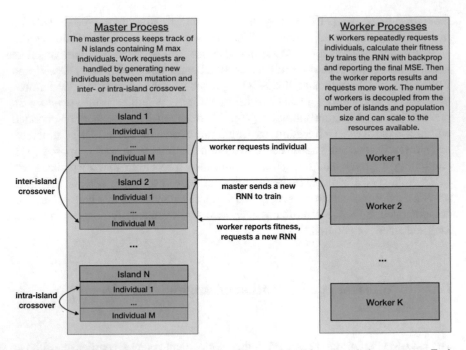

**Fig. 10.2** EXAMM is designed using an asynchronous island-based evolution strategy. Each worker process requests and trains an RNN independently, reports its fitness (mean squared error (MSE) on validation data after training), and requests new work as necessary. The master process sends new RNNs generated by mutation (intra- or inter-island crossover), created from islands in a round robin fashion in order to fulfill work requests. New RNNs are inserted into their island populations if they are better than the worst performing RNN in the population (which is then removed). The number of workers is completely independent of the number islands and their population size

The EXAMM codebase has a multithreaded implementation for multicore CPUs as well as an MPI [57] implementation that allows EXAMM to operate using high performance computing resources.

## 10.3.1 Mutation and Recombination Operations

RNNs are evolved with both edge-level operations, as is done in NEAT, and with novel high level node mutations, as originally proposed in EXALT and EXACT. Whereas NEAT only requires innovation numbers for new edges, EXAMM requires innovation numbers for both new nodes, new edges, and new recurrent edges. The master process keeps track of all node, edge, and recurrent edge innovations made, which are required to perform the crossover operation in linear time without a graph matching algorithm. Figures 10.3, 10.4 10.5, 10.6, 10.7, 10.8, 10.9, 10.10, 10.11

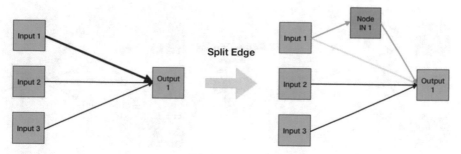

**Fig. 10.3  Split Edge:** The edge between Input 1 and Output 1 is selected to be split. A new node with innovation number (IN) 1 is created

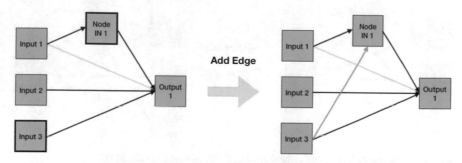

**Fig. 10.4  Add Edge:** Input 3 and Node IN 1 are selected to have an edge between them created and added

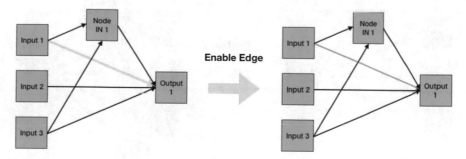

**Fig. 10.5  Enable Edge:** The edge between Input 3 and Output 1 is enabled

and 10.12 display a visual walkthrough of all the mutation operations used by EXAMM. Figure 10.13 provides a visual example of the crossover operation itself. Nodes and edges selected for modification are highlighted (new elements to the RNN are shown in green). Edge innovation numbers are not shown for clarity. Enabled edges are in black while disabled edges are in gray.

It should be noted that for the operations described in the next section, whenever an edge is added, unless otherwise specified, it is probabilistically selected to be a recurrent connection with the following recurrent probability: $p = \frac{n_{re}}{n_{ff}+n_{re}}$, where

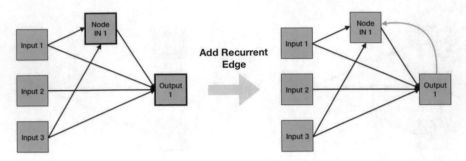

**Fig. 10.6 Add Recurrent Edge:** A recurrent edge is added between Output 1 and Node IN 1

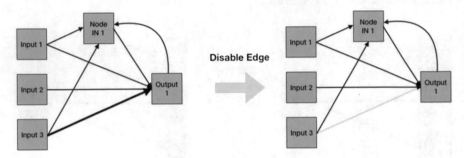

**Fig. 10.7 Disable Edge:** The edge between Input 3 and Output 1 is disabled

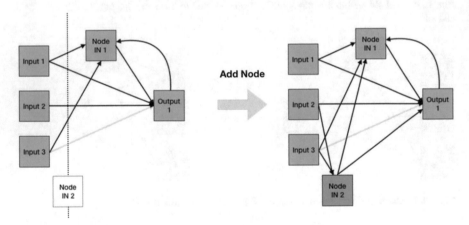

**Fig. 10.8 Add Node:** A node with IN 2 is selected to be added at a depth which is between the inputs and Node IN 1. Edges are randomly added to Input 2 and 3, and Node IN 1 and Output 1

$n_{re}$ is the number of enabled recurrent edges and $n_{ff}$ is the number of enabled feedforward edges in the parent RNN. A recurrent connection will span a randomly selected number of time steps with the bound specified as a search parameter (in this work, we allow between 1 and 10 time steps), allowing for recurrent connections of variable time spans. Any newly created node is selected uniformly at random as

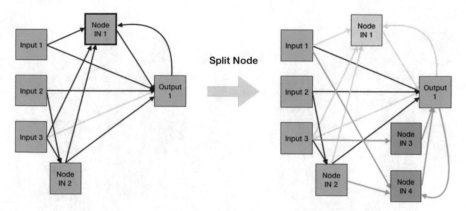

**Fig. 10.9  Split Node:** Node IN 1 is selected to be split. It is disabled along with its input/output edges. The node is split into Nodes IN 3 and 4, which get half of the inputs. Both have an output edge to Output 1 since there was only one output from Node IN 1

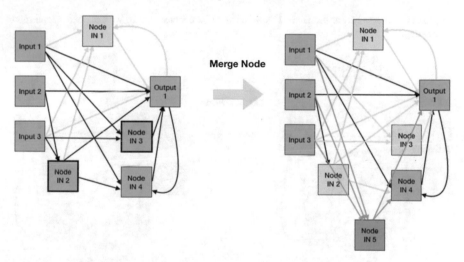

**Fig. 10.10  Merge Node:** Nodes IN 2 and 3 are selected for a merger (note that input/output edges are disabled). Node IN 5 is created with edges to all of their inputs/outputs

a simple neuron or from the memory cell types specified by the EXAMM input parameters.

### 10.3.1.1  Edge Mutations

- **Split Edge** (Fig. 10.3): This operation selects an enabled edge at random and disables it. It creates a new node and two new edges, and connects the input node of the split edge to the new node, and the new node to the output node of the split

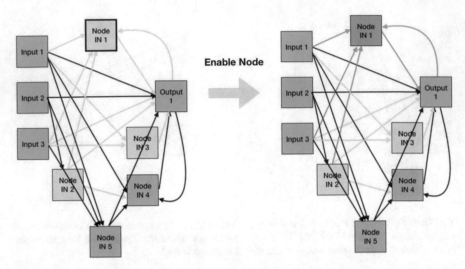

**Fig. 10.11 Enable Node:** Node IN 1 is selected to be enabled, along with all of its input and output edges

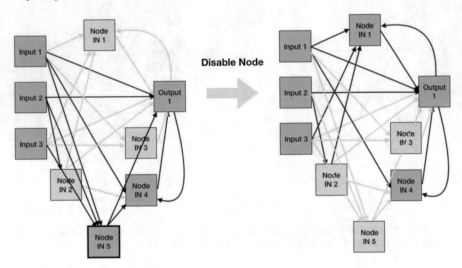

**Fig. 10.12 Disable Node:** Node IN 5 is selected to be disabled, along with all of its input and output edges

edge. The depth of this node is the average of the input and output node of the split edge. If the split edge was recurrent, the new edges will also be recurrent (with the same time skip); otherwise they will be feedforward.

- **Add Edge** (Fig. 10.4): This operation selects two nodes $n_1$ and $n_2$ within the RNN genome at random, such that $depth_{n_1} < depth_{n_2}$ and such that there is not already an edge between those nodes in this RNN genome. Then an edge is added from $n_1$ to $n_2$.

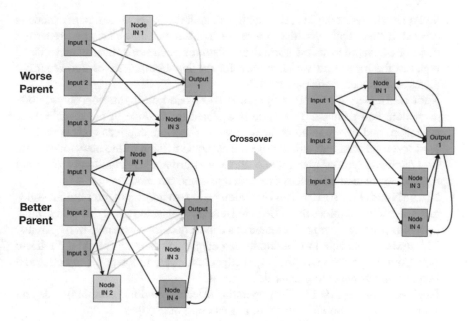

**Fig. 10.13  Crossover:** Two parent RNNs are selected, either from the same island (for intra-island crossover) or between islands (for inter-island crossover). All reachable nodes and edges are combined within the child RNN

- **Enable Edge** (Fig. 10.5): If there are any disabled edges or recurrent edges in the RNN genome, this operation selects a disabled edge or recurrent edge at random and enables it.
- **Add Recurrent Edge** (Fig. 10.6): This operation selects two nodes $n_1$ and $n_2$ within the RNN genome at random and then adds a recurrent edge from $n_1$ to $n_2$ (selecting a time span as described before). The same two nodes can be connected with multiple recurrent connections, each spanning different lengths of time. However, the edge will not create a duplicate recurrent connection with the same time span.
- **Disable Edge** (Fig. 10.7): This operation randomly selects an enabled edge or recurrent edge in an RNN genome and disables it so that it is not used. The edge remains in the genome. Given that the *disable edge* operation can potentially make an output node unreachable, after all mutation operations have been performed to generate a child RNN genome, if any output node is unreachable, then that RNN genome is discarded without training.

### 10.3.1.2   Node Mutations

- **Add Node** (Fig. 10.8): This operation selects a random depth between 0 and 1, noninclusive. Given that the input nodes are always at depth 0 and the output nodes are always at depth 1, this depth will split the RNN in two. A new node is created at that depth, and the number of input and output edges and recurrent edges are generated using normal distributions each with mean and variance equal to the mean and variance used for the input/output edges and recurrent edges of all nodes in the parent RNN.

- **Split Node** (Fig. 10.9): This operation takes one non-input, non-output node at random and splits it. This node is disabled (as is done in the disable node operation) and two new nodes are created at the same depth as their parent. At least one input and one output edge are assigned to each of the new nodes, both of a duplicate type of the parent. Other edges are assigned randomly, ensuring that the newly created nodes have both inputs and outputs.
- **Merge Node** (Fig. 10.10): This operation takes two non-input, non-output nodes at random and combines them. Selected nodes are disabled (as in the disable node operation) and a new node is created at a depth equal to the average of its parents. This node is connected to the inputs and outputs of its parents with a duplicate type from the parent given that input edges connected to lower depth nodes and output edges connect to greater depth nodes.
- **Enable Node** (Fig. 10.11): This operation selects a random (disabled) node and enables it along with all of its incoming and outgoing edges.
- **Disable Node** (Fig. 10.12): This operation selects a random non-output node and disables it along with all of its incoming and outgoing edges. Note that this allows for input nodes to be dropped, which can prove to be useful when it is not previously known which input parameters are correlated to the output.

### 10.3.1.3    Other Operations

- **Crossover** (Fig. 10.13): This operation creates a child RNN using all reachable nodes and edges from two parents. A node or edge is reachable if there is a path of enabled nodes and edges from an input node to itself, as well as a path of enabled nodes and edges from itself to an output node. In other words, a node or edge is reachable if and only if it actually affects the RNN. The crossover operator can be performed either within an island (*intra-island*) or between islands (*inter-island*). Inter-island crossover selects a random parent in the target island as well as the best RNN from the other islands.
- **Clone:** This operation creates a copy of the parent genome (initializing weights with the same values as the parent). This allows a particular genome to continue training in cases where further training may be more beneficial than simply performing a mutation or crossover operation.

## 10.3.2 Lamarckian Weight Initialization

For RNNs generated during population initialization, weights are initialized uniformly at random between $-0.5$ and $0.5$. Biases and weights for new nodes and edges are initialized from a normal distribution based on the average, $\mu$, and variance, $\sigma^2$, of the parents' weights. However, RNNs generated through mutation or crossover re-use parental weights. This allows RNNs to train from where the parents left off, i.e., a *"Lamarckian" weight initialization* scheme.

During crossover, in the case where an edge or node exists in both parents, the child's weights are generated by recombining the parents' weights. Given a random number $-0.5 <= r <= 1.5$, a child's weight $w_c$ is set to $w_c = r(w_{p2}-w_{p1})+w_{p1}$, where $w_{p1}$ is the weight from the more fit parent, and $w_{p2}$ is the weight taken from the less fit parent. This allows a child RNN's weights to be set along a gradient calculated from the weights of its two parents.

This weight initialization strategy is particularly important since, within this scheme, newly generated RNNs do not need to be completely retrained from scratch. In fact, the RNNs only need to be trained for a few epochs to investigate the benefits of newly added structures. In this work, the generated RNNs are only trained for 10 epochs (see Sect. 10.5.2). In contrast, training a static, hand-crafted RNN structure from scratch often requires hundreds or even thousands of epochs to achieve good generalization.

## 10.4  Datasets

This work utilizes two datasets to benchmark the memory cells and RNNs evolved by EXAMM. The first comes from a selection of 10 flights worth of data from the National General Aviation Flight Information Database (NGAFID).[2] The other comes from data selected from 12 burners of a coal-fired power plant (which has requested to remain anonymous). Both datasets are multivariate (26 and 12 parameters, respectively), non-seasonal, and the parameter recordings are not independent. Furthermore, the series samples are very long—the aviation time series ranges from 1 to 3 h worth of per-second data, while the power plant data consists of 10 days worth of per-minute readings. These datasets are provided openly through the EXAMM GitHub repository,[3] in part for reproducibility, but also to provide a valuable resource to the field. To the authors' knowledge, real world time series datasets of this size and at this scale are not freely available.

---

[2] https://ngafid.org.

[3] https://github.com/travisdesell/exact.

### 10.4.1  Aviation Flight Recorder Data

With permission, data from 10 flights was extracted from the NGAFID. Each of the 10 flight data files last over an hour and consists of per-second data recordings from 26 parameters:

1. Altitude Above Ground Level (Alt-AGL)
2. Engine 1 Cylinder Head Temperature 1 (E1 CHT1)
3. Engine 1 Cylinder Head Temperature 2 (E1 CHT2)
4. Engine 1 Cylinder Head Temperature 3 (E1 CHT3)
5. Engine 1 Cylinder Head Temperature 4 (E1 CHT4)
6. Engine 1 Exhaust Gas Temperature 1 (E1 EGT1)
7. Engine 1 Exhaust Gas Temperature 2 (E1 EGT2)
8. Engine 1 Exhaust Gas Temperature 3 (E1 EGT3)
9. Engine 1 Exhaust Gas Temperature 4 (E1 EGT4)
10. Engine 1 Oil Pressure (E1 OilP)
11. Engine 1 Oil Temperature (E1 OilT)
12. Engine 1 Rotations Per minute (E1 RPM)
13. Fuel Quantity Left (FQtyL)
14. Fuel Quantity Right (FQtyR)
15. GndSpd - Ground Speed (GndSpd)
16. Indicated Air Speed (IAS)
17. Lateral Acceleration (LatAc)
18. Normal Acceleration (NormAc)
19. Outside Air Temperature (OAT)
20. Pitch
21. Roll
22. True Airspeed (TAS)
23. Voltage 1 (volt1)
24. Voltage 2 (volt2)
25. Vertical Speed (VSpd)
26. Vertical Speed Gs (VSpdG)

These files had identifying information (fleet identifier, tail number, date and time, as well as latitude/longitude coordinates) which was removed in order to protect the identity of the pilots. The data is provided unnormalized.

*RPM* and *pitch* were selected as prediction parameters from the aviation data since RPM is a product of engine activity, with other engine-related parameters being correlated. Pitch itself is directly influenced by pilot controls. As a result, both of these target variables are particularly challenging to predict. Figure 10.14 provides an example of the RPM and pitch time series from Flight 8 of this dataset. In addition, the pitch parameter represents how many degrees above or below horizontal the aircraft is angled. As a result, the parameter typically remains steady around a value of 0; however, it increases or decreases depending on whether or not the aircraft is angled to fly upward or downward, based on pilot controls and external conditions. On the other hand, RPM will mostly vary between an idling speed, i.e., if the plane is on the ground, and a flight speed, with some variation between takeoff and landing. Since the majority of the flights in NGAFID (and, by extension, all of the flights in the provided sample) are student training flights, multiple practice takeoffs and landings can be found in the data. This results in two different types of time series, both of which are dependent on the other flight parameters but each sporting highly different characteristics—creating excellent time series benchmarks for evaluating RNNs.

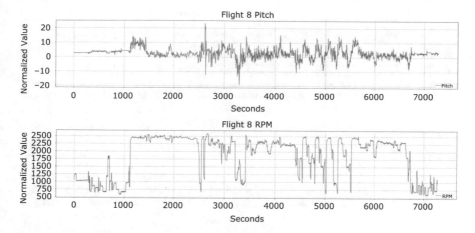

**Fig. 10.14** Example parameters of Flight 8 from the NGAFID dataset

## 10.4.2   Coal-Fired Power Plant Data

This dataset consists of 10 days of per-minute data readings extracted from 12 of the plant's burners. Each of these 12 data files has 12 parameters of time series data:

1. Conditioner Inlet Temp
2. Conditioner Outlet Temp
3. Coal Feeder Rate
4. Primary Air Flow
5. Primary Air Split
6. System Secondary Air Flow Total
7. Secondary Air Flow
8. Secondary Air Split
9. Tertiary Air Split
10. Total Combined Air Flow
11. Supplementary Fuel Flow
12. Main Flame Intensity

In order to protect the confidentiality of the power plant which provided the data (along with any other sensitive data elements) all identifying data has been scrubbed from the datasets (such as dates, times, locations, and facility names). This data was normalized to lie in the range [0, 1] which serves as one more additional data anonymization step. So while the data cannot be reverse engineered to identify the originating power plant or actual parameter values—it still provide an extremely valuable test platform for times series data prediction given that it consists of "in-the-wild" data from a highly complex system composed of interdependent data streams.

For the coal plant data, *main flame intensity* and *supplementary fuel flow* were selected as parameters of interest. Figure 10.15 provides examples of these two parameters from Burner # 2 found in the dataset. Main flame intensity is mostly a product of conditions within the burner and parameters related to coal quality which causes it to vary over time. However, sometimes planned outages occur or conditions in the burner deteriorate so badly that the burner is temporarily shut down. In these cases, sharp spikes occur during the shutdown, which last for an

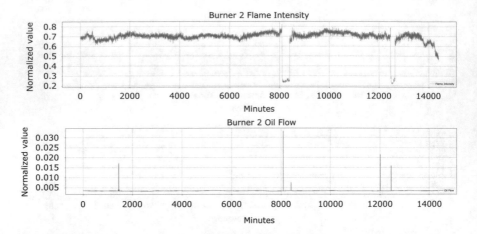

**Fig. 10.15** Example parameters of Burner 2 from the coal dataset

unspecified period of time before the burner turns back on again (and the parameter value sharply increases). The burners can also potentially operate at different output levels, depending on power generation needs. As a result, step-wise behavior is observed.

On the other hand, supplementary fuel flow remains fairly constant. Nonetheless, it yields sudden and drastic spikes in response to decisions made by plant operators. When conditions in the burners become poor due to coal quality or other effects, the operator may need to provide supplementary fuel to prevent the burner from going into shutdown. Of particular interest is to see if an RNN can successfully learn to detect these spikes given the conditions of the other parameters. Similar to the key parameters (RPM and pitch) selected in the NGAFID data, main flame intensity itself is mostly a product of conditions within the (coal) burner, while supplementary fuel flow is more directly controlled by human operators. Despite these similarities, the characteristics of these time series are different from each other as well as from the NGAFID flight data, providing additional, unique benchmark prediction challenges.

## 10.5 Results

### 10.5.1 Experiments

The first set of (5) experiments only permitted the use of a single memory cell type, i.e., exclusively Δ-RNN, GRU, LSTM, MGU, or UGRNN (one experiment per type), and no simple neurons. All of these experiments only allowed the generation of feedforward connections between cell nodes (these experiments were denoted as *delta*, *gru*, *lstm*, *mgu*, or *ugrnn*). The second set of (2) experiments were conducted where the first one only permitted the use of simple neurons and feedforward connections (denoted as *simple*), while the second permitted EXAMM to make use of feedforward connections and simple neurons as well as the choice of any memory cell type (denoted as *all*). The next set of experiments (5) were identical to the first set with the key exception that EXAMM could choose either between simple neurons and one specified specific memory cell type (these experiments are appended with a *+simple*, i.e., *lstm+simple*). The final set of (12) experiments consisted of taking the setting of each of the prior 12 (5 + 2 + 5) runs and re-ran them but with the modification that EXAMM was permitted to generate deep recurrent connections of varying time delays (these runs are appended with a *+rec*). For example, *lstm* from the previous set of experiments is *lstm+rec* in this set of experiments, and similarly *lstm+simple* in the previous set of experiments is *lstm+simple+rec* in this set of experiments.

This full set of (24) experiments was conducted for each of the four prediction parameters, i.e., RPM, pitch, main flame intensity, and supplementary fuel flow. $K$-fold cross validation was carried out for each prediction parameter, with a fold size of 2. This resulted in 5 folds for the NGAFID data (as it had 10 flight data files), and 6 folds for the coal plant data (as it has 12 burner data files). Each fold and EXAMM experiment was repeated 10 times. In total, each of the 24 EXAMM experiments was conducted 220 times (50 times each for the NGAFID parameter $k$-fold validation and 60 times each for the coal data parameter $k$-fold validation), for a grand total of 5280 separate EXAMM experiments/simulations.

## 10.5.2   EXAMM and Backpropagation Hyperparameters

All RNNs were locally trained with backpropagation through time (BPTT) [58] and stochastic gradient descent (SGD) using the same hyperparameters. SGD was run with a learning rate of $\eta = 0.001$, utilizing Nesterov momentum with $mu = 0.9$. No dropout regularization was used since, in prior work, it has been shown to result in worse performance when training RNNs for time series prediction [26]. For the LSTM cells that EXAMM could make use of, the forget gate bias had a value of 1.0 added to it since [59] has shown that doing so improves (LSTM) training time significantly. Otherwise, RNN weights were initialized via EXAMM's Lamarckian strategy.

To control for exploding and vanishing gradients, we apply re-scaling to the full gradient of the RNN, **g**, which is one single vector of all the partial derivatives of the cost function with respect to the individual weights (in terms of a standard RNN, this amounts to flattening and concatenating all of the individual derivative matrices into one single gradient vector). Re-scaling was done in this way due to the unstructured/unlayered RNNs evolved by EXAMM. Computing the stabilized gradient proceeds formally as follows:

$$
\mathbf{g} = \begin{cases} \mathbf{g} * \frac{t_h}{\|\mathbf{g}\|_2}, & \text{if } \|\mathbf{g}\|_2 > t_h \\ \mathbf{g} * \frac{t_l}{\|\mathbf{g}\|_2}, & \text{if } \|\mathbf{g}\|_2 < t_l \\ \mathbf{g} & \text{otherwise} \end{cases}
$$

noting that $\| \cdot \|_2$ is the Euclidean norm operator. $t_h$ is the (high) threshold for preventing diverging gradient values, while $t_l$ is the (low) threshold for preventing shrinking gradient values. In essence, the above formula is composed of two types of gradient re-scaling. The first part re-projects the gradient to a unit Gaussian ball ("gradient clipping" as prescribed by Pascanu et al. [60]) when the gradient norm exceeds a threshold $t_h = 1.0$. The second part, on the other hand, is a trick we propose called "gradient boosting," where, when the norm of the gradient falls below a threshold $t_l = 0.05$, we up-scale it by the factor $\frac{t_l}{\|\mathbf{g}\|_2}$.

For EXAMM, each neuro-evolution run consisted of 10 islands, each with a population size of 5. New RNNs were generated via intra-island crossover (at a rate of 20%), mutation at a rate 70%, and inter-island crossover at 10% rate. All of the EXAMM's mutation operations (except for the *split edge* operator) were utilized, each chosen with a uniform 10% chance. The experiments labeled *all* were able to select any type of memory cell or simple (Elman) neurons at random, each with an equal probability. Each EXAMM run generated 2000 RNNs, with each RNN being trained locally (using the BPTT settings described above) for 10 epochs. Recurrent connections that could span a time skip between 1 and 10 could be chosen (selected uniformly at random). These runs were performed utilizing 20 processors in parallel, and, on average, required approximately 0.5 compute hours. In total, the results we report come from training 10,560,000 RNNs which required ~52,800 CPU hours of compute time.

### 10.5.3 Experimental Results

Figure 10.16 shows the range of the fitness values of the best found RNNs across all of the EXAMM experiments. This combines the results from all folds and all trial repeats—each box in the box plots represents 110 different fitness values. The box plots are ordered according to mean fitness (calculated as mean absolute error, or MAE) of the RNNs for that experiment/setting (across all folds), with the top being the highest average MAE, i.e., the worst performing simulation setting, and the bottom containing the lowest average MAE, i.e., the best performing setting. Means are represented by green triangles and medians by orange bars. Run type names with deep recurrent connections are highlighted in red.

Experimental runs were also analyzed by calculating the mean and standard deviation of all best evolved fitness scores from each repeated experiment across each fold. This was done since each fold of the test data had a different range of

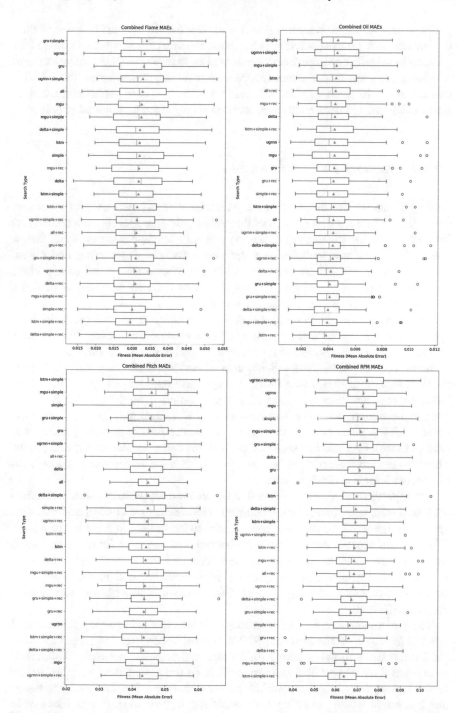

**Fig. 10.16** Consolidated range of fitness (mean absolute error) of the best found neural networks for the flame dataset, fuel flow, pitch, and RPM prediction parameters for each of the 24 experiments across all 6 folds, with 10 repeats per fold. Run types are ordered top-down by mean

potential best results. It was then possible to rank the simulations in terms of their deviation from the mean (providing a less biased metric of improvement). Table 10.1 presents how well each experiment performed as an average of how many standard deviations each was from the mean in their average case performance. Table 10.2 is constructed in the same way but is instead based on best case performance. Search types which utilized deep recurrent connections (+*rec*) are highlighted in bold.

### 10.5.4   Memory Cell Performance

Table 10.3 shows the frequency of a particular memory cell experiment/setting appearing in the three best (Top 3) or three worst (Bottom 3) slots (in a ranked list) for each prediction parameter for the experiment's average and best RNN fitness score. The *simple* row includes only the *simple* and *simple+rec* runs and the *all* row includes the *all* and *all+rec* runs, while the other memory cell rows include the +*simple* and +*rec* versions (e.g., the *delta* row includes occurrences of *delta*, *delta+simple*, *delta+rec*, and *delta+simple+rec*).

Based on these count results, the Δ-RNN memory cells performed the best, appearing in the top 3 for the average and best cases 4 times each. Furthermore, it did not appear in the bottom 3 for the average and best cases *at all*, making these particular memory cells more reliable than others. MGU memory cells performed the next best, appearing twice in the average case and 3 times in the best case. However, they also showed up 4 times in the bottom, 3 on the average case, and once in the bottom 3 for the best case networks. Interestingly enough, while the popular LSTM memory cells showed up frequently for the average performance case (3 times), they did not show up at all in the top 3 for the best found RNNs. They also occurred once in the bottom three for average performance and twice in the bottom 3 for best performance.

The experimental results indicate that simple neurons perform rather well when combined with deep recurrent connections. The *simple+rec* configuration showed up once in the top 3 for the average case, and twice in the top 3 for the best case. When simple neurons did appear in the bottom 3, it was only for the experiments when no deep recurrent connections were permitted. As a result, aside from the Δ and MGU memory cells, *simple+rec* performed better than the other more complicated memory cells, e.g., LSTM, GRU, and was, furthermore, more reliable than the MGU cells.

The performance of the GRU memory cells was intriguing—they showed up 3 times in the top 3 for the best case RNNs, 0 times for the top 3 average case runs, but 2 and 5 times in the bottom 3 for average and best case networks. This seems to indicate that, while GRU memory cells have the potential to find performant networks, they are *highly unreliable* for these time series datasets. We hypothesize that this might be due to either high sensitivity to initialization conditions or to unknown limitations in the way they gate/carry temporal information.

**Table 10.1** EXAMM experiments ranked by how their average fitness performed in how many standard deviations from the mean of all experiments

| (a) Flame Std Devs: Avg MAE | | (b) Fuel Flow Std Devs: Avg MAE | |
|---|---|---|---|
| Type | Devs from mean | Type | Devs from mean |
| gru+simple | 0.45466 | simple | 0.21337 |
| gru | 0.38537 | mgu+simple | 0.18593 |
| mgu+simple | 0.27395 | ugrnn+simple | 0.17120 |
| ugrnn+simple | 0.26383 | **all+rec** | 0.12609 |
| ugrnn | 0.24459 | lstm | 0.11196 |
| all | 0.22218 | delta | 0.08607 |
| mgu | 0.21171 | **lstm+simple+rec** | 0.07886 |
| lstm+simple | 0.05836 | **mgu+rec** | 0.07732 |
| delta+simple | 0.04819 | gru | 0.04823 |
| lstm | 0.03707 | ugrnn | 0.04216 |
| simple | 0.01420 | **gru+rec** | 0.03564 |
| delta | 0.00741 | **simple+rec** | 0.01922 |
| **ugrnn+simple+rec** | −0.04944 | lstm+simple | 0.01803 |
| **mgu+rec** | −0.05957 | mgu | 0.00344 |
| **lstm+rec** | −0.09531 | **ugrnn+simple+rec** | −0.00456 |
| **all+rec** | −0.10439 | all | −0.00649 |
| **delta+rec** | −0.13052 | **ugrnn+rec** | −0.02236 |
| **gru+rec** | −0.15598 | delta+simple | −0.03849 |
| **gru+simple+rec** | −0.16988 | **delta+rec** | −0.08420 |
| **ugrnn+rec** | −0.18138 | **gru+simple+rec** | −0.12883 |
| **mgu+simple+rec** | −0.20289 | gru+simple | −0.13948 |
| **simple+rec** | −0.30881 | **delta+simple+rec** | −0.22414 |
| **lstm+simple+rec** | −0.33808 | **lstm+rec** | −0.28345 |
| **delta+simple+rec** | −0.42524 | **mgu+simple+rec** | −0.28552 |

(continued)

**Table 10.1** (continued)

| (c) Pitch Std Devs: Avg MAE | | (d) RPM Std Devs: Avg MAE | |
|---|---|---|---|
| Type | Devs from mean | Type | Devs from mean |
| lstm+simple | 0.23909 | ugrnn+simple | 0.49582 |
| mgu+simple | 0.20029 | ugrnn | 0.32738 |
| ugrnn+simple | 0.14481 | mgu | 0.30650 |
| simple | 0.14432 | simple | 0.26625 |
| gru+simple | 0.13351 | mgu+simple | 0.23401 |
| gru | 0.13152 | gru+simple | 0.20177 |
| delta | 0.10322 | gru | 0.19246 |
| **all+rec** | 0.09201 | delta | 0.17695 |
| all | 0.06990 | all | 0.10472 |
| delta+simple | 0.05982 | lstm | 0.02485 |
| **simple+rec** | 0.03136 | lstm+simple | 0.01335 |
| **ugrnn+rec** | 0.00369 | delta+simple | 0.00523 |
| **lstm+rec** | $-0.03673$ | **ugrnn+simple+rec** | $-0.00490$ |
| lstm | $-0.03996$ | **lstm+rec** | $-0.02911$ |
| **delta+rec** | $-0.06630$ | **mgu+rec** | $-0.06871$ |
| **mgu+rec** | $-0.08360$ | **all+rec** | $-0.08791$ |
| **mgu+simple+rec** | $-0.08452$ | **ugrnn+rec** | $-0.09029$ |
| **gru+rec** | $-0.11290$ | **delta+simple+rec** | $-0.15970$ |
| **gru+simple+rec** | $-0.11699$ | **gru+simple+rec** | $-0.17044$ |
| ugrnn | $-0.13281$ | **simple+rec** | $-0.21730$ |
| **lstm+simple+rec** | $-0.14307$ | **gru+rec** | $-0.27053$ |
| **delta+simple+rec** | $-0.15425$ | **delta+rec** | $-0.36819$ |
| **ugrnn+simple+rec** | $-0.18647$ | **mgu+simple+rec** | $-0.40717$ |
| mgu | $-0.19593$ | **lstm+simple+rec** | $-0.47505$ |

Lower values indicate better performance

Lastly, UGRNN memory cells performed the worst overall. They only appeared once in the top 3 for the average case and not at all in the top 3 for the best case. At the same time they occurred 4 and 3 times in the bottom 3 for the average and best case performance rankings.

The *all* configurations did not show up at all in the top 3 or bottom 3, most likely due to the significant size of its particular search space. Given the additional option to select from a large pool of different memory cell types, the EXAMM neuro-evolution procedure might just simply require far more time to decide which cell types would yield better/top performing networks.

**Table 10.2** EXAMM experiments ranked by how their best found fitness scores performed with respect to how many standard deviations from the mean of all experiments they fall

| (a) Flame Std Devs: Best MAE | | (b) Fuel Flow Std Devs: Best MAE | |
|---|---|---|---|
| Type | Devs from mean | Type | Devs from mean |
| gru+simple | −1.02844 | **gru+rec** | −1.10116 |
| **mgu+rec** | −1.15701 | ugrnn+simple | −1.18567 |
| ugrnn+simple | −1.21079 | lstm+simple | −1.18625 |
| mgu | −1.24655 | **mgu+rec** | −1.18778 |
| mgu+simple | −1.26880 | **lstm+simple+rec** | −1.21500 |
| gru | −1.29390 | mgu+simple | −1.21509 |
| simple | −1.30901 | all | −1.22138 |
| lstm+simple | −1.35475 | gru+simple | −1.27796 |
| lstm | −1.35496 | **gru+simple+rec** | −1.29070 |
| delta+simple | −1.37473 | **simple+rec** | −1.29699 |
| **ugrnn+rec** | −1.42362 | simple | −1.30479 |
| ugrnn | −1.43371 | ugrnn | −1.30559 |
| delta | −1.48912 | **ugrnn+simple+rec** | −1.31366 |
| **mgu+simple+rec** | −1.55717 | delta | −1.33034 |
| **gru+simple+rec** | −1.58618 | **delta+rec** | −1.35481 |
| **lstm+simple+rec** | −1.63655 | **all+rec** | −1.37338 |
| **all+rec** | −1.64301 | lstm | −1.38003 |
| all | −1.66893 | delta+simple | −1.38368 |
| **lstm+rec** | −1.70057 | **lstm+rec** | −1.38510 |
| **ugrnn+simple+rec** | −1.71172 | ugrnn+rec | −1.42369 |
| **gru+rec** | −1.73098 | mgu | −1.45259 |
| **delta+rec** | −1.95685 | **mgu+simple+rec** | −1.50962 |
| **simple+rec** | −1.97756 | gru | −1.53812 |
| **delta+simple+rec** | −2.08205 | **delta+simple+rec** | −1.54667 |

(continued)

**Table 10.2** (continued)

| (c) Pitch Std Devs: Best MAE | |
|---|---|
| Type | Devs from mean |
| ugrnn+simple | −0.99073 |
| gru | −1.01889 |
| lstm+simple | −1.09707 |
| gru+simple | −1.10143 |
| delta | −1.19651 |
| lstm | −1.24966 |
| all | −1.25872 |
| **delta+rec** | −1.42943 |
| mgu+simple | −1.48976 |
| **all+rec** | −1.55755 |
| **ugrnn+rec** | −1.58235 |
| **mgu+rec** | −1.60397 |
| **lstm+rec** | −1.63888 |
| **ugrnn+simple+rec** | −1.64192 |
| mgu | −1.67690 |
| ugrnn | −1.70299 |
| **delta+simple+rec** | −1.77567 |
| delta+simple | −1.78042 |
| **gru+rec** | −1.81352 |
| **lstm+simple+rec** | −1.89858 |
| simple | −2.05128 |
| **mgu+simple+rec** | −2.09451 |
| **gru+simple+rec** | −2.09545 |
| **simple+rec** | −2.24764 |

| (d) RPM Std Devs: Best MAE | |
|---|---|
| Type | Devs from mean |
| gru | −0.94516 |
| simple | −0.99991 |
| gru+simple | −1.08121 |
| mgu | −1.17371 |
| ugrnn+simple | −1.19714 |
| **all+rec** | −1.34347 |
| ugrnn | −1.36917 |
| **ugrnn+simple+rec** | −1.44366 |
| **gru+simple+rec** | −1.49508 |
| mgu+simple | −1.49991 |
| lstm | −1.50167 |
| **delta+simple+rec** | −1.51271 |
| delta+simple | −1.51795 |
| **mgu+rec** | −1.52494 |
| delta | −1.57259 |
| lstm+simple | −1.64965 |
| all | −1.69526 |
| **lstm+simple+rec** | −1.71450 |
| **ugrnn+rec** | −1.72680 |
| **lstm+rec** | −1.74024 |
| **simple+rec** | −1.74335 |
| **gru+rec** | −1.88070 |
| **mgu+simple+rec** | −1.89718 |
| **delta+rec** | −2.05063 |

Lower values mean better performance

**Table 10.3** How often a memory cell type appeared in the top 3 or bottom 3 experiments in the best and average cases

| Memory cell | Top 3 | | Bottom 3 | |
|---|---|---|---|---|
| | Avg | Best | Avg | Best |
| all | 0 | 0 | 0 | 0 |
| simple | 1 | 2 | 1 | 1 |
| delta | 4 | 4 | 0 | 0 |
| gru | 0 | 3 | 2 | 5 |
| lstm | 3 | 0 | 1 | 2 |
| mgu | 2 | 3 | 4 | 1 |
| ugrnn | 1 | 0 | 4 | 3 |

**Table 10.4** Performance improvement (in std. dev. from the mean) for adding simple neurons

| Type | Dev for avg | Dev for best |
|------|-------------|--------------|
| delta | −0.07663 | −0.07420 |
| gru | −0.02369 | 0.04575 |
| lstm | −0.02973 | 0.02485 |
| mgu | −0.03463 | −0.18857 |
| ugrnn | 0.07991 | 0.15908 |
| overall | −0.02365 | −0.02018 |

**Table 10.5** Performance improvement (in std. dev. from the mean) for adding deep recurrent connections

| Type | Dev for avg | Dev for best |
|------|-------------|--------------|
| all | −0.09113 | −0.01828 |
| simple | −0.27842 | −0.40014 |
| delta | −0.25571 | −0.30079 |
| gru | −0.31534 | −0.43257 |
| lstm | −0.14463 | −0.24462 |
| mgu | −0.11507 | 0.01901 |
| ugrnn | −0.19291 | −0.08625 |
| overall | −0.19903 | −0.20909 |

## 10.5.5   Effects of Simple Neurons

Table 10.4 provides measurements for how the addition of simple neurons changed the performance of the varying memory cell types. In it, we show how many standard deviations from the mean the average case moved when averaging the differences of *mgu* to *mgu+simple* and *mgu+rec* to *mgu+simple+rec* (over all four prediction parameters). In the average case, adding simple neurons did appear to yield a modest improvement, improving deviations from the mean by −0.02 overall, and improving deviation from the mean for all memory cells except for UGRNNs. Adding simple neurons had a similar overall improvement for the best found RNNs; however, this incurred a much wider variance. Nonetheless, despite this variance, 2 of the 3 best found networks had *+simple* as an option, with a third being *simple+rec*. This seems to indicate that most memory cell types could either benefit by mixing or combining them with simple neurons.

## 10.5.6   Effects of Deep Recurrent Connections

Table 10.5 provides similar measurements for EXAMM settings that permitted the addition of deep recurrent edges to the varying memory cell types, as well as the *all* and *simple* runs. Compared to adding *+simple*, the *+rec* setting showed an order of magnitude difference, improving deviations from the mean by −0.2 overall. In addition, for each of the prediction parameters, the best found RNN utilized deep recurrent connections. Looking at the top 3 best and top 3 average case RNNs, 11

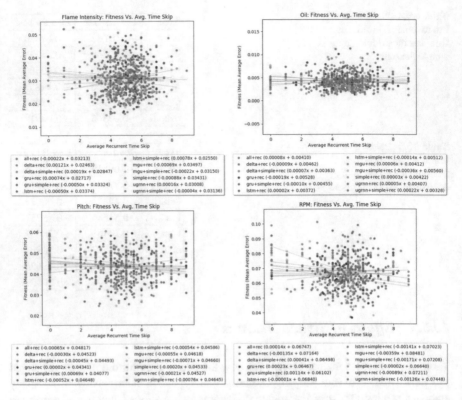

**Fig. 10.17** Scatter plots of the fitness of the best found RNNs for the flame, oil, pitch, and RPM datasets against the average time skip of their recurrent connections (with fit trendlines)

out of 12 utilized deep recurrent connections. Similarly, in the bottom 3 best, +*rec* occurs twice and does not appear at all in the bottom 3 average case run types. For the Flame and RPM parameters, on the average case, even the worst performing run type with +*rec* performs better than any of the simulations/experiments without it.

Figures 10.17 plots the average time skip of all recurrent connections (not counting the recurrent connections within the memory cells) in the best found RNN of each EXAMM run against their fitness. RNNs that had a 0 average (measurement) were those without any time skip recurrent connections. While the average number of time skip connections for all these RNNs were centered around 5.5 (which makes sense as the depth of the time skips ranged from 1 to 10, and were selected uniformly at random), many of the search types exhibit negative trendlines indicating that deeper time skips resulted in more accurate RNNs, with RPM prediction demonstrating particularly strong trendlines.

## 10.6   Discussion

The results presented in this chapter provide some significant insights to the domain of recurrent neural network-based time series data prediction. The main findings of this study include:

- *Deep Time Skip Connections:* The most significant improvements in recurrent neural network (RNN) performance were shown with networks that included deep time skip connections, and in some experiments were more important than the use of memory cells, i.e., the *simple+rec* search types performing quite strongly. For all four benchmark datasets, the best found RNNs included deep recurrent connections. Overall, adding deep recurrent connections to the evolutionary process resulted in large shifts of improvement in terms of the standard deviations from mean measurements. These results are particularly significant given that the common story told is that LSTM and other memory cells are the only effective means of capturing long-term time dependencies when their internal connections only go back a single time step.
- *Strong simple+rec Performance:* Another very interesting finding was that only using simple neurons and deep time skip connections without any memory cells (the *simple+rec* search type) could perform quite well, finding the best RNN in the case of the Pitch dataset, second best on the Flame dataset, and fourth best on the RPM dataset. This shows that, at least in some cases, it might be more important to effectively utilize deep temporal/recurrent connections than to use more complicated memory cells.
- *Strong Δ-RNN Memory Cell Performance:* While there is no free lunch in statistical learning, the newer Δ-RNN memory cell did stand out as performing better than the other memory cells. In three out of the four datasets, runs based on it found the best performing RNN, and for the average case performance, these runs made it into the top 3 search types across all four datasets. Furthermore, unlike the other memory cells search types, the Δ-RNN based searches did not perform in the bottom 3 for any of the search types, either in the average or best cases. The only other search type to boast top 3 best performance and no bottom 3 performance was the *simple+rec* search type. However, this did not perform as well in the average case, only appearing in the top 3 twice. Though it is a rather newer memory cell, these results indicate that the Δ-RNN cell warrants strong consideration when designing and developing RNN-based predictors in the future.

## 10.7   Future Work

This work raises a number of points that warrant of further study. Admittedly, the choice of selecting time skip depths uniformly at random between the hyper-parameters of 1 and 10 was somewhat arbitrarily chosen. Given that some of the experiments demonstrated strong improvements in fitness as the average time skip depth of recurrent connections increased, further investigation exploring even deeper time skips/delays could be an avenue to further predictive accuracy improvement. In addition, an adaptive approach for selecting the span of the time skips based on previously well performing connections might provide better accuracy while removing the need for any external tuning of these hyperparameters. Perhaps the most interesting direction to pursue is to extend gated neural structures and memory cells to utilize synaptic connections that go back (explicitly) in time farther than the last time step. This might yield improved memory retention and generalization, potentially even alleviate the need for evolutionary-driven selection of additional deep time skip connections (which EXAMM in this study found necessary to do).

The general question of hyperparameter optimization is also of particular interest given that there are a large number of parameters that could be adjusted to improve performance of EXAMM, e.g., the probabilities at which the various mutation and crossover operations are selected and the rates at which different memory cell types are selected. These rates could be adapted at run-time to better generate new neural networks (incrementally). However, there are challenges in doing this as noted in prior work—local selections of memory cell types based on how well they improve the evolving populations does not necessarily result in the selection of the best memory cell types for the problem at hand [43]. Additionally, investigating the co-evolution of the (local) backprop hyperparameters jointly with the network topology/structure be particularly valuable as indicated by results (in other work) that show to evolve more efficient convolutional neural networks [61].

The strong performance of the *simple+rec* search type might also suggest that generating and training RNNs using an evolutionary process with Lamarckian weight initialization may make training RNNs with non-gated recurrent connections easier since RNNs that use backprop to adjust weights progress to poor regions of the search space, often due to the vanishing and exploding gradient problems. In neuro-evolution, these RNNs will be discarded and thus not be added to the populations used to generate children—this means the evolutionary process will tend to preserve RNNs which have been training well. In future work, this will be examined by retraining the best found architectures from scratch and comparing their performance.

More recently, work by Camero et al. has shown that a mean absolute error (MAE) random sampling strategy can provide good estimates of RNN performance [62] and has successfully used it to speed up the neuro-evolution of LSTM RNNs [63]. By determining the sampled MAEs of randomly initializations of an RNN, performance can be estimated in a manner that is potentially faster than using backprop (locally) itself. In this work, EXAMM trained RNNs for 10 epochs and Camaro et al.'s work utilized 100 forward pass samples, so this estimation method is more computationally complex (assuming one backprop epoch is approximately equal to two sampling forward passes). However, it is of interest to see if the estimation could be a potentially more accurate estimation of an RNN's performance, given that an uninformed stochastic backprop could potentially traverse a poor path through the search space, discarding RNNs that may otherwise be well performing.

One of the central motivations behind using EXAMM as an analysis tool is that it could potentially evolve better RNNs than methods using a singular memory cell type (since EXAMM chooses from a pool of different types, mixing them with complex recurrent connectivity structures) while being more robust. While the *all* and *all+rec* search types did not perform poorly in any case—they avoided the bottom 3 for the best and average cases—they also did not find top-performing RNNs and did not appear in the top 3 for best or average cases either. This may, in part, be due to the fact that these search types induced a much larger search space, where the *add node* mutation was selecting from seven possible neurons (each memory cell and simple neurons) as opposed to only one or two. This additional complexity can slow down the evolutionary process and, thus, it is worth investigating to see if *all* or *all+rec* can outperform the other search types if more computation time is allocated.

Another potential avenue for enhancing EXAMM would utilize layer-based mutations. Previous work has shown that adding node-level mutations (described in Sect. 10.3.1.2) increased the speed of evolution allowing neuro-evolution meta-heuristic approaches to find better performant ANN structures [28, 61]. This would entail adding an even higher level type of mutation where layers of memory cell types are added at once. Incorporating such a mechanism might provide model generalization improvements, while at the same time further improving EXAMM's ability to break out of local minima when searching complex spaces.

## 10.8   Conclusions

While most work in the field of neuro-evolution focuses on the evolution of artificial neural networks that can outperform hand-crafted architectures, this work showcases the potential of neuro-evolution for a different purpose: a robust analysis and investigation of the performance capabilities of different neural network components (in this chapter's case, recurrent memory cells). Rigorously investigating a new neural network component can be quite challenging since its performance can be tied to the architectures it is used within. For most work, new architectural components or strategies are typically only investigated using a few select architectures. This may not necessarily represent how well such a new component would perform given the much wider range of possible architectures. Neuro-evolution can help alleviate this issue by allowing the most successful networks with given components to guide how the architectures are designed, providing a more fair comparison. In short, neuro-evolution metaheuristics might be using in developing an automated and strong experimental methodology for investigating various machine learning methodologies, especially those based on complex, black-box models such as recurrent neural networks.

**Acknowledgments** This material is in part supported by the U.S. Department of Energy, Office of Science, Office of Advanced Combustion Systems under Award Number #FE0031547 and by the Federal Aviation Administration National General Aviation Flight Information Database (NGAFID) award. We also thank Microbeam Technologies, Inc., as well as Mark Dusenbury, James Higgins, Brandon Wild at the University of North Dakota for their help in collecting and preparing the coal-fired power plant and NGAFID data, respectively.

# References

1. Krizhevsky, A., Sutskever, I., Hinton, G.E.: ImageNet classification with deep convolutional neural networks. In: Pereira, F., Burges, C., Bottou, L., Weinberger, K. (eds.) Advances in Neural Information Processing Systems 25, pp. 1097–1105. Curran Associates, Inc., Red Hook (2012). Available: http://papers.nips.cc/paper/4824-imagenet-classification-with-deep-convolutional-neural-networks.pdf
2. LeCun, Y., Bottou, L., Bengio, Y., Haffner, P.: Gradient-based learning applied to document recognition. Proc. IEEE **86**(11), 2278–2324 (1998)
3. Karpathy, A., Toderici, G., Shetty, S., Leung, T., Sukthankar, R., Fei-Fei, L.: Large-scale video classification with convolutional neural networks. In: Proceedings of the IEEE Conference on Computer Vision and Pattern Recognition, pp. 1725–1732 (2014)
4. Kim, Y.: Convolutional neural networks for sentence classification. Preprint, arXiv:1408.5882 (2014)
5. Hinton, G., Deng, L., Yu, D., Dahl, G.E., Mohamed, A.-R., Jaitly, N., Senior, A., Vanhoucke, V., Nguyen, P., Sainath, T.N., et al.: Deep neural networks for acoustic modeling in speech recognition: the shared views of four research groups. IEEE Signal Process. Mag. **29**(6), 82–97 (2012)
6. Simonyan, K., Zisserman, A.: Very deep convolutional networks for large-scale image recognition. Preprint, arXiv:1409.1556 (2014)
7. Szegedy, C., Liu, W., Jia, Y., Sermanet, P., Reed, S., Anguelov, D., Erhan, D., Vanhoucke, V., Rabinovich, A.: Going deeper with convolutions. In: Proceedings of the IEEE Conference on Computer Vision and Pattern Recognition, pp. 1–9 (2015)
8. He, K., Zhang, X., Ren, S., Sun, J.: Deep residual learning for image recognition. In: Proceedings of the IEEE Conference on Computer Vision and Pattern Recognition, pp. 770–778 (2016)
9. Ororbia, A.G. II, Mikolov, T., Reitter, D.: Learning simpler language models with the differential state framework. Neural Comput. **0**(0), 1–26 (2017). PMID: 28957029. Available: https://doi.org/10.1162/neco_a_01017
10. Chung, J., Gulcehre, C., Cho, K., Bengio, Y.: Empirical evaluation of gated recurrent neural networks on sequence modeling. Preprint, arXiv:1412.3555 (2014)
11. Hochreiter, S., Schmidhuber, J.: Long short-term memory. Neural Comput. **9**(8), 1735–1780 (1997)
12. Zhou, G.-B., Wu, J., Zhang, C.-L., Zhou, Z.-H.: Minimal gated unit for recurrent neural networks. Int. J. Autom. Comput. **13**(3), 226–234 (2016)
13. Collins, J., Sohl-Dickstein, J., Sussillo, D.: Capacity and trainability in recurrent neural networks. Preprint, arXiv:1611.09913 (2016)
14. Gomez, F., Schmidhuber, J., Miikkulainen, R.: Accelerated neural evolution through cooperatively coevolved synapses. J. Mach. Learn. Res. **9**, 937–965 (2008)
15. Salama, K., Abdelbar, A.M.: A novel ant colony algorithm for building neural network topologies. In: Swarm Intelligence, pp. 1–12. Springer, Berlin (2014)
16. Xie, L., Yuille, A.: Genetic CNN. Preprint, arXiv:1703.01513 (2017)

17. Suganuma, M., Shirakawa, S., Nagao, T.: A genetic programming approach to designing convolutional neural network architectures. In: Proceedings of the Genetic and Evolutionary Computation Conference, GECCO '17, pp. 497–504. ACM, New York (2017). Available: http://doi.acm.org/10.1145/3071178.3071229

18. Sun, Y., Xue, B., Zhang, M.: Evolving deep convolutional neural networks for image classification. CoRR, vol. abs/1710.10741 (2017). Available: http://arxiv.org/abs/1710.10741

19. Miikkulainen, R., Liang, J., Meyerson, E., Rawal, A., Fink, D., Francon, O., Raju, B., Shahrzad, H., Navruzyan, A., Duffy, N., Hodjat, B.: Evolving deep neural networks. Preprint, arXiv:1703.00548 (2017)

20. Real, E., Moore, S., Selle, A., Saxena, S., Suematsu, Y.L., Le, Q., Kurakin, A.: Large-scale evolution of image classifiers. Preprint, arXiv:1703.01041 (2017)

21. Stanley, K., Miikkulainen, R.: Evolving neural networks through augmenting topologies. Evol. Comput. **10**(2), 99–127 (2002)

22. Stanley, K.O., D'Ambrosio, D.B., Gauci, J.: A hypercube-based encoding for evolving large-scale neural networks. Artif. Life **15**(2), 185–212 (2009)

23. Rawal, A., Miikkulainen, R.: Evolving deep LSTM-based memory networks using an information maximization objective. In: Proceedings of the Genetic and Evolutionary Computation Conference 2016, pp. 501–508. ACM, New York (2016)

24. Rawal, A., Miikkulainen, R.: From nodes to networks: evolving recurrent neural networks. CoRR, vol. abs/1803.04439 (2018). Available: http://arxiv.org/abs/1803.04439

25. Desell, T., Clachar, S., Higgins, J., Wild, B.: Evolving deep recurrent neural networks using ant colony optimization. In: European Conference on Evolutionary Computation in Combinatorial Optimization, pp. 86–98. Springer, Berlin (2015)

26. ElSaid, A., El Jamiy, F., Higgins, J., Wild, B., Desell, T.: Optimizing long short-term memory recurrent neural networks using ant colony optimization to predict turbine engine vibration. Appl. Soft Comput. **73**, 969–991 (2018)

27. ElSaid, A., Jamiy, F.E., Higgins, J., Wild, B., Desell, T.: Using ant colony optimization to optimize long short-term memory recurrent neural networks. In: Proceedings of the Genetic and Evolutionary Computation Conference, pp. 13–20. ACM, New York (2018)

28. Desell, T.: Accelerating the evolution of convolutional neural networks with node-level mutations and epigenetic weight initialization. In: Proceedings of the Genetic and Evolutionary Computation Conference Companion, pp. 157–158. ACM, New York (2018)

29. ElSaid, A., Benson, S., Patwardhan, S., Stadem, D., Travis, D.: Evolving recurrent neural networks for time series data prediction of coal plant parameters. In: The 22nd International Conference on the Applications of Evolutionary Computation, Leipzig, April 2019

30. Elsken, T., Metzen, J.H., Hutter, F.: Neural architecture search: a survey. Preprint, arXiv:1808.05377 (2018)

31. Real, E., Aggarwal, A., Huang, Y., Le, Q.V.: Regularized evolution for image classifier architecture search. In: Proceedings of the AAAI Conference on Artificial Intelligence, vol. 33, pp. 4780–4789 (2019)

32. Pham, H., Guan, M.Y., Zoph, B., Le, Q.V., Dean, J.: Efficient neural architecture search via parameter sharing. Preprint, arXiv:1802.03268 (2018)

33. Liu, C., Zoph, B., Neumann, M., Shlens, J., Hua, W., Li, L.-J., Fei-Fei, L., Yuille, A., Huang, J., Murphy, K.: Progressive neural architecture search. In: Proceedings of the European Conference on Computer Vision (ECCV), pp. 19–34 (2018)

34. Zoph, B., Le, Q.V.: Neural architecture search with reinforcement learning. Preprint, arXiv:1611.01578 (2016)

35. McClelland, J.L., Rumelhart, D.E., P. R. Group, et al.: Parallel distributed processing, vol. 2. MIT Press, Cambridge (1987)

36. Lin, T., Horne, B.G., Tino, P., Giles, C.L.: Learning long-term dependencies in NARX recurrent neural networks. IEEE Trans. Neural Netw. **7**(6), 1329–1338 (1996)

37. Lin, T., Horne, B.G., Giles, C.L.: How embedded memory in recurrent neural network architectures helps learning long-term temporal dependencies. Neural Netw. **11**(5), 861–868 (1998)

38. Lin, T., Horne, B.G., Giles, C.L., Kung, S.-Y.: What to remember: how memory order affects the performance of NARX neural networks. In: 1998 IEEE International Joint Conference on Neural Networks Proceedings. IEEE World Congress on Computational Intelligence (Cat. No. 98CH36227), vol. 2, pp. 1051–1056. IEEE, Piscataway (1998)

39. Giles, C.L., Lin, T., Horne, B.G., Kung, S.-Y.: The past is important: a method for determining memory structure in NARX neural networks. In: 1998 IEEE International Joint Conference on Neural Networks Proceedings. IEEE World Congress on Computational Intelligence (Cat. No. 98CH36227), vol. 3, pp. 1834–1839. IEEE, Piscataway (1998)

40. Diaconescu, E.: The use of NARX neural networks to predict chaotic time series. WSEAS Trans. Comput. Res. **3**(3), 182–191 (2008)

41. Chen, J., Chaudhari, N.S.: Segmented-memory recurrent neural networks. IEEE Trans. Neural Netw. **20**(8), 1267–1280 (2009)

42. ElSaid, A., Wild, B., Higgins, J., Desell, T.: Using LSTM recurrent neural networks to predict excess vibration events in aircraft engines. In: 2016 IEEE 12th International Conference on e-Science (e-Science), pp. 260–269. IEEE, Piscataway (2016)

43. Ororbia, A., ElSaid, A., Desell, T.: Investigating recurrent neural network memory structures using neuro-evolution. In: Proceedings of the Genetic and Evolutionary Computation Conference, GECCO '19, pp. 446–455. ACM, New York (2019). Available: http://doi.acm.org/10.1145/3321707.3321795

44. Elman, J.L.: Finding structure in time. Cogn. Sci. **14**(2), 179–211 (1990)

45. Jordan, M.I.: Serial order: a parallel distributed processing approach. Adv. Psychol. **121**, 471–495 (1997)

46. Ororbia, I., Alexander, G., Linder, F., Snoke, J.: Using neural generative models to release synthetic twitter corpora with reduced stylometric identifiability of users. Preprint, arXiv:1606.01151 (2018)

47. Ororbia, A.G., Mali, A., Wu, J., O'Connell, S., Miller, D., Giles, C.L.: Learned iterative decoding for lossy image compression systems. In: Data Compression Conference. IEEE, Piscataway (2019)

48. El Hihi, S., Bengio, Y.: Hierarchical recurrent neural networks for long-term dependencies. In: Advances in Neural Information Processing Systems, pp. 493–499. MIT Press, Cambridge (1996)

49. Kalinli, A., Sagiroglu, S.: Elman network with embedded memory for system identification. J. Inf. Sci. Eng. **22**(6), 1555–1568 (2006)

50. Bengio, Y., Simard, P., Frasconi, P.: Learning long-term dependencies with gradient descent is difficult. IEEE Trans. Neural Netw. **5**(2), 157–166 (1994)

51. Williams, R.J., Zipser, D.: A learning algorithm for continually running fully recurrent neural networks. Neural Comput. **1**(2), 270–280 (1989)

52. Glüge, S., Böck, R., Palm, G., Wendemuth, A.: Learning long-term dependencies in segmented-memory recurrent neural networks with backpropagation of error. Neurocomputing **141**, 54–64 (2014)

53. Desell, T.: Asynchronous global optimization for massive scale computing. Ph.D. dissertation, Rensselaer Polytechnic Institute (2009)

54. ElSaid, A.A., Ororbia, A.G., Desell, T.J.: The ant swarm neuro-evolution procedure for optimizing recurrent networks. Preprint, arXiv:1909.11849 (2019)

55. Desell, T.: Large scale evolution of convolutional neural networks using volunteer computing. CoRR, vol. abs/1703.05422 (2017). Available: http://arxiv.org/abs/1703.05422

56. Alba, E., Tomassini, M.: Parallelism and evolutionary algorithms. IEEE Trans. Evol. Comput. **6**(5), 443–462 (2002)

57. Message Passing Interface Forum: MPI: a message-passing interface standard. Int. J. Supercomput. Appl. High Perform. Comput. **8**(3/4), 159–416 (Fall/Winter 1994)

58. Werbos, P.J.: Backpropagation through time: what it does and how to do it. Proc. IEEE **78**(10), 1550–1560 (1990)

59. Jozefowicz, R., Zaremba, W., Sutskever, I.: An empirical exploration of recurrent network architectures. In: International Conference on Machine Learning, pp. 2342–2350 (2015)

60. Pascanu, R., Mikolov, T., Bengio, Y.: On the difficulty of training recurrent neural networks. In: International Conference on Machine Learning, pp. 1310–1318 (2013)
61. Desell, T.: Developing a volunteer computing project to evolve convolutional neural networks and their hyperparameters. In: The 13th IEEE International Conference on eScience (eScience 2017), pp. 19–28, Oct 2017
62. Camero, A., Toutouh, J., Alba, E.: Low-cost recurrent neural network expected performance evaluation. Preprint, arXiv:1805.07159 (2018)
63. Camero, A., Toutouh, J., Alba, E.: A specialized evolutionary strategy using mean absolute error random sampling to design recurrent neural networks. Preprint, arXiv:1909.02425 (2019)

# Chapter 11
# Neuroevolution of Generative Adversarial Networks

Victor Costa, Nuno Lourenço, João Correia, and Penousal Machado

**Abstract** Generative Adversarial Networks (GAN) is an adversarial model that became relevant in the last years, displaying impressive results in generative tasks. A GAN combines two neural networks, a discriminator and a generator, trained in an adversarial way. The discriminator learns to distinguish between real samples of an input dataset and fake samples. The generator creates fake samples aiming to fool the discriminator. The training progresses iteratively, leading to the production of realistic samples that can mislead the discriminator. Despite the impressive results, GANs are hard to train, and a trial-and-error approach is generally used to obtain consistent results. Since the original GAN proposal, research has been conducted not only to improve the quality of the generated results but also to overcome the training issues and provide a robust training process. However, even with the advances in the GAN model, stability issues are still present in the training of GANs. Neuroevolution, the application of evolutionary algorithms in neural networks, was recently proposed as a strategy to train and evolve GANs. These proposals use the evolutionary pressure to guide the training of GANs to build robust models, leveraging the quality of results, and providing a more stable training. Furthermore, these proposals can automatically provide useful architectural definitions, avoiding the manual discovery of suitable models for GANs. We show the current advances in the use of evolutionary algorithms and GANs, presenting the state-of-the-art proposals related to this context. Finally, we discuss perspectives and possible directions for further advances in the use of evolutionary algorithms and GANs.

## 11.1 Introduction

Generative Adversarial Networks (GAN) [16] is an adversarial model that makes use of neural networks to produce samples based on an input distribution. GANs

V. Costa (✉) · N. Lourenço · J. Correia · P. Machado
CISUC, Department of Informatics Engineering, University of Coimbra, Coimbra, Portugal
e-mail: vfc@dei.uc.pt; naml@dei.uc.pt; jncor@dei.uc.pt; machado@dei.uc.pt

© Springer Nature Singapore Pte Ltd. 2020
H. Iba, N. Noman (eds.), *Deep Neural Evolution*, Natural Computing Series,
https://doi.org/10.1007/978-981-15-3685-4_11

293

can be applied in several contexts, for example, in the generation of image, video, sound, and text, being able to produce impressive results concerning the quality of the created samples. This model gained a lot of relevance in recent years, leveraging the interest of the community on improving the original proposal.

Despite the fact that GANs can be used as a generative component to produce samples in a variety of areas, applications in the image domain are more frequently reported by the production of realistic samples, representing significant advances when compared to other methods [3, 21, 51]. Therefore, the focus of this chapter is on the applications of GANs to the image domain. Nevertheless, the techniques presented here can be extended and adapted to other contexts.

Although GANs have attained incredible results, their training is challenging, and the presence of problems such as the vanishing gradient and the mode collapse is common [7, 13]. The balance between the discriminator and the generator is frequently the cause of these problems. In the case of the vanishing gradient, the discriminator becomes so powerful that it can distinguish almost perfectly between samples created by the generator and real samples. After this, because of the training approach used in GANs, the process stagnates. Regarding the mode collapse, the problem occurs when the generator fails to capture the entire representation of the distribution used as input to the discriminator. This is an undesired behavior, as we want not only to reproduce realistic samples but also to reproduce the diversity of the input distribution. Although there is a diversity of strategies and techniques to minimize the effect of these problems, they are still affecting the GAN training [17, 39]. Most of the proposed solutions appeal to mathematical models to deal with these problems, such as the use of more robust loss functions and stable neural network layers [3, 5, 27, 51]. Other proposals also worked on the architecture of the neural networks in order to avoid these issues [31, 35].

In spite of these issues, research was also conducted to improve the original GAN model with respect to the quality of the results, leveraging it to impressive levels [3, 21, 27]. Other researches also proposed changes on the model to introduce a conditional input [20, 29, 32, 37]. Thus, a relevant effort is being made to improve GANs, not only to overcome the difficulties on the original model but also to extend the initial concept to different objectives.[1]

In GANs, the adversarial characteristics and the necessity of an equilibrium between the generator and the discriminator make the design of the network crucial for the quality of the results. Therefore, the topology and hyperparameters that compose the neural networks of the generator and the discriminator are important to keep the balance between them in the training process. If one component becomes more powerful than the other, the GAN training will probably become unstable and may not produce the desired outcome. In this case, the design of the neural network is paramount to achieve convergence on training.

---

[1]A list of proposals related to GANs can be found at https://github.com/hindupuravinash/the-gan-zoo.

The design of a neural network is usually defined by hand in an empirical process, based on expert knowledge, which requires spending human time in repetitive tasks, such as experimentation and fine-tuning [7]. Experiments are used to validate and fine-tune the model, aiming to find efficient architectures to produce a neural network for a specific problem. However, some approaches can be used to automatize this process. In the field of evolutionary computation, neuroevolution can be used to design and optimize neural networks [28, 42, 50]. An evolutionary algorithm (EA) is based on the evolutionary mechanism found in nature, using it to evolve a population of potential solutions, producing better outcomes for a given problem [41]. In neuroevolution, this concept is adapted to the context of neural networks. In this case, the population is composed of individuals encoded through a genotype that represents, in some level of abstraction, neural networks. The genotype is used in a transformation procedure that creates the phenotype of an individual, which expresses the concrete implementation of a neural network. As in a regular EA, the phenotypes are used to evaluate and select individuals for reproduction to form the next generations of potentially better solutions.

Neuroevolution can be applied to evolve both the network architecture (e.g., topology, hyperparameters, and optimization method) and the internal parameters (e.g., weights) [50]. NeuroEvolution of Augmented Topologies (NEAT) [42] is a well-known neuroevolution method that evolves the weights and topologies of neural networks. A further proposal originated DeepNEAT [28], a modification of the model that expands NEAT to larger search spaces, such as in deep neural networks.

Although neuroevolution is usually applied to standalone neural networks, the concepts can also be applied in the context of GANs. Furthermore, in the mechanics of the GAN model, the generator and discriminator are competing in a zero-sum game in the task of creating and discriminating fake and real samples. Therefore, a competitive model can be suitable to represent populations of individuals in GANs. In EAs, coevolution is the simultaneous evolution of at least two distinct species [19, 36, 43]. In competitive coevolution, individuals of these species are competing together, and their fitness function directly represents this competition. Thus, the applicability of a competitive coevolution environment in an EA to train GANs can also be evaluated [9, 10, 14, 46].

In recent years, researchers have been applying the concepts of EAs to improve the performance of GANs with different strategies [1, 9, 10, 14, 46, 47]. The authors found advances not only in the quality of the outcome but also regarding the stability issues in the training of GANs. We present in this chapter the state-of-the-art of these proposals, discussing their main advantages and drawbacks, and presenting further directions for improvements. The following proposals will be described in this chapter: E-GAN [47], Pareto GAN [14], Lipizzaner [1], Mustangs [46], and COEGAN [9, 10].

The remainder of this chapter is organized as follows. Section 11.2 introduces the concepts of GANs, presenting the challenges and advances in this field. Section 11.3 summarizes the possibilities regarding the application of EAs in the context of GANs. Section 11.4 presents the current proposals that use EAs with

GANs. Section 11.5 discusses the application of EAs in GANs, drawing particular attention to the drawbacks and advantages of each approach, presenting directions for further improvements. Finally, Sect. 11.6 concludes this chapter with the final considerations about the subject.

## 11.2 Generative Adversarial Networks

Generative Adversarial Networks (GAN) [16] is an adversarial model that became relevant mostly for the performance achieved in generative tasks on the image domain, representing significant improvements over other generative methods. We present in this section the model definition, the common issues found when training a GAN, and how to evaluate and compare GANs using the state-of-the-art metrics.

### 11.2.1 Definition

A GAN combines two neural networks in a unified training algorithm: a discriminator $D$ and a generator $G$. The discriminator $D$ aims to distinguish between real and fake examples. The generator $G$ outputs fake samples, attempting to capture the input distribution used in the training of $D$.

Both the discriminator and generator use backpropagation and gradient descent in the GAN training. Thus, different loss functions are used in the GAN components. The loss function of the discriminator is defined as follows:

$$J^{(D)}(D, G) = -\mathbb{E}_{x \sim p_{data}}[\log D(x)] - \mathbb{E}_{z \sim p_z}[\log(1 - D(G(z)))]. \qquad (11.1)$$

For the generator, the non-saturating version of the loss function is defined by

$$J^{(G)}(G) = -\mathbb{E}_{z \sim p_z}[\log(D(G(z)))]. \qquad (11.2)$$

In Eq. (11.1), $p_{data}$ represents the dataset used as input to the discriminator. In Eqs. (11.1) and (11.2), $z$ is the latent space used as input to the generator, $p_z$ is the latent distribution, $G$ is the generator, and $D$ represents the discriminator.

GANs are hard to train, and training stability is an issue that systematically affects the results. So, to achieve good outcomes in training, a trial-and-error approach is frequently used. Some works developed a set of techniques to train GANs to improve the probability to achieve convergence [35, 39]. However, these strategies only minimize the effect of the problems that usually happen in the training process. Several other variations of the original GAN model were proposed to improve the effect of these problems [3, 17, 21, 27, 31]. In Sect. 11.2.2, we describe some of these problems regarding the training of GANs.

## 11.2.2  Common Problems in GAN Training

The vanishing gradient and the mode collapse are among the most common problems affecting the stability when training GANs. They are widespread and represent a significant challenge to obtain useful representations for applying GANs in different domains. These issues are often part of a bigger problem: the balance between the discriminator and the generator during the training. Although several approaches tried to minimize those obstacles, they still affect the training and remain unsolved [3, 17, 39]. Following we describe the mode collapse and the vanishing gradient issues, presenting how they affect the training of GANs.

### 11.2.2.1  Mode Collapse

In the mode collapse problem, the generator captures only a small portion of the dataset distribution provided as input to the discriminator. This diminished representation is not desirable since it is expected that a generative model reproduces the whole distribution of the data to achieve variability on the output samples.

Figure 11.1 represents images created by a generator after a failed training of a GAN using the MNIST dataset [24]. The effects of the mode collapse can be clearly seen in these images. We can see in the samples on the left of Fig. 11.1 that only the digits 9 and 7 are represented. However, in the samples on the right, the digits cannot be identified correctly. The generator creates only a superposed combination of digits. The lack of variability demonstrated in these examples characterizes the problem as mode collapse.

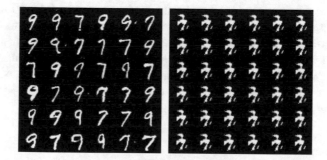

**Fig. 11.1** Samples created by a GAN after training that resulted in the mode collapse issue. Note that the GAN was trained using the MNIST dataset, which contains digits from 0 to 9. However, on the left, the generator can only create samples related to the digits 7 and 9. In the right, the generator failed to create a real digit, outputting the same unrealistic pattern

#### 11.2.2.2 Vanishing Gradient

The vanishing gradient occurs when one of the GAN components, i.e., the discriminator or the generator, becomes powerful enough to harm the balance required on the training. For example, the discriminator can become too strong and not be fooled anymore by the generator when distinguishing between fake and real samples. Hence, the loss function is too small, the gradient does not flow through the neural network of the generator, and the GAN progress stagnates. In the GAN training, the equilibrium between the discriminator and generator is essential to the training convergence. The vanishing gradient problem happens when this equilibrium is violated in an irreversible way.

Figure 11.2 presents an example of a GAN training that suffers from the vanishing gradient problem. We can see in this figure the progression of losses of the generator and discriminator through iterations. Note that when the discriminator loss becomes zero (marked by the dashed vertical line), the generator stops to improve and stagnates until the end of the training. As such, the quality of samples created by the generator will not improve anymore. It is important to note that the divergence between the generator and discriminator, expressed by the losses, does not need to always decrease [13]. Even when the loss increases, the training can reach a good solution in the end. Therefore, regarding the vanishing gradient, the problem only occurs when the loss approximates to zero. The GAN model tolerates steps with a reduction in the loss without losing convergence capabilities.

### 11.2.3 Evaluation Metrics

Several metrics can be used to quantify the performance of a GAN [6, 49]. As the generators are commonly the most relevant component of a GAN, these metrics usually target them. However, the measurement of the performance when executing generative tasks is a relevant problem and there is not a consensus yet in the community about the best metric to use. We highlight here two of the most commonly reported metrics for GANs in the literature: the Inception Score and the Fréchet Inception Distance (FID) score.

**Fig. 11.2** Losses of the generator and discriminator of a training experiment with the vanishing gradient issue. As the loss of the discriminator approximates to zero, the loss of generator stagnates

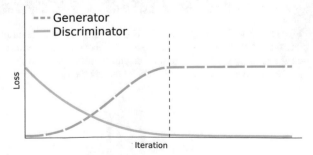

Other metrics, such as the skill rating [33], were evaluated and obtained relevant results. Despite this, they are still not widely used by the community, becoming hard to use them in a comparison study to evaluate a proposal with other works. However, they can still be useful to use in the context of EAs. They can be used not only as comparison criteria between the solutions but also as fitness functions to guide the evolution.

### 11.2.3.1 Inception Score

The Inception Score (IS) [39] is an automatic metric to evaluate synthetic image samples that were created based on an input dataset. This method uses the Inception Network [44, 45] to get the conditional label distribution of the images created by a generative algorithm, such as a GAN. This network should be previously trained using a dataset, usually the ImageNet dataset [38]. Therefore, the Inception Score is defined as:

$$IS(x, y) = \exp(\mathbb{E}_x KL(p(y|x)||p(y))), \tag{11.3}$$

where $x$ is the input data, $y$ is the label of the data, $p(y)$ is the label distribution, $p(y|x)$ is the conditional label distribution, and KL is the Kullback–Leibler divergence between the distributions $p(y|x)$ and $p(y)$. It is recommended to evaluate the IS metric on a large number of samples, such as 50,000, in order to provide enough diversity to the score [39].

The IS metric has some drawbacks, such as the sensitivity to the weights of the Inception Network used in the calculation [4]. Moreover, the network used in the Inception Score, which was trained in the ImageNet dataset, may not be applicable with consistent performance to other datasets.

### 11.2.3.2 Fréchet Inception Distance

Fréchet Inception Distance (FID) [18] is the state-of-the-art metric to compare the generative components of GANs. The FID score outperforms other metrics, such as the Inception Score, with respect to diversity and quality [26]. As in the Inception Score, FID also uses a trained Inception Network in the computation process. In the FID score, a hidden layer of Inception Net (also usually trained on ImageNet) is used in the transformation of images into the feature space, which is interpreted as a continuous multivariate Gaussian. This transformation is applied to a subset of the real dataset and samples created by the generative method. The mean and covariance of the two resulting Gaussians are estimated and the Fréchet distance between these Gaussians is given by

$$FID(x, g) = ||\mu_x - \mu_g||_2^2 + Tr(\Sigma_x + \Sigma_g - 2(\Sigma_x \Sigma_g)^{1/2}). \tag{11.4}$$

In Eq. (11.4), $\mu_x$, $\Sigma_x$, $\mu_g$, and $\Sigma_g$ represent the mean and covariance estimated for the real dataset $x$ and fake samples $g$, respectively. In summary, the FID score is given by the norm of the means and the trace of the covariances between real and fake samples.

## 11.3 Exploring the Evolution of GANs

Several aspects that compose the GAN model can be actively used as evolvable components in an evolutionary algorithm. However, it is important to keep in mind that the EA should preserve the balance of these components in order to tackle the issues listed in Sect. 11.2.2. We discuss in this section the possibilities for the application of EAs to the GAN model. The options related to neuroevolution and the aspects of GANs will be presented as possible choices to design an algorithm.

### 11.3.1 Neuroevolution

Neuroevolution is the application of EAs in the evolution of a neural network. It can be applied to evolve weights, topology, and hyperparameters of a neural network [50]. When used to discover the network topology, a substantial benefit is the automation of the architecture design and parameter decision, transforming a manual human effort into an automatic procedure. This automation is even more critical with the rise of deep learning, which is producing deeper models and increasing the search space [28]. However, the increase in the search space is also a challenge for neuroevolution. These methods have high time-consuming executions that may turn their application unfeasible.

Neuroevolution can be fully applied in the context of GANs. The evolution of the topologies of the discriminator and the generator should take into account that the equilibrium between them is paramount to the convergence of the training process. Not only the structure (i.e., the number of layers and the connections between them) but also the internal characteristics of each layer composing a neural network can be the subject of evolution. For example, the type of a layer (e.g., convolution or fully connected), the number of output features, and the activation function (e.g., ReLU, ELU, Tanh). Other aspects relevant to the network can also be a variable of the individual, such as the choice for the optimizer used in the training, the learning rate, the batch size, and the number of the training iterations.

We can also make use of other techniques regarding evolutionary computation in neuroevolution, such as coevolution. Coevolution is the simultaneous evolution of at least two distinct populations (also denominated species) [19, 36]. There are two types of coevolution algorithms: cooperative and competitive. In cooperative coevolution, individuals of different species cooperate in the search for efficient solutions, and the fitness function of each species is designed to reward this cooperation. In competitive coevolution, individuals of different species are competing between them in the search for better solutions. Here, their fitness function directly represents this competition in a way that scores between species are inversely related. For example, NEAT was successfully applied to a competitive coevolution environment [43].

The coevolutionary approach used in an EA can lead to some issues, such as intransitivity and disengagement [2, 30]. The intransitivity occurs when a solution $a$ is better than $b$ and $b$ is better than $c$, but this does not guarantee that $a$ is better than $c$. This issue can lead to cycling between these solutions during the evolutionary process, preventing the progress of individuals toward optimal solutions. Disengagement occurs when the equilibrium between the populations is broken. In this case, individuals from one population are much better than individuals from the other, leading to ineffective progression.

GANs can be modeled as a competitive coevolution problem. We can consider a population of discriminators as competitors to a population of generators. Therefore, an EA can make use of competitive coevolution concepts to match individuals from these two populations at the evaluation phase. Furthermore, we can relate problems that frequently affect the training of GANs (Sect. 11.2.2) to coevolution problems. For example, the vanishing gradient can be linked to the disengagement issue. Thus, the use of coevolution can be explored in combination with other techniques (e.g., neuroevolution) to solve the challenges of the GAN training process.

## 11.3.2 Variations of GANs

Several advances over the original GAN model were recently proposed. These proposals focused not only on the improvement of the quality of the created samples but also on the improvement of the training stability. These proposals can be divided into two main categories: architecture improvements and alternative loss functions [34, 48].

In the category of architecture improvements, we have DCGAN [35], a set of constraints and rules that guide the design of the components of a GAN. DCGAN became a reference architecture for the discriminator and the generator in GANs. Some of these rules are:

- Use batch normalization in the generator and discriminator;
- Use the ReLU activation function in all hidden layers of the generator;
- Use LeakyReLU in all layers of the discriminator.

In the experiments presented with DCGAN, the training stability was improved, but there are still issues such as the mode collapse problem in some executions [35].

Other proposals introduced different aspects into the original GAN model [5, 7, 11, 12, 15, 21, 22, 51]. We can use some of these strategies as inspiration for an EA. For example, the method described in [21] uses a predefined strategy to grow a GAN during the training procedure. The main idea is to grow the model progressively, increasing layers in both discriminator and generator. This mechanism will make the model more complex while the training procedure runs, resulting in the generation of higher resolution images at each phase. However, these layers are added progressively in a preconfigured way, i.e., they are not produced by a stochastic procedure. These concepts can be expanded to be used in an EA. Instead of a predefined grow, the progression of the discriminator and the generator can be guided by evolution, using a fitness function that can prevent and discard unfitted individuals.

Other approaches use multiple components instead of only a single generator and a single discriminator. For example, GMAN [11] proposed a model that uses multiple discriminators in the training algorithm. On the other hand, MAD-GAN [15] explored the use of multiple generators in the GAN training. An EA can be aligned with these concepts with the proposal of a solution that contains two entirely different populations of discriminators and generators.

Another strategy to overcome the training issues and improve the original GAN model is the use of alternative loss functions. A variety of alternative loss functions were proposed to minimize the problems and leverage the quality of the results, such as WGAN [3], LSGAN [27], and SN-GAN [31]. WGAN proposes the use of the Wasserstein distance to model the loss functions. LSGAN uses the least-squares function as the loss for the discriminator. SN-GAN proposes the use of spectral normalization to improve the training of the discriminator. An EA can take advantage of these variations and use the loss function as an interchangeable component.

## 11.4   Current Proposals

We present in this section the state-of-the-art on the application of evolutionary algorithms in GANs. These proposals are aligned with the possibilities presented in Sect. 11.3, presenting solutions to apply them and improve the GAN training process. To the best of our knowledge, these are the proposals that use EAs in the context of GANs: E-GAN [47], Pareto GAN [14], Lipizzaner [1], Mustangs [46], and COEGAN [9, 10]. In this section we describe these solutions, focusing on the choices concerning the aspects of the EA and the characteristics of GANs. Therefore, we report the characteristics of the algorithms concerning the selection method, fitness functions, variation operators, evaluation, and experiments.

## 11.4.1   E-GAN

A model called E-GAN[2] was proposed to use EAs in GANs [47]. The approach applies an EA to GANs using a mutation operator that can only switch the loss function of the generator. Therefore, the evolution occurs only in the generator, and a single-fixed discriminator is used as the adversarial for the population of generators. The network architectures for the generator and the discriminator are fixed and based on DCGAN [35].

The population of generators contains individuals that have different loss functions. The mutation operator used in the process can change the loss function of the individual to another one selected from a predefined set. Each loss function in the predefined set focused on an objective to help in the GAN learning process. A minimal population of individuals is used to capture all possibilities of the predefined losses and provide an adaptive objective for the training. In this case, the population of generators is composed of three individuals, each one representing one of the possible losses.

The possibilities for losses are implemented through three mutation operators: minimax, heuristic, and least-squares mutation. The minimax mutation follows the original GAN objective given by Eq. (11.2), minimizing the probability of the discriminator to detect fake samples. On the other hand, the heuristic mutation aims to maximize the probability of the discriminator to make mistakes regarding fake samples. The least-squares mutation is based on the objective function used in LSGAN [27]. Only these operations are available and crossover is not used in the E-GAN algorithm.

Two criteria were used as fitness in the evaluation phase of the algorithm. The first, called quality fitness score, is defined as:

$$F_q = \mathbb{E}_z[(D(G(z)))], \tag{11.5}$$

that is similar to the loss function used in the generator of the original GAN model (Eq. (11.1)). The second criteria, called the diversity fitness score, is defined as:

$$F_d = -\log \|\nabla_D - \mathbb{E}_x[\log(D(x))] - \mathbb{E}_z[\log(1 - D(G(z)))]\|. \tag{11.6}$$

In Eqs. (11.5) and (11.6), $z$, $G$, and $D$ represent the latent space, the generator, and the discriminator, respectively. These two fitness criteria are combined as follows:

$$F = F_q + \gamma F_d, \tag{11.7}$$

where the $\gamma$ parameter is used to regulate the influence of the diversity criteria on the final fitness.

---

[2]Code available at https://github.com/WANG-Chaoyue/EvolutionaryGAN.

At each generation, individuals are evaluated following their specific loss function, and only the best-fitted generator survives for the next steps. In the next generation, the survivor individual is used to train the discriminator and to generate the three children for the next evaluation.

The E-GAN model was evaluated on the CIFAR-10, LSUN, and CelebA datasets. The Inception Score was used as the metric to analyze the results. As specified in E-GAN, the population used in the experiments consist of a single discriminator and three generators. The authors concluded that E-GAN improved the training stability and achieved satisfactory performance, outperforming other methods in some scenarios.

### 11.4.2  Pareto GAN

A neuroevolution approach for training GANs was proposed in [14]. Although not named by the authors, we refer to this solution as Pareto GAN.[3] The proposal uses a genetic algorithm to evolve the architecture of the neural networks used for both the generator and the discriminator. A single individual $(G_i, D_i)$ is used to represent both the generator and the discriminator in the EA.

The crossover operator combines two parents exchanging the discriminator and the generator between them. For example, a crossover between the individuals $(G_1, D_1)$ and $(G_2, D_2)$ produces the children $(G_1, D_2)$ and $(G_2, D_1)$. The crossover operator does not change the internal state of the generator and the discriminator in each individual. To accomplish this, a set of possible mutations is applied to individuals when creating a new generation.

Regarding the architecture of the neural networks, the mutation can change, add, or remove a layer. Mutation can also change the internal state of a layer, such as the weights or the activation function. Some mutation operators also work on the GAN algorithm level. There is an operator to change the loss function used in the GAN algorithm by using a predefined set of possibilities. Another possibility is to change the characteristics of the algorithm. Here, it is possible to change the number of iterations for the generator and the discriminator when applying the GAN training algorithm to an individual.

A benchmark for GANs based on the problem of Pareto set approximations was also proposed [14]. The comparison between the Pareto front of a solution and the real front is used to assess the quality of the samples and can also identify issues, such as the mode collapse problem. Therefore, the inverted generational distance (IGD) [8] was used as fitness to drive the EA. The IGD measures the

---

[3]Code available at https://github.com/unaigarciarena/GAN_Evolution.

smallest distance between points in the true Pareto front and in the Pareto front approximation and is given by

$$\text{IGD} = \frac{1}{|R|} \left( \sum_{r \in R} \min_{a \in A} d(r, a)^p \right)^{\frac{1}{p}}, \quad d(r, a) = \left( \sum_{k=1}^{m} (r_k - a_k)^2 \right)^{\frac{1}{2}}, \quad (11.8)$$

where $R$ is the real Pareto front, $A$ is the Pareto approximation, and $m$ is the number of vectors in $R$.

The evaluation phase will transform each individual $(G_i, D_i)$ into a concrete GAN, composed of a discriminator and a generator, that will be trained according to the regular GAN algorithm. The fitness is calculated, and the selection uses the Pareto dominance to compose the offspring that will form the next generation.

The proposed solution was evaluated using bi-objective functions as the input data, each one with 10 input variables. A population of 20 individuals, evaluated for 500 generations, was used in the experiments. The authors concluded that the algorithm was able to discover architectures that improved the Pareto set approximation for discriminators and generators. The experiments do not include evaluations with image datasets. However, experiments using the same data dimension as the MNIST dataset, i.e., with 784 input variables, were also conducted. The authors demonstrated that the solution is scalable to this dimension, as the results showed that useful architectures were also found in this case.

### 11.4.3   Lipizzaner

A model called Lipizzaner[4] defines a coevolutionary framework to train and evolve GANs [1]. In Lipizzaner, the evolution occurs only on the internal parameters of the generator and discriminator, such as the weights of their neural networks. Thus, the network architecture used in both the discriminator and generator is fixed and defined a priori. The architecture varies with the dataset used in the experiments. For MNIST, an MLP network composed of four layers and 700 neurons was used. On the other hand, an architecture based on DCGAN was used for the experiments with the CelebA dataset.

The fitness used in Lipizzaner for the generators and discriminators is based on the GAN objective function, defined as:

$$\mathcal{L}(u, v) = \mathbb{E}_{x \sim p_{data}}[\phi(D_v(x))] + \mathbb{E}_{x \sim G_u}[\phi(1 - D_v(x))], \quad (11.9)$$

---

[4]Code available at https://github.com/ALFA-group/lipizzaner-gan.

where $\phi$ is a concave function in the interval $[0, 1]$, $p_{data}$ is the input dataset, $G_u$ is the generator with the parameters $u$, and $D_v$ represents the discriminator with its parameters $v$.

At the evaluation step, $\mathcal{L}(u_i, v_j)$ is calculated for each pair $(G_i, D_j)$, and the fitness values are updated as $f_{u_i} \mathrel{-}= \mathcal{L}(u_i, v_j)$ and $f_{v_j} \mathrel{+}= \mathcal{L}(u_i, v_j)$ for generators and discriminators, respectively.

Spatial coevolution was used to design the algorithm that trains and evolve the generators and discriminators. Individuals are distributed over a two-dimensional toroidal grid, where each cell contains individuals from the generator and discriminator populations. In the evaluation phase, the EA matches individuals in neighbor cells following a coevolutionary pairing approach. A five-cell neighborhood was used to determine these interactions. Figure 11.3 displays an example of a $3 \times 3$ grid with the spatial coevolution strategy used in Lipizzaner. The generator is determined as a mixture of generators in this neighborhood.

Lipizzaner uses two mutation operators. The first operator mutates the learning rates of the optimization method used in the generator and the discriminator. In this case, a normal distribution is used to change the learning rate at small steps at each generation. The second operator is a gradient-based mutation that updates the weights of the individuals in the populations of generators and discriminators. Lipizzaner uses the Adam optimizer [23] to update the weights. Furthermore, an evolution strategy combined with a performance metric (e.g., the Inception Score or FID) is used to update the mixture of weights.

The model was evaluated on the MNIST and CelebA datasets, using a $2 \times 2$ grid, forming a population of 4 generators and 4 discriminators. These populations were evolved through 400 generations. The authors found that Lipizzaner was able to avoid the mode collapse problem in most of the experiments. The model can recover

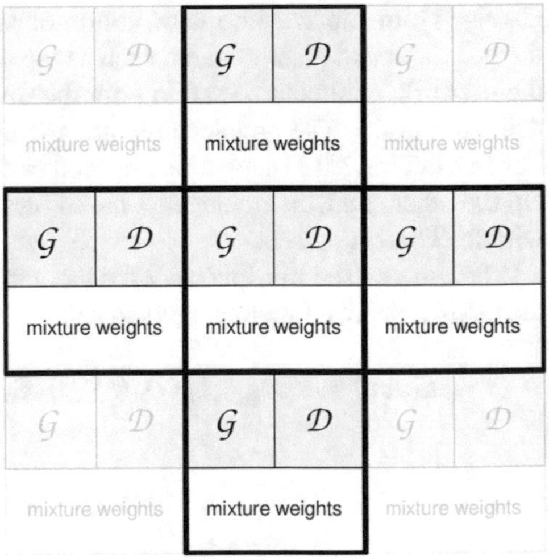

**Fig. 11.3** A $3 \times 3$ grid representing the spatial coevolution mechanism used in Lipizzaner. The neighborhood of the central cell includes the four-highlighted nodes in the grid. Each cell contains one discriminator, one generator, and a mixture of weights

from the mode collapse issue and continue to improve as the training advances through the next generations.

## 11.4.4 Mustangs

The models E-GAN and Lipizzaner were combined in a hybrid approach to train and evolve GANs, called Mutation Spatial GANs (Mustangs)[5] [46]. As in Lipizzaner and E-GAN, the topologies of the generator and discriminator are fixed during the algorithm, i.e., the architectures are not a target of the EA.

Mustangs combines the mutation operators used in E-GAN and the spatial coevolution mechanism used in Lipizzaner. The goal is to increase the diversity of genomes in the population. Thus, the loss function of generators can be modified by the mutation operator, as in E-GAN. As in Lipizzaner, the match between individuals occurs in a toroidal grid, and the internal weights of the neural networks are calculated based on the neighborhood.

The Mustangs model uses the same fitness strategy used in Lipizzaner, i.e., the fitness is based on the GAN objective function $\mathcal{L}(u, v)$, defined by Eq. (11.9). Thus, at the evaluation step, the value $\mathcal{L}(u_i, v_j)$ is also calculated for each pair $(G_i, D_j)$, and the fitness values are also updated as $f_{u_i} \mathrel{-}= \mathcal{L}(u_i, v_j)$ and $f_{v_j} \mathrel{+}= \mathcal{L}(u_i, v_j)$ for generators and discriminators, respectively.

The operators used in Mustangs are a combination of the ones used in Lipizzaner and E-GAN. Therefore, as in E-GAN, the loss function of the individuals can be changed. However, the strategy used here is to randomly select one of the three possibilities for the loss function, instead of evaluating the individuals using all losses. The mutation operators used in Lipizzaner are also applied for Mustangs. Mustangs also applies an evolution strategy to update the weights. Crossover is not used in this proposal.

The evaluation phase follows the same proposal of Lipizzaner. Mustangs uses spatial coevolution to pair discriminators and generators, using a toroidal grid to spatially distribute the individuals. Therefore, individuals are matched using the grid neighborhood to calculate the fitness and evaluate each individual. As in Lipizzaner, the generator is determined as a mixture of generators in this neighborhood.

Mustangs was evaluated with the MNIST and the CelebA datasets. As the architectures of the neural networks that compose a GAN are fixed and predefined, the authors chose different topologies according to the dataset used in the experiments. A four-layer MLP network with 700 neurons and a DCGAN-based architecture were used for the experiments with the MNIST and the CelebA dataset, respectively. For MNIST, a grid size of $3 \times 3$ was used with a time limit of 9 h. For CelebA, the experiments were executed with a $2 \times 2$ grid for 20 epochs. A comparison between standard GAN, E-GAN, Lipizzaner, and Mustangs was presented. The

---

[5]Code available at https://github.com/mustang-gan/mustang.

authors found that Mustangs is able to generate the best results concerning the FID score. They also concluded that spatial coevolution is an efficient way to model the population of generators and discriminators to train GANs.

### 11.4.5  COEGAN

Coevolutionary Generative Adversarial Networks (COEGAN),[6] a proposal combining neuroevolution and coevolution to train and evolve GANs, was proposed by us in [9, 10]. This approach took inspiration on DeepNEAT [28], adapting and extending the EA to the context of GANs.

An array of genes compose the genome of COEGAN. The genotype-phenotype mapping transforms this array into a sequence of layers to compose a neural network. Each gene represents either a linear, convolution, or transpose convolution layer (also known as deconvolution layer). Moreover, each gene also has some common internal parameters, such as the activation function, chosen from the following set: ReLU, Leaky ReLU, ELU, Sigmoid, and Tanh. The genes representing a convolution or transpose convolution layer only have the number of output channels as a variable parameter. The number of input channels is calculated dynamically, based on the setup of the previous layer. The stride and the kernel size are previously defined but are dynamically adjusted to fit the output size of a layer. Similarly, the linear layer only has the number of output features as a variable parameter. The previous layer is also used to calculate the number of input features. Thus, the parameters subject to variation operations are the activation function, the number of output features, and the number of output channels.

Figure 11.4 illustrates examples of the genotypes of a discriminator and a generator. The genotype of the discriminator is composed of a convolutional section and followed by a linear section (composed of fully connected layers). As in the original GAN model, the discriminator outputs the probability that the input sample is a real sample, drawn from the dataset. Similarly, the genotype of the generator is composed of a linear section and followed by a transpose convolutional section (also known as convolutional section). The generator outputs a fake sample, with the same characteristics (i.e., dimension and channels) of a real sample.

Competitive coevolution was used to model the algorithm. Therefore, COEGAN is composed of two separated subpopulations: a population of generators, where each $G_i$ represents a generator; and a population of discriminators, where each $D_j$ represents a discriminator. A speciation mechanism, inspired by the strategy used in NEAT, was used in each subpopulation to promote innovation. The speciation mechanism ensures that recently modified individuals will have the chance to survive for enough generations to be as powerful as individuals from previous generations. For this, each population is divided into species based on a similarity

---

[6]Code available at https://github.com/vfcosta/coegan.

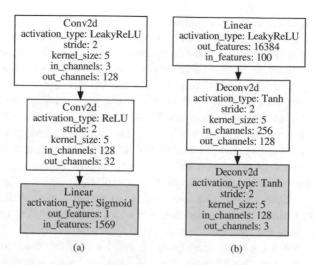

(a)                                          (b)

**Fig. 11.4** Genotypes of a discriminator (**a**) and a generator (**b**). In both cases, the genotype is composed of three genes: two convolutions and one linear for the discriminator, one linear and two deconvolutions for the generator. The phenotype transformation creates a network with three layers in the same linear sequence as displayed in the genomes. For the discriminator, the output layer is represented by the linear gene and outputs the probability of the input sample to be real or fake. For the generator the final gene represents a deconvolution layer that outputs the samples created by the generator

function (used to group similar individuals). Thus, the innovation, represented by the addition of new genes into a genome, may cause the creation of new species in order to fit the individuals containing these new genes. The individuals belonging to new species will have a higher chance to survive because they will not directly compete with more powerful individuals from other species.

COEGAN is only interested in the evolution of the neural network architectures. Thus, the parameters of the layers in the phenotype (e.g., weights and bias) are not part of the evolution, being modified by the training with a gradient descent method. The variation operators are focused on the evolution of the network topology.

Different fitness functions for the generator and the discriminator were used in COEGAN. For discriminators, the fitness is based on the loss function of the original GAN model, i.e., the fitness is equivalent to Eq. (11.1) (Sect. 11.2). The same approach was tested on the generator using Eq. (11.2) (Sect. 11.2), but preliminary results presented instabilities when using this strategy, making it not suitable to be used as fitness. Thus, the generator uses the FID score [18] as fitness, i.e., the fitness is represented by Eq. (11.4) (Sect. 11.2.3). FID is the state-of-the-art metric to compare GANs and outperforms other metrics, such as the Inception Score [39]. The use of the FID score as fitness puts selection pressure in COEGAN and directs the evolution of the population towards the creation of better generators, and consequently better discriminators.

Only mutations are used as variation operators for COEGAN. The mutation process is composed of three operations: add a new layer, remove a layer, and change an existing layer. In the addition operation, a new layer is randomly selected from the set of possible layers (linear or convolution for discriminators and linear or transpose convolution for generators). The remove operation randomly selects an existing layer and excludes it from the genotype, adjusting the connections between the previous and the next layers. The change operation acts on the activation function and the specific attributes of a layer. The activation function is randomly drawn from the set of possibilities. The number of output features and the number of output channels can be mutated for the linear and convolution layers, respectively. These attributes are mutated using a uniform distribution with a predefined range to limit the possible values. Crossover was also experimented and evaluated in preliminary experiments but it was discarded as it promotes instability, decreasing the performance of the system.

COEGAN keeps the parameters (weights and bias) of the genes involved in a mutation operator when possible. So, the new individual will carry the information from previous generations and the training continues from the last state, simulating the transfer learning mechanism used in deep neural networks. However, in some cases these parameters cannot be kept, such as when the change occurs in the parameters of a linear or a convolution layer. In these cases, the new setup of the layer is incompatible with the previous configuration, and the new layer will be trained from the beginning.

In the evaluation step of the EA, individuals from the populations of discriminators and generators must be paired to be trained and to calculate the fitness for the individuals. The pairing strategy is crucial to coevolution, and some challenges can be related to the issues occurred in the GAN training (see Sect. 11.3.1). Two pairing strategies were used to evaluate COEGAN: *all vs. all* and *all vs. k-best*.

In *all vs. all*, each discriminator is paired with each generator, resulting in all possible matches. In this case, the fitness for discriminators is the average of the losses obtained by the training with each generator. As the FID score does not use the discriminator in the calculation, the pairing strategy does not affect the fitness for generators. The *all vs. all* strategy is important to promote diversity in the GAN training and improve the variability of the environment for both discriminators and generators. However, the trade-off is the time to execute this approach. The *all vs. all* approach was used in the experiments presented in [9].

In *all vs. k-best*, $k$ individuals are selected from one population to be matched against all individuals in the other population. Therefore, each generator is paired with $k$ best discriminators from the previous generation and, similarly, each discriminator with $k$ best generators. For the first generation, a random approach is used, i.e., $k$ random individuals are selected for pairing in the initial evaluation. This approach provides less variability in the training but is more efficient, as fewer matches will be executed per generation. The *all vs. k-best* approach with $k = 3$ was used in the experiments presented in [10]. The *all vs. all* strategy achieved better results than *all vs. k-best*, presenting a more stable training for COEGAN [9].

For the selection phase, COEGAN uses a strategy based on NEAT [42]. The populations of generators and discriminators are divided into subpopulations using a speciation strategy based on the one used in NEAT. Each species is composed of individuals with similar genomes, i.e., similar network structures. Therefore, the similarity between individuals is based only on the parameters of each gene composing the genome, excluding the weights of the similarity calculation. The distance $\delta$ between two genomes $i$ and $j$ is defined as the number of genes that exist only in $i$ or $j$. The speciation approach uses the distance to cluster individuals based on a $\delta_t$ threshold. This threshold is calculated at each generator in order to fit the previously chosen number of species. Fitness sharing is used to adjust the fitness of individuals inside each species. Individuals are selected in proportion to the average fitness of the species they belong to. Besides this process, a tournament between $k_t$ individuals is applied in each species to finally select the individuals to breed and compose the next population.

To evaluate the COEGAN proposal, experiments using the MNIST and the Fashion MNIST datasets were presented. These experiments compare COEGAN, a DCGAN-based solution, and a random search method using the FID Score, Inception Score, and the root mean square error (RMSE) metrics. The size of the genome was limited to six layers. The probabilities for the variation operators are 20%, 10%, and 10% for the add, remove, and change mutations, respectively. The number of output features and channels follows a uniform distribution, delimited by the interval [32, 1024] and [16, 128], respectively. The experiments ran for 50 generations, using 10 individuals for the populations of generators and discriminators.

Figure 11.5 presents the results of the FID score on the MNIST dataset (lower is better). We can see that COEGAN outperforms the other approaches. The random

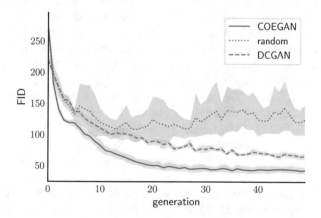

**Fig. 11.5** The FID score on the MNIST dataset comparing COEGAN, the DCGAN-based architecture, and the random search method. Note that, as expected, the random search does not achieve good results and presents high variability on the FID score. The DCGAN-based result shows the convergence of the GAN training. However, COEGAN presents the best results and a smooth decreasing pattern on the FID score

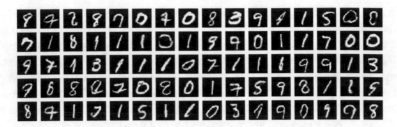

**Fig. 11.6** Samples generated by COEGAN when training on the MNIST dataset

approach presented high variability and worse results in terms of this metric. This is evidence that the choices for the fitness functions for COEGAN provide enough evolutionary pressure to guide the evolution to better outcomes.

Figure 11.6 displays the samples created by the generator after the COEGAN training process. We can see the samples in this figure resembling the data in MNIST. No evidence of the vanishing gradient was found in the experiments with COEGAN, and the mode collapse occurred only partially in some executions. COEGAN avoids these issues by using the evolutionary pressure to discard failed individuals from the population. As these individuals will perform worse than others, they will eventually not be selected, and their issues will not persist through generations. The diversity provided by the population of generators and discriminators is also a factor that prevents these issues from happening. The variability of the training with multiple instances of generators and discriminators, instead of a single generator and discriminator, can be a way to provide a stronger and stable training for GANs.

In order to assess the applicability of the solution in complex datasets, we expand the experiments with COEGAN to include the results with the CelebA dataset [25]. For this, we use an experimental setup similar to the one applied in [9]. However, for the sake of simplicity, we only use convolution and transpose convolution layers when adding a new gene, excluding the linear layer from the set of possibilities. Furthermore, we allow only ReLU and Leaky ReLU as possible activation functions in the mutation operators. The populations of generators and discriminators contain 10 individuals each, divided into three species. The *all vs. all* pairing strategy was applied, using 100 batches of 64 images to train each pair. The images from the CelebA dataset were rescaled to 64 × 64. Each experiment was repeated three times, and the presented results are the average of these executions with a confidence interval of 95%.

Figure 11.7 presents the FID score for COEGAN through generations. As expected, we can note the decreasing behavior of the FID score, resembling the behavior presented in the MNIST results (Fig. 11.5). This is an indication of the generalization ability of COEGAN to effectively work with more complex datasets like CelebA. The average FID score achieved by COEGAN at the last generation is $89.8 \pm 17.2$. No evidence of the vanishing gradient and mode collapse was found in the experiments.

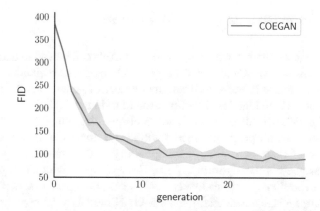

**Fig. 11.7** The FID score of COEGAN on the CelebA dataset. COEGAN achieves a FID score of $89.8 \pm 17.2$ at the last generation

**Fig. 11.8** Samples generated by COEGAN when training on the CelebA dataset

Figure 11.8 displays samples created by COEGAN at the final generation of one experiment. We can clearly see the formation of faces in each created sample, with elements coherently positioned in each face. The variety achieved on samples also demonstrates that COEGAN achieved convergence when training, avoiding problems such as the mode collapse. However, the produced samples are not perfect. Undesired artifacts can be seen in some samples, affecting the quality of the outcome.

## 11.5   Discussion

Section 11.4 presented the current proposals that apply evolutionary algorithms in the context of GANs. We can see that a variety of techniques frequently used in EAs, and introduced here in Sect. 11.3, were used in these proposals. Following we present and discuss these characteristics regarding the aspects of the GAN model used in the proposals, the choices concerning the EA, and the experimental results.

## 11.5.1  Characteristics of the GAN Model

Table 11.1 presents choices with respect to the GAN model used in each proposal. These proposals are compared under the perspective of four attributes: the number of discriminators used in the algorithm, the number of generators, the architecture of each component, and the loss function used to train the GAN.

Except for E-GAN, all proposals used multiple discriminators in their model. For the generators, all proposals used multiple generators, with E-GAN using a fixed number of three generators, corresponding to the number of possible loss functions designed in the algorithm. Thus, E-GAN works with small populations, limiting the evolutionary options that can emerge through generations. On the other hand, Mustangs adapted successfully the E-GAN model in the context of a larger population, using the spatial coevolution approach of Lipizzaner to handle the individuals.

Regarding the architecture, only the Pareto GAN and COEGAN used an evolvable approach. The other proposals used a predefined and fixed architecture for the neural networks of generators and discriminators. Therefore, Pareto GAN and COEGAN work with larger search spaces, as the architectures that can emerge from the EA have a high number of possibilities. They are also potentially able to enhance the balance between generators and discriminators, as the complexity of the architecture is determined by the algorithm.

Lipizzaner and COEGAN use a fixed loss function for the GAN training. E-GAN, Pareto GAN, and Mustangs use an evolvable approach to the loss function. This approach uses a set of predefined possibilities to select and attribute a loss function to an individual. A more flexible approach can also be used instead of a predefined set, using genetic programming to discover better loss functions for GANs. However, the proposals analyzed in this chapter did not explore this approach.

**Table 11.1**  Aspects of the GAN used in the evaluated proposals

| Algorithm | Discriminator | Generator | Architecture | Loss function |
|-----------|---------------|-----------|--------------|---------------|
| E-GAN | Single-fixed | Three | DCGAN-based | Evolvable[a] |
| Pareto GAN | Many | Many | Evolvable | Evolvable[a] |
| Lipizzaner | Many | Many | MLP and DCGAN-based[b] | Original GAN |
| Mustangs | Many | Many | MLP and DCGAN-based[b] | Evolvable[a] |
| COEGAN | Many | Many | Evolvable | Original GAN |

[a]The loss function is selected using a predefined set of possibilities
[b]The DCGAN-based architecture was used with the CelebA dataset and a simpler approach was applied with the MNIST dataset (see Sects. 11.4.3 and 11.4.4)

## 11.5.2  Aspects of the Evolutionary Algorithm

Table 11.2 presents a comparison between the solutions presented in Sect. 11.4, focusing on the aspects of the evolutionary algorithm. Four aspects of the EA were analyzed: the pairing approach, the variation operators, the fitness function, and the selection method.

As multiple generators and/or discriminators are used in all proposals, and the GAN training occurs using generators and discriminators as adversarial, an approach has to be used to pair the individuals. With the exception of Pareto GAN, all other solutions use separated individuals to represent discriminators and generators. In E-GAN, as there are only a single discriminator and three generators, the policy for pairing is to use the discriminator to evaluate all three generators. In COEGAN, the *all vs. all* and *all vs. k-best* were used. Lipizzaner and Mustangs use the same spatial coevolution strategy to match generators and discriminators. It is important to note that the spatial coevolution mechanism applied in Lipizzaner and Mustangs uses the mixture of weights from the neighborhood to compose the weights of the generator in each cell, taking advantage of multiple individuals to produce a single model. The other solutions do not apply an analogous mechanism to combine weights from different individuals. COEGAN and Pareto GAN have individuals with diverse architectural characteristics in the population, preventing the use of the mixture mechanism designed for Lipizzaner and Mustangs.

The variation operators are paramount to provide diversity in the search for good solutions in an EA. Pareto GAN uses crossover and mutation as operators. It is also the solution that provides the most variability regarding the elements that can be evolved through generations in the EA. As Pareto GAN models its individual as a representation of the entire GAN, i.e., encoding both the discriminator and the generator into the genotype, the crossover works exchanging the generator and the discriminator between two parents to form the offspring. The other solutions modeled the GAN with independent genotypes to represent the generator and the discriminator. Therefore, this approach is not applicable to them. COEGAN also evaluated a strategy to apply crossover, using a cut point to share parts of the neural network between parents. However, this strategy proved to be not efficient for the method.

**Table 11.2**  Aspects of the evolutionary algorithm used in the evaluated proposals

| Algorithm | Pairing | Variation operators | Fitness | Selection |
|---|---|---|---|---|
| E-GAN | One-vs-three | Mutation (loss) | Custom | Best individual |
| Pareto GAN | – | Crossover and mutation | IGD | Pareto dominance |
| Lipizzaner | Spatial coevolution | Mutation (weights) | GAN objective | Spatial |
| Mustangs | Spatial coevolution | Mutation (weights, loss) | GAN objective[a] | Spatial |
| COEGAN | all vs. (all | k-best)[b] | mutation (architecture) | FID and loss | NEAT-based |

[a]The FID score is used as the performance metric to evolve the mixture of weights in Mustangs
[b]COEGAN presented experiments using both the *all vs. all* and the *all vs. k-best* approaches

COEGAN and Pareto GAN are the only solutions that have evolvable neural network architectures. The mutation operator is used to provide small changes in these architectures that are built through generations to produce strong discriminators and generators. E-GAN, Lipizzaner, and Mustangs use a restricted mutation strategy. In E-GAN, only the loss function can be switched. In Lipizzaner, gradient-based mutations are applied to update the weights of generators and discriminators. Furthermore, Lipizzaner uses an evolution strategy to update the mixture of weights used for generators. Mustangs combines the operators of E-GAN and Lipizzaner. Different from Lipizzaner and Mustangs, COEGAN does not apply a mutation operator directly to the weights. However, this option can be explored to develop a hybrid approach that evolves the weights when the gradient descent training stagnates for a number of generations.

The choice for fitness is diverse among the proposals. E-GAN uses a custom function that represents the quality and diversity of the created samples. As only the generator is subject to evolution, the discriminator does not have a fitness associated. Pareto GAN based its fitness on the concepts of the Pareto front, using the inverted generational distance (IGD) to represent the fitness value. Lipizzaner and Mustangs use the GAN objective function to calculate the fitness for the individuals. In addition, the FID score was used as the performance metric to evolve the mixture of weights in [46]. COEGAN follows a distinct approach for the fitness function. The loss function of discriminators of the original GAN model is used as fitness for them. In the generator, the FID score is used as fitness. COEGAN takes advantage of the capabilities in the FID distance to represent the diversity and quality of the created samples. As the FID is commonly used by researchers to compare GANs, the implementation of this metric into an EA is a way to provide automatic insight about the solutions produced by the method.

The selection method used in E-GAN is based on the choice of the best generator. As E-GAN has only three generators, each one with a specific loss function, the fitness guides the evolution by selecting the function that fits the best generator for the current environment. The switches between functions through generations give to E-GAN sufficient training diversity to achieve convergence. In Pareto GAN, Pareto dominance is used as the strategy to select individuals to form the next generation. Lipizzaner and Mustangs have a selection strategy based on the spatial coevolution mechanism used in the evaluation phase. The neighborhood is used to evaluate and replace the individual in the center of a neighborhood according to the fitness. COEGAN uses an approach based on classical NEAT selection. Therefore, speciation is used to ensure that individuals from different species will have the opportunity to develop the skills needed to survive. Some of these strategies can be combined into a single solution to build a stronger algorithm. For example, the mechanism that guides the selection for Lipizzaner and Mustangs can be applied in COEGAN to reduce the complexity of the evaluation phase and bring the advantages given by spatial coevolution.

## 11.5.3  Experiments and Results

Table 11.3 compares the proposals under the perspective of the experimental setup used to assess the contributions of each solution. Four experimental attributes are presented: the dataset used in the training, the number of generators and discriminators in the populations, the number of generations used in training, and the metric used to evaluate the results.

Except for Pareto GAN, all proposals used image datasets in the experiments. Pareto GAN uses bi-objective functions to validate the model, also including a function that simulates the data dimension of the MNIST dataset. In the category of images, MNIST is a simple dataset and should be used carefully to draw generic conclusions about the performance of a solution. The CelebA dataset is perhaps the most commonly used data to validate GANs. Therefore, it would be important to assess the performance of Pareto GAN in this dataset.

The populations used in the experiments vary a lot among the proposals. Except for E-GAN, the solutions used multiple individuals for both populations in the experiments. Although it is possible to use more individuals in E-GAN, the experiments used only a single discriminator and three possibilities for generators (representing each possible loss function). In Pareto GAN, one individual completely represents a GAN. Therefore, 20 individuals were used, meaning that 20 independent GANs with their own generator and discriminator was trained through generations. Lipizzaner and Mustangs use spatial coevolution to distribute the individuals in a grid of $2 \times 2$ for the MNIST dataset. For CelebA, Mustangs used a grid of $3 \times 3$. As these grids hold a single generator and discriminator in each cell, the population is composed of 4 and 9 individuals for the $2 \times 2$ and $3 \times 3$ setups, respectively. As a five-cell neighborhood is applied, spatial coevolution reduces the number of iterations needed to evaluate the individuals. Thus, a larger number of individuals can be used to evaluate Lipizzaner and Mustangs. Besides, COEGAN can adopt the spatial coevolution approach to reduce the training time and also increase the number of individuals in the experiments.

**Table 11.3** Comparison of the experiments presented in the proposals

| Algorithm | Dataset | Population ($D \times G$) | Generations | Metric |
|---|---|---|---|---|
| E-GAN | CIFAR-10, LSUN, CelebA | $1 \times 3$ | 200,000 | Inception Score |
| Pareto GAN | Bi-objective functions | 20[a] | 500 | IGD |
| Lipizzaner | MNIST, CelebA | $4 \times 4$ | 400 | – |
| Mustangs | MNIST, CelebA | $4 \times 4, 9 \times 9$ | Time-limited, 20 | FID score |
| COEGAN | MNIST, Fashion MNIST, CelebA | $10 \times 10$ | 30, 50 | FID score |

[a]In Pareto GAN one individual completely represents a GAN, i.e., it contains both a generator and a discriminator

The number of generations used to evaluate each approach also presents high variability. Each approach adapted the experiments to use a number of generations respecting their internal characteristics. For example, as E-GAN works with smaller populations, the number of generations needed to converge is much higher than the others. On the other hand, COEGAN used only 50 generations on the experiments with MNIST and Fashion MNIST. For the experiments with CelebA (Sect. 11.4.5), COEGAN ran for 30 generations. Mustangs used a time-limited strategy of 9 h for the experiments with MNIST and a limit of 20 generations for experiments with CelebA. The time-limited approach used in the MNIST experiments corresponds to more than 150 generations.

Because COEGAN uses a population of 10 individuals for generators and discriminators with the *all vs. all* pairing approach, each individual will execute the training process for ten times at each generation. Furthermore, COEGAN uses multiple batches when training a pair of generators and discriminators at each generation. Mustangs uses a five-cell neighborhood to train the individuals, having a lower number of samples in each training step when compared to COEGAN. However, it is important to note that COEGAN also evolves the architecture of the neural networks, requiring more samples per training step to achieve convergence. On the other hand, the architectures of generators and discriminators in Mustangs are fixed. Therefore, Mustangs is more efficient with respect to the number of samples used at each training step, but COEGAN also provides neural architecture search for discriminators and generators in the solution.

A metric is commonly used to evaluate the samples created by the generator. COEGAN and Mustangs use the FID score to report and analyze the results. As discussed in Sect. 11.2.3, the FID score is currently the state-of-the-art metric used to evaluate and compare GANs. The Inception Score, the former most used metric for GANs, was applied in the E-GAN experiments. Pareto GAN adopted the IGD as the metric, that is adequate to its approach that is based on the Pareto set approximations. Lipizzaner analyzed the results through visual inspections and does not present an evaluation with respect to some objective measurement.

As the proposals use different metrics, we cannot directly compare the results between all proposals. Only COEGAN and Mustangs share the same metric in the evaluation of the results. The average FID for experiments with MNIST reported by COEGAN [10] and Mustangs [46] are 49.2 and 42.235, respectively. Further experiments for COEGAN [9] achieved an average of 42.6 for the FID score. However, the difference between the average FID scores of COEGAN and Mustangs is small and experiments with equal conditions should be made to better compare these solutions.

For the CelebA dataset, the FID score reported in experiments with Mustangs was 36.148, outperforming the FID score of 89.8 obtained by COEGAN. However, this difference is not evident in a visual inspection of the samples produced by both solutions.

## 11.6   Conclusions

We present in this chapter the state-of-the-art of evolutionary algorithms applied to Generative Adversarial Networks (GANs). An overview of GANs introduces the challenges of the training method and how the common problems affect the resulting performance. We also explore the applicability of concepts related to evolutionary computation in the context of GANs, showing components that can be evolved and participate actively in an EA. These concepts are materialized into the state-of-the-art proposals of EAs applied to GANs that can be found in the literature. We discuss the characteristics of these proposals, demonstrating the drawbacks and possible improvements for further research.

Despite the recent advances in GANs, it is possible to see that there are still open problems. The stability of training remains a challenge, being tackled by researches using different approaches, such as the proposal of new loss functions and/or alternative architectures. GAN is a relatively new model, and the use of EAs in this context is in its early years. With the rise of the computational power and new methods to apply EAs with robust machine learning techniques (e.g., deep learning), EAs can be viewed as a strong way to train and evolve GANs. In this way, the proposals presented in this chapter showed advantages in the union between EAs and GANs. A set of different techniques was used by them, with different choices concerning the GAN model and the EA. The diversity of strategies present in GANs and also in evolutionary computation composes a large number of open possibilities for exploration.

As future work, the techniques used in the proposals presented in this chapter can be combined in the development of new solutions. For example, the spatial coevolution strategy used in Mustangs and Lipizzaner can be adapted to the other proposals. On the other hand, the neuroevolution techniques used in Pareto GAN and COEGAN can also be evaluated in the other solutions. Besides, the proposed solutions can be explored in larger experiments. The algorithms can run on a larger number of generations and, when possible, with a larger population of generators and discriminators. These experiments can make possible to evaluate the quality of the outcome and also the scalability of the proposals. Complex datasets can also be used to assess the robustness of the proposed solutions. Different techniques related to GANs can also be incorporated into the algorithm. For example, the use of alternative loss functions (as in WGAN [3]), spectral normalization [31], or the self-attention module for GANs [51]. Concerning neural networks, other techniques can also be experimented, such as the recently proposed competitive gradient descent algorithm [40]. Alternative fitness functions can also be investigated to better guide the progress of GANs in an EA. For example, the skill rating metric [33] uses the mechanism that classifies the skill of players in a game to quantify the performance of generators and discriminators in GANs. The adversarial characteristics of GANs and a competitive coevolution environment can leverage the advantage with the use of this metric, providing an efficient evaluation of individuals in the population of generators and discriminators.

# References

1. Al-Dujaili, A., Schmiedlechner, T., Hemberg, E., O'Reilly, U.M.: Towards distributed coevolutionary GANs. In: AAAI 2018 Fall Symposium (2018)
2. Antonio, L.M., Coello, C.A.C.: Coevolutionary multiobjective evolutionary algorithms: survey of the state-of-the-art. IEEE Trans. Evol. Comput. **22**(6), 851–865 (2018)
3. Arjovsky, M., Chintala, S., Bottou, L.: Wasserstein generative adversarial networks. In: International Conference on Machine Learning, pp. 214–223 (2017)
4. Barratt, S., Sharma, R.: A note on the inception score. Preprint, arXiv:1801.01973 (2018)
5. Berthelot, D., Schumm, T., Metz, L.: BEGAN: Boundary equilibrium generative adversarial networks. Preprint, arXiv:1703.10717 (2017)
6. Borji, A.: Pros and cons of GAN evaluation measures. Comput. Vis. Image Underst. **179**, 41–65 (2019)
7. Brock, A., Donahue, J., Simonyan, K.: Large scale GAN training for high fidelity natural image synthesis. In: 7th International Conference on Learning Representations, ICLR 2019, New Orleans, 6–9 May 2019
8. Coello, C.A.C., Sierra, M.R.: A study of the parallelization of a coevolutionary multi-objective evolutionary algorithm. In: Mexican International Conference on Artificial Intelligence, pp. 688–697. Springer, Berlin (2004)
9. Costa, V., Lourenço, N., Correia, J., Machado, P.: COEGAN: evaluating the coevolution effect in generative adversarial networks. In: Proceedings of the Genetic and Evolutionary Computation Conference, pp. 374–382. ACM, New York (2019)
10. Costa, V., Lourenço, N., Machado, P.: Coevolution of generative adversarial networks. In: International Conference on the Applications of Evolutionary Computation (Part of EvoStar), pp. 473–487. Springer, Berlin (2019)
11. Durugkar, I., Gemp, I., Mahadevan, S.: Generative multi-adversarial networks. Preprint, arXiv:1611.01673 (2016)
12. Elgammal, A., Liu, B., Elhoseiny, M., Mazzone, M.: CAN: Creative adversarial networks, generating "art" by learning about styles and deviating from style norms. Preprint, arXiv:1706.07068 (2017)
13. Fedus, W., Rosca, M., Lakshminarayanan, B., Dai, A.M., Mohamed, S., Goodfellow, I.: Many paths to equilibrium: GANs do not need to decrease a divergence at every step. In: 6th International Conference on Learning Representations, ICLR 2018, Vancouver, 30 April–3 May 2018, Conference Track Proceedings (2018)
14. Garciarena, U., Santana, R., Mendiburu, A.: Evolved GANs for generating Pareto set approximations. In: Proceedings of the Genetic and Evolutionary Computation Conference, GECCO '18, pp. 434–441. ACM, New York (2018)
15. Ghosh, A., Kulharia, V., Namboodiri, V.P., Torr, P.H., Dokania, P.K.: Multi-agent diverse generative adversarial networks. In: Proceedings of the IEEE Conference on Computer Vision and Pattern Recognition, pp. 8513–8521 (2018)
16. Goodfellow, I., Pouget-Abadie, J., Mirza, M., Xu, B., Warde-Farley, D., Ozair, S., Courville, A., Bengio, Y.: Generative adversarial nets. In: Advances in Neural Information Processing Systems, pp. 2672–2680 (2014)
17. Gulrajani, I., Ahmed, F., Arjovsky, M., Dumoulin, V., Courville, A.C.: Improved training of Wasserstein GANs. In: Advances in Neural Information Processing Systems, pp. 5769–5779 (2017)
18. Heusel, M., Ramsauer, H., Unterthiner, T., Nessler, B., Hochreiter, S.: GANs trained by a two time-scale update rule converge to a local NASH equilibrium. In: Advances in Neural Information Processing Systems, pp. 6629–6640 (2017)
19. Hillis, W.D.: Co-evolving parasites improve simulated evolution as an optimization procedure. Physica D **42**(1–3), 228–234 (1990)

20. Isola, P., Zhu, J.Y., Zhou, T., Efros, A.A.: Image-to-image translation with conditional adversarial networks. In: Proceedings of the IEEE Conference on Computer Vision and Pattern Recognition, pp. 1125–1134 (2017)
21. Karras, T., Aila, T., Laine, S., Lehtinen, J.: Progressive growing of GANs for improved quality, stability, and variation. In: 6th International Conference on Learning Representations, ICLR 2018, Vancouver, 30 April–3 May 2018, Conference Track Proceedings (2018)
22. Karras, T., Laine, S., Aila, T.: A style-based generator architecture for generative adversarial networks. Preprint, arXiv:1812.04948 (2018)
23. Kingma, D.P., Ba, J.: Adam: A method for stochastic optimization. In: 3rd International Conference on Learning Representations, ICLR 2015, San Diego, 7–9 May 2015, Conference Track Proceedings (2015)
24. LeCun, Y.: The MNIST database of handwritten digits. http://yann.lecun.com/exdb/mnist/ (1998)
25. Liu, Z., Luo, P., Wang, X., Tang, X.: Deep learning face attributes in the wild. In: Proceedings of the IEEE International Conference on Computer Vision, pp. 3730–3738 (2015)
26. Lucic, M., Kurach, K., Michalski, M., Gelly, S., Bousquet, O.: Are GANs created equal? A large-scale study. Preprint, arXiv:1711.10337 (2017)
27. Mao, X., Li, Q., Xie, H., Lau, R.Y., Wang, Z., Smolley, S.P.: Least squares generative adversarial networks. In: 2017 IEEE International Conference on Computer Vision (ICCV), pp. 2813–2821. IEEE, Piscataway (2017)
28. Miikkulainen, R., Liang, J., Meyerson, E., Rawal, A., Fink, D., Francon, O., Raju, B., Navruzyan, A., Duffy, N., Hodjat, B.: Evolving deep neural networks. Preprint, arXiv:1703.00548 (2017)
29. Mirza, M., Osindero, S.: Conditional generative adversarial nets. Preprint, arXiv:1411.1784 (2014)
30. Mitchell, M.: Coevolutionary learning with spatially distributed populations. In: Computational Intelligence: Principles and Practice, pp. 137–154. IEEE Computational Intelligence Society, Piscataway (2006)
31. Miyato, T., Kataoka, T., Koyama, M., Yoshida, Y.: Spectral normalization for generative adversarial networks. In: 6th International Conference on Learning Representations, ICLR 2018, Vancouver, 30 April–3 May 2018, Conference Track Proceedings (2018)
32. Odena, A., Olah, C., Shlens, J.: Conditional image synthesis with auxiliary classifier GANs. In: Proceedings of the 34th International Conference on Machine Learning-Volume 70, pp. 2642–2651. JMLR.org (2017)
33. Olsson, C., Bhupatiraju, S., Brown, T., Odena, A., Goodfellow, I.: Skill rating for generative models. Preprint, arXiv:1808.04888 (2018)
34. Pan, Z., Yu, W., Yi, X., Khan, A., Yuan, F., Zheng, Y.: Recent progress on generative adversarial networks (GANs): A survey. IEEE Access 7, 36322–36333 (2019)
35. Radford, A., Metz, L., Chintala, S.: Unsupervised representation learning with deep convolutional generative adversarial networks. Preprint, arXiv:1511.06434 (2015)
36. Rawal, A., Rajagopalan, P., Miikkulainen, R.: Constructing competitive and cooperative agent behavior using coevolution. In: 2010 IEEE Symposium on Computational Intelligence and Games (CIG), pp. 107–114 (2010)
37. Reed, S., Akata, Z., Yan, X., Logeswaran, L., Schiele, B., Lee, H.: Generative adversarial text to image synthesis. Preprint, arXiv:1605.05396 (2016)
38. Russakovsky, O., Deng, J., Su, H., Krause, J., Satheesh, S., Ma, S., Huang, Z., Karpathy, A., Khosla, A., Bernstein, M., et al.: ImageNet large scale visual recognition challenge. Int. J. Comput. Vis. 115(3), 211–252 (2015)
39. Salimans, T., Goodfellow, I., Zaremba, W., Cheung, V., Radford, A., Chen, X.: Improved techniques for training GANs. In: Advances in Neural Information Processing Systems, pp. 2234–2242 (2016)
40. Schäfer, F., Anandkumar, A.: Competitive gradient descent. Preprint, arXiv:1905.12103 (2019)
41. Sims, K.: Evolving 3d morphology and behavior by competition. Artif. Life 1(4), 353–372 (1994)

42. Stanley, K.O., Miikkulainen, R.: Evolving neural networks through augmenting topologies. Evol. Comput. **10**(2), 99–127 (2002)
43. Stanley, K.O., Miikkulainen, R.: Competitive coevolution through evolutionary complexification. J. Artif. Intell. Res. **21**, 63–100 (2004)
44. Szegedy, C., Liu, W., Jia, Y., Sermanet, P., Reed, S., Anguelov, D., Erhan, D., Vanhoucke, V., Rabinovich, A.: Going deeper with convolutions. In: Proceedings of the IEEE Conference on Computer Vision and Pattern Recognition, pp. 1–9 (2015)
45. Szegedy, C., Vanhoucke, V., Ioffe, S., Shlens, J., Wojna, Z.: Rethinking the inception architecture for computer vision. In: Proceedings of the IEEE Conference on Computer Vision and Pattern Recognition, pp. 2818–2826 (2016)
46. Toutouh, J., Hemberg, E., O'Reilly, U.M.: Spatial evolutionary generative adversarial networks. Preprint, arXiv:1905.12702 (2019)
47. Wang, C., Xu, C., Yao, X., Tao, D.: Evolutionary generative adversarial networks. Preprint, arXiv:1803.00657 (2018)
48. Wang, Z., She, Q., Ward, T.E.: Generative adversarial networks: a survey and taxonomy. Preprint, arXiv:1906.01529 (2019)
49. Xu, Q., Huang, G., Yuan, Y., Guo, C., Sun, Y., Wu, F., Weinberger, K.: An empirical study on evaluation metrics of generative adversarial networks. Preprint, arXiv:1806.07755 (2018)
50. Yao, X.: Evolving artificial neural networks. Proc. IEEE **87**(9), 1423–1447 (1999)
51. Zhang, H., Goodfellow, I., Metaxas, D., Odena, A.: Self-attention generative adversarial networks. Preprint, arXiv:1805.08318 (2018)

# Part V
# Applications and Others

# Chapter 12
# Evolving Deep Neural Networks for X-ray Based Detection of Dangerous Objects

Ryotaro Tsukada, Lekang Zou, and Hitoshi Iba

**Abstract** In recent years, neural networks with an additional convolutional layer, referred to as convolutional neural networks (CNN), have widely been recognized as being effective in the field of image recognition. In the majority of these previous researches, the structures of networks were designed by hand, and were based on experience. However, there is no established theory explaining how to build networks with higher learning abilities. In this chapter, we propose a framework on automatically obtaining network structures with the highest learning ability for image recognition, through the combination of the various core technologies. We employ EC (evolutionary computation) for the automatic extraction and synthesis of network structures. Additionally, we attempt to perform an effective search in a larger parameter space by gradually increasing the number of training epochs during the generation change process. In order to show the effectiveness of our approach, we apply the proposed method to the task of detecting dangerous objects in an X-ray image data set. Compared with the previous results, we have achieved an improvement in the mAP value. We can also find several by-passes in the structures that were actually obtained.

## 12.1 Introduction

In recent years, machine learning methods using neural networks have significantly outperformed traditional methods in areas such as image recognition [5], speech recognition [2], and natural language processing [1].

Neural networks with an additional convolutional layer, referred to as convolutional neural networks (CNN), are widely recognized as being effective in the field of image recognition. In ILSVRC 2012, a worldwide image recognition contest conducted in 2012, Hinton et al. [5] used a method of image recognition

R. Tsukada · L. Zou · H. Iba (✉)
Graduate School of Information Science and Technology, University of Tokyo, Tokyo, Japan
e-mail: tsukada@iba.t.u-tokyo.ac.jp; zou@iba.t.u-tokyo.ac.jp; iba@iba.t.u-tokyo.ac.jp

© Springer Nature Singapore Pte Ltd. 2020
H. Iba, N. Noman (eds.), *Deep Neural Evolution*, Natural Computing Series,
https://doi.org/10.1007/978-981-15-3685-4_12

325

based on convolutional neural networks and achieved more than 10% significant improvement (compared with previous methods) in terms of recognition accuracy. Subsequently, this achievement has triggered a lot of proactive research on image recognition based on convolutional neural networks; this has resulted in the numerous proposals and performance enhancement of neural networks with more complex and varied structures such as the GoogLeNet [18], ResNet [4], and YOLO [12].

Image recognition is the process of extracting features from images obtained in the real world in order to recognize objects such as characters, symbols, people's faces, and animals that may appear in an image, and it has a wide range of applications. Hence, building systems with increasingly higher recognition accuracies is necessary.

A characteristic of convolutional neural networks is that networks can be constructed by combining layers that perform specific functions, just like blocks that are put together. In the majority of the previous research on image recognition using convolutional neural networks, the structures of networks were designed by hand, and were based on experience. In this context, the improvement in the learning ability of various networks from various core technologies such as dropout, batch normalization, GoogLeNet's inception module [18], and residual learning introduced in ResNet [4] has been empirically verified. However, there is no established theory explaining how to combine these technologies to build networks with higher learning abilities. In fact, networks that currently exhibit the highest learning levels contain a huge number of parameters and are deep and complex. Therefore, specialists must perform a lot of trial-and-error and craftwork in order to yield the highest learning ability on a specific data set. Consequently, there is ongoing research on the automatic design of network structures using genetic programming (GP) [9, 17], and network structure search methods using neural networks [11].

The present research focuses on automatically obtaining network structures with the highest learning ability for image recognition, through the combination of the various core technologies itemized above. We used genetic algorithms (GA) for the automatic extraction and synthesis of network structures. The advantage of genetic algorithms in the present research is that a simple network can gradually evolve into a complex network during the search process with very little prior input from a human. Additionally, we attempt to perform a search in a larger parameter space by gradually increasing the number of training epochs that evaluate each individual during the generation change process.

## 12.2 Related Research

### 12.2.1 Neuro-Evolution

Neuro-evolution in the broad sense is an attempt to generate neural networks by the use of evolutionary computation methods [9]. In the present research, our goal is to optimize the structure of the convolutional neural network by using GA. This process can be likened to an evolutionary computation; thus, this approach can be considered as a form of neuro-evolution.

The NEAT (NeuroEvolution of Augmented Topologies [16]) method is an example of neuro-evolution. The method is characterized by the growth of small structures into larger structures as they get optimized. Using GA, the network undergoes evolution by crossover and mutation, which, respectively, results in better structures and changes in the connectional relationship between nodes. An example of structural change resulting from a mutation in NEAT is illustrated in Fig. 12.1.

### 12.2.2 Genetic CNN

Genetic convolutional neural network (Genetic CNN) [19] is a proposed example of a convolutional neural network structure search using GA. In Genetic CNN, a stage composed of multiple convolutional layers and subsequent pooling layers is repeated multiple times. The convolutional layers in each stage are connected in the form of directed acyclic graphs. As shown in Fig. 12.2, the binary values 0 or 1 are

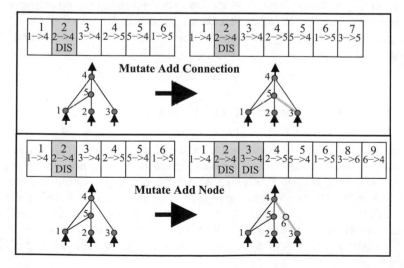

**Fig. 12.1** Example of structure change due to mutation in NEAT [16]

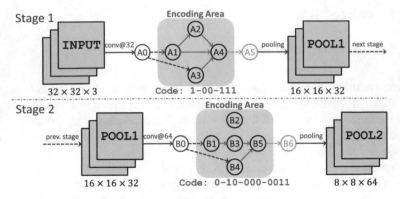

**Fig. 12.2** Example of network structure encoding in Genetic CNN [19]

encoded as a sequence, specifying which previous convolutional layer is connected to itself. These codes are viewed as "genes," and the process of searching for the network structures with higher learning abilities using GA is referred to as Genetic CNN. The number of stages and the number of convolutional layers in each stage are preset; therefore, the length of the sequence that makes up the gene is also fixed.

To represent the fitness of GA, we use the recognition accuracy yielded by the network pertaining to the individual based on a specific data set used for learning. Crossover is achieved by exchanging bits between sequences, and mutations are performed by bit reversal. For the selection, a roulette wheel selection process is adopted where the probability of selection is proportional to the difference in fitness with respect to the individual with the lowest fitness. Therefore, individuals having networks with a structure that yields higher accuracy are more likely to survive into the next generation. This way, trial-and-error involving the combinations of network structures and mutations eventually results in the output of individuals having networks with the highest accuracy. The Genetic CNN algorithm is described in Algorithm 1:

## 12.2.3 Aggressive Selection and Mutation

In Genetic CNN, training is performed from scratch to evaluate the individuals. This process is repeated multiple times for all individuals of the generation during the generation change process. Hence, if the final number of generations is $T$ and the number of individuals in each generation is $N$, training is repeated $T \times N$ times until the end of the generation change process. However, the training of a convolutional neural network normally takes a considerable amount of time for just one round. Consequently, a drawback of these repetitions is the considerable amount of time required.

---

**Algorithm 1** Genetic CNN

---

1: **Input:** data set $\mathcal{D}$, final generation number $T$, number of individuals in each generation $N$, probabilities of crossover and mutation $p_C$ and $p_M$, parameters $q_C, q_M$ related to crossover and mutation

2: **Initialization:** the initial generation is formed, containing $N$ individuals. Each individual consists of a structure where bits 0 and 1 are selected at random. The individual fitness is evaluated (more on this later).

3: **for** $t = 1, 2, \ldots, T$ **do**

4:    **Selection:** considering individuals from generation $t - 1$, $N$ individuals are selected in a roulette wheel scheme where the selection probability is proportional to the fitness difference with regard to the individual with the lowest (worst) fitness; selection of the same individual multiple times is allowed.

5:    **Crossover:** for each pair of selected neighboring individuals in the same generation $t$, the crossover is performed with probability $p_C$ (each bit is exchanged with probability $q_C$).

6:    **Mutation:** for individuals that did not undergo crossover as above, mutation is carried out with probability $p_M$ (each bit is reversed with a probability $q_M$).

7:    **Evaluation:** the network of an individual is trained on data set $\mathcal{D}$ and tested using test images to obtain fitness, which is the accuracy obtained.

8: **end for**

9: **Output:** individuals and their recognition accuracies in the final generation.

---

Additionally, due to the utilization of a roulette wheel selection scheme where the selection probability is proportional to the difference with respect to the fitness of the individual with the lowest fitness, it is possible that individuals having networks with weak structures that are not expected to yield further improvements in accuracy could avoid elimination from selection and survive.

Furthermore, elements not related to connections in the convolutional layer (hyperparameters such as the layout of the pooling layer, number of channels in each layer, filter size, and stride.) must be previously determined, resulting in a small search space for the parameters subject to search.

In [7], the above problems are addressed in the following ways:

(i) evaluation of individuals is accelerated by roughly assessing the fitness of an individual by training using a small number of epochs;

(ii) by introducing a selection and mutation scheme called "aggressive selection and mutation," weak individuals are eliminated early so that new individuals based on strong ones can be born more easily;

(iii) the space of parameters to be searched is enlarged by increasing the elements subject to mutation.

The training of convolutional neural networks is performed by repeatedly feeding the same training data to the networks. The number of such repetitions is called "number of epochs." Usually, a sufficient level of recognition accuracy is not achieved if the network is trained with a small number of epochs. However, to estimate fitness as a reference for generational change in GA, a rough estimation based on a small number of epochs is expected to be enough.

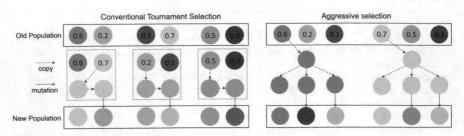

**Fig. 12.3** Different strategies. Aggressive selection and mutation for $N = 6$ and $k = 2$ are shown in the right [7]

The aggressive selection and mutation method involve selecting only $k(\ll N)$ individuals with high fitness from the parents' generation, complementing the lacking part with clones, and applying mutation to those clones. The method is similar to the random mutation hill-climbing method because it searches for a solution based only on mutations. An example of aggressive selection and mutation for $N = 6$ and $k = 2$ is illustrated in Fig. 12.3.

In Genetic CNN, only connectional relations in the convolutional layer are set as a target for optimization. However, in aggressive selection and mutation, various types of mutation operations are available, such as adding or deleting layers other than convolutional layers or changing preset values of the convolutional layer itself (hyperparameters). With this method, the parameter search space gets significantly expanded.

Consequently, the time required to find the best individual is significantly reduced and the recognition accuracy of the best individual is improved dramatically.

The aggressive selection and mutation algorithm is described in Algorithm 2.

---

**Algorithm 2** Aggressive selection and mutation

---

1: **Input:** data set $\mathcal{D}$, final generation number $T$, number of individuals per generation $N$, number of elite individuals $k$ to be added to the next generation, threshold $d$ for the distance between individuals

2: **Initialization:** the 0-th generation is formed by $N$ individuals having a fixed initial structure. The fitness of each individual is evaluated (to be explained later).

3: **for** $t = 1, 2, \ldots, T$ **do**

4:    **Selection:** $k$ individuals with high fitness are selected sequentially from generation $t - 1$. The individual is added to generation $t$ unless the distance with respect to the individuals already added to generation $t$ is less than $d$. A total of $N - k$ clones of the $k$ added individuals are added.

5:    **Mutation:** a mutation operation is selected and applied to each of the $N - k$ cloned individuals. Nothing is done to the remaining $k$ individuals.

6:    **Evaluation:** the network of each individual is trained on data set $\mathcal{D}$, and the recognition accuracy on test images is stored as the fitness value.

7: **end for**

8: **Output:** individuals in the final generation and the recognition accuracy for these individuals.

Details on the initial structure and methods of mutation are explained in Sect. 12.4.

## 12.2.4 YOLO

YOLO is one of single-shot object detection CNN models. YOLO first resizes the input image into a square and divides it into equal-sized regions using $S \times S$ grids. Each grid will predict $B$ bounding boxes and probability values for $C$ categories. An example for $S = 7, B = 2$, and $C = 20$ is illustrated in Fig. 12.4 [13]. Each bounding box needs to be represented by five parameters: the coordinate of the center point on the $x$ and $y$ axes, the height $h$ and width $w$ of the bounding box, and the confidence value $c$. Finally, YOLO selects the prediction box with the highest confidence value as the detection result. In order to avoid multiple detections of the same object, YOLO uses the non-maximum suppression method to ensure that each object is detected only once.

As shown in Fig. 12.4, the entire YOLO network is composed of convolutional layers and fully connected layers without any sub-network structure. Here, the output dimension of YOLO is $7 \times 7 \times 30$. This is because when $S = 7, B = 2$, and $C = 20$, each grid predicts two bounding boxes, probability values for 20 categories, and each bounding box needs five parameters $\{x, y, h, w, c\}$ in total requiring $7 \times 7 \times (5 \times 2 + 20) = 7 \times 7 \times 30$ parameters.

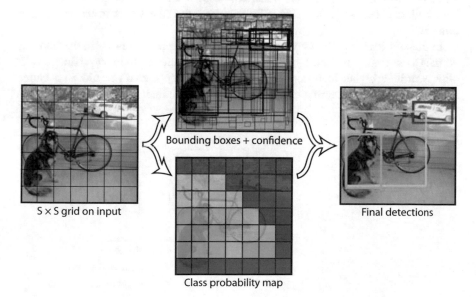

Bounding boxes + confidence

$S \times S$ grid on input

Class probability map

Final detections

**Fig. 12.4** The system model of YOLO [13]

## 12.2.5  Transfer Learning

Transfer learning [10] is a framework in machine learning where data on related problems and derived knowledge are used to effectively and efficiently solve the target problem. The sender and the receiver of knowledge to be transferred are called the source domain and target domain, respectively.

Humans learn various things from transfer learning. For instance, when someone who can play the piano starts to learn the electronic organ at the same time as someone who cannot play the piano, the former can learn to play the electronic organ better and in a shorter amount time than the latter. In the framework of transfer learning, the piano is the source domain, the electronic organ is the target domain, and proficiency in playing the piano assists in the learning of the electronic organ. Transfer learning is applied in various fields, including natural language processing, voice recognition, and image processing.

Figure 12.5 shows a rough flow of transfer learning. In this example, the source task concerns training on female speech and the target task is to recognize speech from males. Learning is carried out using data and knowledge associated with problems in the source and target domains to ultimately answer problems in the target domain efficiently and with high precision. Usually it is assumed that the source and the target domains have some structural relationship.

Transfer learning is very effective when there is little training data in the target domain, but substantial data in the source domain. Moreover, transferring knowledge from a domain that is highly similar to the target domain results in more efficient learning. In contrast, the transfer of knowledge from a source domain with low similarity results in a decrease in learning performance, which is called negative transfer.

In transfer learning, the maximum limit of learning performance in the training domain is normally limited by the learning performance in the source domain. In other words, a higher learning precision in the source domain results in a better chance of improving learning efficiency in the target domain.

**Fig. 12.5**  Image of transfer learning

An important problem in transfer learning is determining what knowledge to transfer. If the data from the source domain is necessary, the most obvious approach is to appropriately map the data and use it when learning in the target domain. On the other hand, transferring feature values or parameters that exist both in the source and target domains is also possible. What knowledge can be transferred and what knowledge successfully works depends on each domain, and determining which to apply is difficult.

## 12.3   Proposed Method

In Sect. 12.2, we have explained the aggressive selection and mutation scheme [7]. In this section, we further introduce an extended method, which we call ASM+, where generation alternation is performed by increasing the number of epochs according to the generation number during the evaluation training of the individuals. Let $n_{min}$, $n_{max}$, and $T$ be the minimum number of epochs, the maximum number of epochs, and the number of total generations, respectively. In our method ASM+, the number of epochs $n(t)$ related to generation $t$ is defined by the following equation:

$$n(t) = \frac{(T-t) \cdot n_{min} + t \cdot n_{max}}{T}$$

Thus, the interval between $n(0) = n_{min}$ and $n(T) = n_{max}$ is uniform and depends on the final generation number $T$.

Due to the slope with respect to the epoch number, the generation change cycle is faster at the beginning of evolution, thereby enabling an evaluation of individuals with various structures over a wide range. Subsequently, individuals exhibiting good structures at the end of evolution are evaluated locally with higher accuracy.

## 12.4   Experiments on Evolutionary Synthesis
## of Convolutional Neural Networks

Experiments were carried out using two types of combinations of data sets and tasks. First, the effectiveness of the proposed method ASM+ is evaluated through a relatively simple handwritten number classification task using MNIST [6] as the data set. Second, we also apply ASM+ to the task of detecting dangerous objects in an X-ray image data set simulating luggage inspection.

## 12.4.1 MNIST Handwritten Number Classification Experiment

For the first experiment, the MNIST [6] data set is used. MNIST is a widely used data set designed for the classification of handwritten numbers. MNIST was selected because of a possible comparison with the previous researches [19] and [7] mentioned in Sect. 12.2. Moreover, the task is simple, and it permits the evaluation of the effectiveness of the method with a small number of computations.

The MNIST data set consists of 60,000 images for training and 10,000 images for testing. Each image is formed by a $28 \times 28$-sized gray scale corresponding to an Arabic number from 0 to 9 which are uniformly drawn on the image.

### 12.4.1.1 Initial Generation

As shown in Fig. 12.6, an individual in the initial generation has a network formed by 3 layers: an input layer, a global max pooling layer, and a fully connected layer. Global max pooling is the process of selecting the maximum value from all channels of the input feature map, with an output to a $1 \times 1$-sized feature map with the same number of channels. For the structure of the initial generation, the accuracy on the MNIST data set is about 11%, which is approximately the same as a network that just outputs random results. This network was selected as the initial structure to confirm that it is unnecessary to introduce restrictions in the search range by including human intervention.

### 12.4.1.2 Mutation Operations

In [7], the 15 types of operations described below were utilized for mutation operations. In the following, "random" refers to the selection of candidates with a uniform probability.

**Fig. 12.6** Network structure of an individual in the initial generation. For details on the notation of the figure, refer to Fig. 12.8

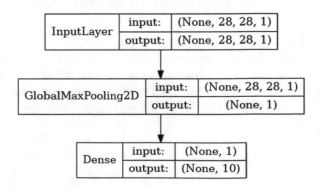

- `add_convolution`: inserts a convolutional layer with 32 channels, one stride, 3 × 3 filter size, and one-pixel padding in a random position. The number of dimensions of the feature maps for the input and output before and after this operation does not change. The activation function used is ReLU.
- `remove_convolution`: selects a single convolutional layer at random and deletes it.
- `alter_channel_number`: selects a single convolutional layer at random, and selects a channel number out of {8, 16, 32, 48, 64, 96, 128} at random for the replacement.
- `alter_filter_size`: selects a convolutional layer at random and selects a filter size out of {1 × 1, 3 × 3 or 5 × 5} for the replacement.
- `alter_stride`: selects a convolutional layer at random and randomly selects a stride out of {1 or 2} for the replacement.
- `add_dropout`: selects a convolutional layer at random and inserts a dropout soon after. The dropout ratio is fixed at 0.5.
- `remove_dropout`: selects a dropout at random for removal.
- `add_pooling`: selects a convolutional layer at random and inserts a pooling layer soon after. Max pooling is adapted and the kernel size is fixed as 2 × 2.
- `remove_pooling`: selects a pooling layer at random and deletes it.
- `add_skip`: inserts a residual network, which was introduced in ResNet [4]. It precisely selects a random pair of layers where the feature maps of the outputs have the same dimension and inserts a layer that has an output formed by the sum of these outputs.
- `remove_skip`: selects at random one of the skip layers above and deletes it.
- `add_concatenate`: like `add_skip`, it selects a random pair of layers where the feature maps of the outputs have the same dimension (however, the numbers of channels may not be the same), and inserts a layer that has an output formed by concatenating the previous outputs.
- `remove_concatenate`: selects one of the concatenate layers above at random and deletes it.
- `add_fully_connected`: selects a random position just after other fully connected layers or the last layer and inserts a fully connected layer. The dimension of the output is selected at random from {50, 100, 150, 200}.
- `remove_fully_connected`: a single fully connected layer is selected at random and deleted.

Selecting and applying one out of the 15 types of operations above is considered a mutation. However, to facilitate the evolution to more complex structures, in our method ASM+, the probability of selecting `add_convolution`, `add_skip`, `add_concatenate`, `alter_stride`, `alter_filter_size`, and `alter_channel_number` is two times higher than other operations.

In some cases, it may be impossible to apply a certain operation. For instance, it is impossible to apply `remove_convolution` to an individual that does not have a convolutional layer. In such cases, the operation is selected again. In other cases, the network structure of a given individual may be considered invalid due to the dimensional mismatch after beginning the actual evaluation training. In such cases, the fitness of such individual is set to 0.

### 12.4.1.3 Experimental Results

Experiments were conducted on a single GPU on Google Colaboratory. Two patterns were defined as follows: with and without batch normalization process inserted just after the convolutional layer.

Other settings were: $T = 30$ (final generation number), $N = 10$ (number of individuals in each generation), $k = 1$ (number of elite individuals to be added to the next generation), $n_{min} = 3$ (minimum number of epochs), and $n_{max} = 12$ (maximum number of epochs).

Figure 12.7 shows how the fitness for the best individual (or image recognition accuracy) evolved along the generations for each pattern. The recognition accuracy obtained for the best individual in the final generation re-evaluated with the maximum number of epochs $n_{max}$ and the time required for the entire evolution process are shown in Table 12.1.

**Fig. 12.7** Changes in fitness value for the best individual in each generation

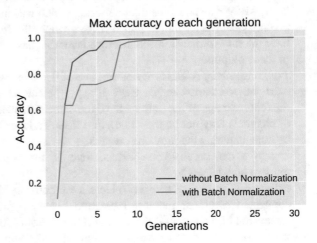

**Table 12.1** Comparison of the recognition accuracy for the best individual in the final generation across different methods

| Method | Recognition accuracy | Computation time |
|---|---|---|
| Genetic CNN [19] | 0.9966 | 48 GPUH |
| Aggressive selection and mutation [7] | 0.9969 | 35 GPUH |
| ASM + (without batch normalization) | 0.9932 | N.A. |
| ASM + (with batch normalization) | 0.9913 | 9 GPUH |

**Fig. 12.8** Network structure for the best individual in the final generation (with batch normalization). InputLayer represents the input layer; Conv2D is a 2-dimensional convolutional layer. BatchNormalization represents batch normalization, and Activation denotes the application of the activation function (ReLU). GlobalMaxPooling2D represents the maximum pooling operation in 2 dimensions across all channels, and Dense represents a fully connected layer. The values in the input and output fields are the dimensions of the input and output, respectively, and individually represent batch size, height, width, and number of channels. "None" indicates that the batch size is arbitrary. Note that fully connected layers have neither height nor width, thus the height and width notations are not shown before and after Dense layer

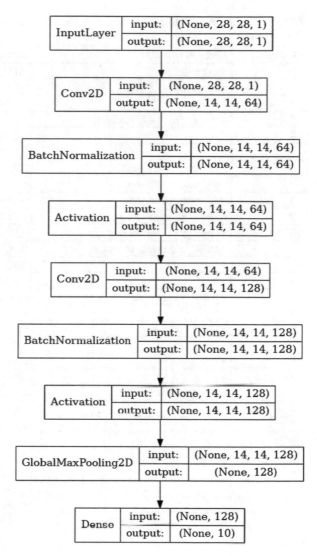

Figure 12.8 shows the network structure for the best individual in the final generation obtained by inserting batch normalization. A list of all the mutation operations that were selected in each generation resulting in that individual is given in Table 12.2. The evolution of the best individual in each generation is illustrated in Fig. 12.9 as a tree diagram.

**Table 12.2** Mutation operations performed on the best individual of each generation (with batch normalization)

| Generation | Mutation operation | Maximum recognition accuracy |
|---|---|---|
| 0 | initialize | 0.1135 |
| 1 | add_convolution | 0.6254 |
| 2 | - | 0.6254 |
| 3 | alter_channel_number | 0.7398 |
| 4 | - | 0.7398 |
| 5 | - | 0.7398 |
| 6 | alter_filter_size | 0.7558 |
| 7 | add_convolution | 0.7695 |
| 8 | add_convolution | 0.9559 |
| 9 | alter_filter_size | 0.9728 |
| 10 | alter_filter_size | 0.9783 |
| 11 | alter_channel_number | 0.9825 |
| 12 | - | 0.9825 |
| 13 | - | 0.9825 |
| 14 | alter_filter_size | 0.9873 |
| 15 | - | 0.9873 |
| 16 | alter_stride | 0.9911 |
| 17 | - | 0.9911 |
| 18 | - | 0.9911 |
| 19 | - | 0.9911 |
| 20 | alter_stride | 0.9916 |
| 21 | - | 0.9916 |
| 22 | - | 0.9916 |
| 23 | - | 0.9916 |
| 24 | alter_channel_number | 0.9922 |
| 25 | - | 0.9922 |
| 26 | - | 0.9922 |
| 27 | - | 0.9922 |
| 28 | alter_channel_number | 0.9933 |
| 29 | - | 0.9933 |
| 30 | - | 0.9933 |

Blank fields denote that no operation took place on the best individual in that generation

## 12.4.2 Experiment on Detecting Dangerous Objects in X-ray Images

The next stage involves experiments that are carried out on a data set of X-ray images created by Zou [20]. This data set contains 6121 X-ray images with 3 types

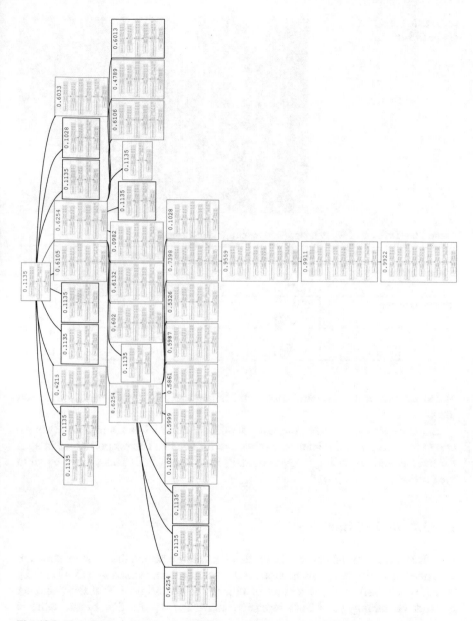

**Fig. 12.9** Tree diagram showing the evolution process of the best individual in each generation (partially omitted). The value accompanying the image of the network structure of an individual represents its fitness. Individuals with a value written in red are those selected for the next generation. Colored arrows indicate the selection flow. Red arrows indicate the occurrence of mutation, and blue arrows indicate cloning

**Fig. 12.10** Portable X-ray
device (NS-100-L)

**Table 12.3** Division of X-ray images into training and test groups

|                  | Training images | Test images | Total      |
| ---------------- | --------------- | ----------- | ---------- |
| Natural images   | 662             | 442         | 1104       |
|                  | (10.82%)        | (7.22%)     | (18.04%)   |
| Synthetic images | 3010            | 2007        | 5017       |
|                  | (49.17%)        | (32.79%)    | (81.96%)   |
| Total            | 3672            | 2449        | 6121       |
|                  | (59.99%)        | (40.01%)    | (100.00%)  |

of dangerous objects (scissors, knives, PET bottles), with annotations related to each image.[1]

Experiments were carried out after dividing the 6121 images into two groups: one to be used for training and the other one for testing. The proportion of images for training and for testing is approximately 6:4 (Table 12.3). This ratio is equal to the previous research [20].[2]

### 12.4.2.1  Initial Generation

Individuals in the initial generation are defined as those having the network structure described in Table 12.4. This network is the same as the one used in YOLOv2 [12], which specialized in the detection of objects. In the original YOLOv2, another network containing the layers indicated above the double line of the table is

---

[1] Instead of multi-view X-ray devices which are expensive and heavy, we use a portable X-ray device (see Fig. 12.10). While multi-view devices need to colorize images, our device can collect single-view X-ray images in real time.

[2] More precisely, the actual images to be used for training and testing are randomly changed each time and the system is run 20 times from Run 1 up to Run 20. Here, we used the set of images corresponding to Run 11, whose results are the most representative of the average.

**Table 12.4** Network structure of individuals in the initial generation

| Type | Filters | Size/Stride | Output |
|---|---|---|---|
| Convolutional | 32 | $3 \times 3$ | $416 \times 416$ |
| Maxpool | | $2 \times 2/2$ | $208 \times 208$ |
| Convolutional | 64 | $3 \times 3$ | $208 \times 208$ |
| Maxpool | | $2 \times 2/2$ | $104 \times 104$ |
| Convolutional | 128 | $3 \times 3$ | $104 \times 104$ |
| Convolutional | 64 | $1 \times 1$ | $104 \times 104$ |
| Convolutional | 128 | $3 \times 3$ | $104 \times 104$ |
| Maxpool | | $2 \times 2/2$ | $52 \times 52$ |
| Convolutional | 256 | $3 \times 3$ | $52 \times 52$ |
| Convolutional | 128 | $1 \times 1$ | $52 \times 52$ |
| Convolutional | 256 | $3 \times 3$ | $52 \times 52$ |
| Maxpool | | $2 \times 2/2$ | $26 \times 26$ |
| Convolutional | 512 | $3 \times 3$ | $26 \times 26$ |
| Convolutional | 256 | $1 \times 1$ | $26 \times 26$ |
| Convolutional | 512 | $3 \times 3$ | $26 \times 26$ |
| Convolutional | 256 | $1 \times 1$ | $26 \times 26$ |
| Convolutional (*) | 512 | $3 \times 3$ | $26 \times 26$ |
| Maxpool | | $2 \times 2/2$ | $13 \times 13$ |
| Convolutional | 1024 | $3 \times 3$ | $13 \times 13$ |
| Convolutional | 512 | $1 \times 1$ | $13 \times 13$ |
| Convolutional | 1024 | $3 \times 3$ | $13 \times 13$ |
| Convolutional | 512 | $1 \times 1$ | $13 \times 13$ |
| Convolutional | 1024 | $3 \times 3$ | $13 \times 13$ |
| Convolutional | 1024 | $3 \times 3$ | $13 \times 13$ |
| Convolutional | 1024 | $3 \times 3$ | $13 \times 13$ |
| Concatenate (**) | $1024 + 256$ | – | $13 \times 13$ |
| Convolutional | 1024 | $3 \times 3$ | $13 \times 13$ |
| Convolutional | 40 | $1 \times 1$ | $13 \times 13$ |

The structure is the same as the network of YOLOv2 [12]. The outputs of convolutional layers marked with (*) are also connected to concatenate layers marked with (**), being combined along the direction of the channel number. Here, the outputs (*) are reduced to 64 channels by 64 filters that perform $1 \times 1$ convolution. Furthermore, to obtain a $13 \times 13$ height and width for the output, they are, respectively, reduced by $\frac{1}{2}$, and the channel number is converted to $64 \times 2^2 = 256$. The by-pass described in Sect. 12.2.4 is realized by this concatenate layer

previously trained on the ImageNet [15] data set where it acquires the ability to extract the features of object recognition. Subsequently, the layers below the double line are replaced by those shown in Table 12.4 and the training for object recognition is repeated, thereby characterizing a transfer learning method (see Fig. 12.11). Using this as a reference, only the layers below the double line becomes the target of evolution. The output is a $13 \times 13 \times 40$ tensor because $B = 5$ and $C = 3$, similar

**Fig. 12.11** Transfer learning for X-ray based detection of dangerous objects

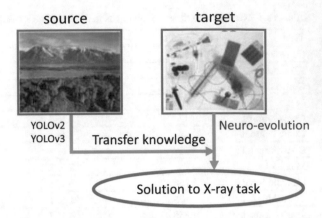

to the case mentioned in Sect. 12.2.4. Here, $C = 3$ represents the three classes of scissors, knives, and PET bottles.

Since evaluating the fitness (mAP, see Sect. 12.4.2.4) of individuals from this initial generation takes a certain amount of time, unlike the initial generation described in Sect. 12.4.1.1 for the MNIST experiment, this process was skipped. Therefore, the group of individuals for the initial generation was just a formality. The experiment essentially began with the group of individuals of the 1st generation, who were obtained by performing mutation operations on the initial generation.

### 12.4.2.2 Mutation Operations

As in Sect. 12.4.1.2, the following nine types of operations based on [7] were chosen as mutation operations. As mentioned in Sect. 12.2.4, the output size of the last layer is fixed in order to output results related to object recognition. Furthermore, to avoid overfitting in YOLO, batch normalization is introduced; thus, dropout is unnecessary [12]. Due to these restrictions, six operations, namely `add_dropout`, `remove_dropout`, `add_skip`, `remove_skip`, `add_fully_connected`, `remove_fully_connected`, were removed. In the following, "random" was defined as the selection of each candidate with uniform probability.

- `add_convolution`: inserts a convolutional layer with 1024 channels, one stride, a $3 \times 3$ filter size, and one-pixel padding in a random position. The numbers of dimensions of the feature maps for the input and output before and after this operation do not change. The activation function used is Leaky ReLU [8].
- `remove_convolution`: selects a single convolutional layer at random and deletes it.
- `alter_channel_number`: selects a single convolutional layer at random and selects a random channel number out of {512, 1024 or 2048} for the replacement.

- `alter_filter_size`: selects a convolutional layer at random and randomly selects a filter size out of {1 × 1, 3 × 3 or 5 × 5} for the replacement.
- `alter_stride`: selects a convolutional layer at random and selects a stride out of {1 or 2} at random for the replacement.
- `add_pooling`: inserts a pooling layer in a random position. Max pooling is adapted, and the kernel size is fixed as 2 × 2.
- `remove_pooling`: selects a pooling layer at random and deletes it.
- `add_concatenate`: selects a layer above and a layer below the double line of Table 12.4. It then inserts a layer with an output formed by concatenating those layers. The insertion point is just after the selected lower layer.
- `remove_concatenate`: selects one of the concatenate layers above at random and deletes it.

Mutation is defined here as the operation of selecting and applying one of the nine operations above. As in Sect. 12.4.1.2, the probability of selecting operations `add_convolution`, `add_concatenate`, `alter_stride`, `alter_filter_size`, and `alter_channel_number` is set as twice as large as other operations to help facilitate the evolution of the network structure of an individual to a more complex one.

### 12.4.2.3  Restructuring of Network Structures Due to Mutation

The results of Sect. 12.4.1.3 indicate that in some cases a mutation operation selected at random cannot be applied due to restrictions related to the number of dimensions of the input and output before and after the layer. For these cases, we introduced a method to enhance the probability that the selected operation can be applied by trying to restructure the network as much as possible to make dimensions match when `add_concatenate` is selected as a mutation operation. The restructuring algorithm is shown in Algorithm 3. This algorithm imitates the operations carried out in the concatenate layer marked with (**) in Table 12.4.

### 12.4.2.4  Fitness (mAP) Calculation Method

In the MNIST handwritten classification task mentioned in Sect. 12.4.1, the classification accuracy was used as a measure of fitness. However, in an object detection task, it is necessary not only to classify the object but also to estimate the position and area where the object exists. Therefore, it is not possible to introduce the concept of correct and incorrect detection results as it is. For this reason, it is necessary to redefine the concept of fitness.

**Algorithm 3** Restructure algorithm for the network structure to apply `add_concatenate`

1: **Input:** input from the immediately preceding layer (height $h_{in}$, width $w_{in}$, number of channels $c_{in}$), input from the source layer (height $h_{source}$, width $w_{source}$, channel $c_{source}$)
2: **if** $h_{source} = m \cdot h_{in}$ **or** $h_{source} = \frac{1}{n} \cdot h_{in}$ $(m, n \in N)$ **then**
3:     **Pointwise convolution:** connects the input from the source layer to the convolutional layer having 64 channels, one stride, $1 \times 1$ filter size, and one-pixel padding. Due to this operation, the number of channels of the input from the source layer is fixed to 64. Hence, the number of parameters is reduced, and the growth in the number of channels for the next operation is suppressed.
4:     **Reorganization:** performs a transformation where the height and width of the input from the source layer are, respectively, reduced by $\frac{1}{m}$ (magnified by $n$ times), and the number of channels is multiplied by $m^2$ (reduction by $\frac{1}{n^2}$). As a result of this operation, we have $h_{source} = h_{in}$, $w_{source} = w_{in}$ and the height and width dimensions of the inputs from the two layers coincide. The number of parameters does not change before and after this operation.
5:     **Concatenation:** the inputs from two layers whose height and width dimensions coincide are concatenated along the channel number direction to form the output of this layer.
6: **else**
7:     **Failure:** since restructuring is impossible, another mutation operation is selected.
8: **end if**
9: **Output:** formed by the concatenation of inputs from two layers along the channel number direction

There are several performance evaluation criteria for the object detection task, but here we consider mAP (mean Average Precision) as a fitness measure. Below is an explanation of the mAP calculation method [3]. For simplicity, the subject of detection is only one type of object.

If the image of the target of object detection is input to a network, several square areas (set of numbers stating the coordinates of the center, width, and height) are obtained. The output also includes the probability that an object exists in each rectangular area (confidence). Of the above, we only consider the rectangular areas where confidence is greater than or equal to a threshold $c$. Three terms are defined below:

**True Positive**   the number of predicted areas that was correctly "detected"
**False Positive**   the number of predicted areas that was incorrectly not "detected"
**False Negative**   the number of areas that should have been "detected" but was not.

Here a predicted area "detects" a correct area if the value of the intersection over union (IoU) is equal to or greater than 0.5. Note that IoU is a measure of overlap between the two areas. Figure 12.12 shows the definition of IoU. Hence, IoU is the ratio of the intersection of two areas to the area of the union between them. If several predicted areas cover a single correct area, only the predicted area of one of them is counted as True Positive; all other predicted areas are counted as False Positive.

**Fig. 12.12** Definition of IoU

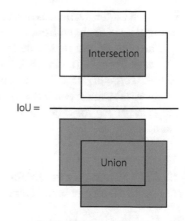

$$IoU = \frac{\quad}{\quad}$$

**Fig. 12.13** Example of PR
curve. The blue-painted
region represents AP

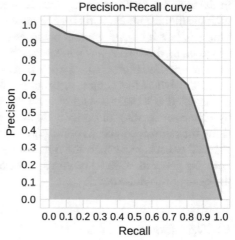

Based on the definition above, the values of Recall and Precision are obtained by counting the number of True Positive (TP), False Positive (FP), and False Negative (FN) for all the given test images, as follows:

$$Recall = \frac{TP}{TP + FN}$$

$$Precision = \frac{TP}{TP + FP}$$

Recall is the ratio of "the number of actually detected objects to the number of all objects to be detected," while Precision is a ratio of "the number of areas where objects actually exist out of the areas where they are predicted to exist." Here, if the threshold value $c$ is changed, the number of predicted areas to be considered also changes. Therefore, the values of TP, FP, and FN also change, and so do the values of Recall and Precision. Thus, a plot such as the one in Fig. 12.13 (PR curve: precision-recall curve) can be obtained by plotting Recall on the horizontal axis,

Precision on the vertical axis, and connecting the points obtained when the value of threshold $c$ is changed from 0 to 1.

The area of the region delimited by the PR curve and the axes is called AP (Average Precision) and assumes a value between 0 and 1. The higher the value of the AP, the higher is the detection accuracy.

Even if the number of classes to be detected is increased, it is possible to calculate the value of AP for each class. mAP is the average value of AP calculated for each class.

### 12.4.2.5 Changing the Evaluation Method for Individuals in the Elite Group of Each Generation

In the experiment described in Sect. 12.4.1, when individuals from the elite group with the highest fitness in their generation are cloned into the next generation, their fitness is just inherited without being re-evaluated. This results in an issue in the proposed method ASM+ where the number of epochs for evaluation increases with each generation. Furthermore, when individuals in the elite group are in a state of overfitting where their fitness does not improve with the number of epochs, they are not updated and enter a state of local stagnation. However, this issue was solved by also re-evaluating those individuals in the elite group just like the other ones using the number of epochs corresponding to the generation. Therefore, unlike the MNIST experiment where the fitness of the best individual improves uniformly with the generation, in some cases the fitness of the best individual decreases as the generation advances.

### 12.4.2.6 Experimental Results

Experiments were conducted on an NVIDIA GeForce GTX 1080 Ti GPU. The parameters used are shown in Table 12.5.

**Table 12.5** Parameters used in the experiment

| Parameter | Value |
|---|---|
| Final generation number $T$ | 4 |
| Number of individuals in each generation $N$ | 4 |
| Number $k$ of elite individuals to be added to the next generation | 1 |
| Batch size $b$ | 64 |
| Number of training images $L_{train}$ (Table 12.3) | 3672 |
| Number of testing images $L_{test}$ (Table 12.3) | 2449 |
| Minimum iteration number $i_{min}$ | 10,000 |
| Maximum iteration number $i_{max}$ | 25,000 |

For reasons related to implementation, instead of defining the minimum number of epochs $n_{min}$ and the maximum number of epochs $n_{max}$, the minimum and maximum number of iterations, $i_{min}$ and $i_{max}$, have, respectively, been defined. Referring to the number of iterations as $i$ and the number of epochs as $n$, the number of epochs is converted to the number of iterations using the equation below:

$$n = \frac{b \cdot i}{L_{train}}$$

Since batch size $b = 64$ and $L_{train} = 3672$, if we set $i_{min} = 10,000$ and $i_{max} = 25,000$, this is almost equivalent to setting $n_{min} = 174$, $n_{max} = 436$.

The method proposed ASM+ in Sect. 12.3 can be applied in the same way even if the number of epochs is replaced by the number of iterations. Hence, in order to split the interval between $i_{min}$ and $i_{max}$ uniformly with respect to generation number $T = 4$, the number of iterations is varied from the first up to the 4th generation in the following order: 10,000, 15,000, 20,000, and 25,000.

Figure 12.14 shows how the fitness (mAP) for an individual changed from generation to generation. Table 12.6 shows the mAP for the best individual obtained by evolution; it also shows the computation time required by the entire process. Table 12.7 shows which mutation operations were selected in each generation.

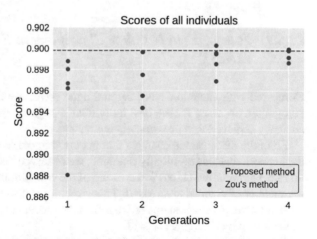

**Fig. 12.14** Change of fitness of all individuals in each generation. The red dot is the result of reference [20] (Exp. 5/Run 11, including data during the training), where the number of epochs upon evaluation is plotted after conversion to the corresponding generation number

**Table 12.6** Comparison of True Positive (TP), False Positive (FP), False Negative (FN), AP, mAP (converted to %), and computation times for each method

| Method | AP | | | | | | | |
|---|---|---|---|---|---|---|---|---|
| | TP | FP | FN | Scissors | Knives | PET bottles | mAP | Computation time |
| Zou [20] | 5474 | 517 | 469 | 90.46% | 88.82% | 90.67% | 89.98% | 7 GPUH |
| ASM+ | 5485 | 459 | 458 | 90.32% | 89.16% | 90.60% | 90.03% | 84 GPUH |

The threshold used for the computation of TP, FP, and FN is $c = 0.25$

**Table 12.7** Mutation operations performed on the best individual of each generation

| Generation | Mutation operation | Maximum mAP |
|---|---|---|
| 0 | `initialize` | – |
| 1 | `add_concatenate` | 0.8988 |
| 2 | - | 0.8997 |
| 3 | `add_convolution` | 0.9003 |
| 4 | `add_concatenate` | 0.8999 |

Blank fields indicate that no update was carried out on a specific individual/generation

The following Figs. 12.15 and 12.16 show the network structures of the individuals exhibiting the best and the second-best fitness during the evolution process. Figure 12.17 is a tree diagram describing the evolution process of the best individual of each generation.

However, note that, in Figs. 12.15, 12.16, and 12.17, batch normalization and Leaky ReLU are applied immediately after all Conv2D (2-dimensional convolutional layer) in all layers except the last one; however, this was abridged in the figures.

## 12.5 Discussion

### 12.5.1 Handwritten Number Classification Experiment Using MNIST

Compared with Genetic CNN [19] and aggressive selection and mutation [7], the recognition accuracy for the best individual is slightly lower; however, the level of accuracy achieved is approximately equivalent.

From the dimensional changes that occurred in the feature map of each layer of the network that was eventually obtained, we observed that the vertical and horizontal dimensions of the output feature map of the convolutional layer decreased, while the number of channels increased. This structure is also seen in classical networks such as LeNet [6] and is supposed to reflect the process of grasping the local features of the image by changing the scale.

However, the final network has an extremely simple structure formed by two convolutional layers. There is no pooling layer, dropout, or by-pass, which could have been introduced by `add_pooling`, `add_dropout`, and `add_skip`, respectively. This could be attributed to the strict conditions for the application of mutations; therefore, the operations involving the addition of such structures were not selected. Moreover, from Fig. 12.9 and Table 12.2, we observed that the addition of a convolutional layer at an early stage of evolution produces a dramatic effect

**Fig. 12.15** Network structure of an individual from the 3rd generation, which exhibited the best fitness among all individuals (fitness = 0.9003). For the notation, refer to Fig. 12.8

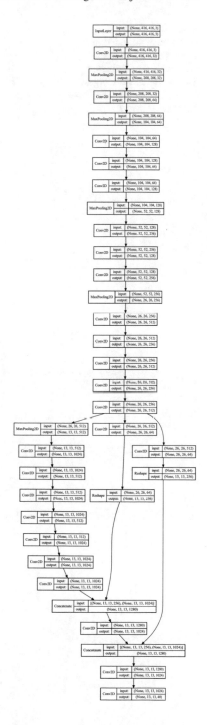

**Fig. 12.16** Network
structure of an individual
from the 4th generation
whose network structure
yielded the second largest
fitness among all individuals
(fitness = 0.8999)

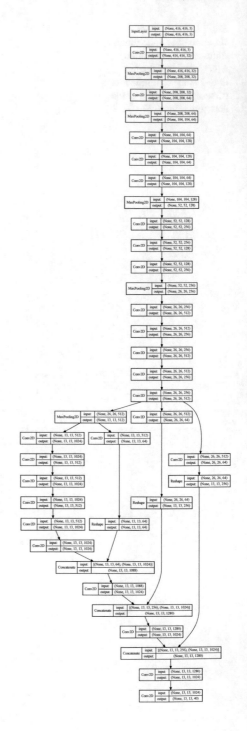

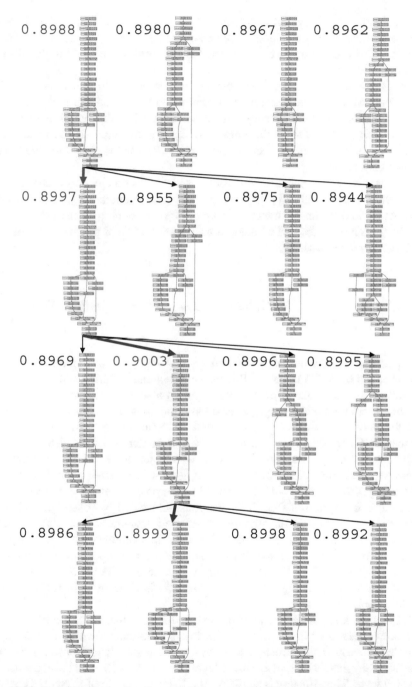

**Fig. 12.17** Tree diagram showing the evolution process of the best individual of each generation. The value accompanying the image of the network structure of an individual represents its fitness (mAP). Individuals with a value written in red are those selected for the next generation. Colored arrows indicate the selection flow. Red arrows indicate the occurrence of mutation, and blue arrows indicate cloning

on recognition accuracy; subsequently, only operations that change the hyperpa-rameters in the convolutional layer (`alter_stride`, `alter_filter_size`, `alter_channel_number`) are selected as operations that result in the birth of better individuals.

Furthermore, from Fig. 12.7, we observe that evolution saturates around the 15th generation. Similarly, the selection of operations to be applied to the individuals of that generation is restricted, resulting in the selection of locally optimized solutions. Another factor is that since the number of epochs increases as the generations advance, overfitting tends to occur in new individuals, and the best individuals end up not being updated.

Conversely, the computation time for the entire evolution process is shorter than the previous results [19] and [7]. This is possibly due to a persisting situation where only networks with simple structures that require short training times were obtained.

The presence or absence of batch normalization did not produce a significant difference in the results.

### 12.5.2  Experiment on Detecting Dangerous Objects in X-ray Images

Compared with the results of the method used by Zou [20], individuals showing an improvement of 0.05% in the mAP value were obtained (Table 12.6). The number of "detected" dangerous objects (True Positive) also increased. Additionally, the number of other objects wrongly detected as dangerous (False Positive) decreased, and the number of dangerous objects that were not detected (overlooked) also decreased (False Negative). Therefore, for the dangerous object detection task, both "Recall" and "Precision" improved, which denotes a definite performance improvement. Figure 12.18 shows an example of a dangerous object that was not detected by Zou's method but was detected by the proposed method ASM+.

In the tree diagram of Fig. 12.17, we can find several by-passes generated by `add_concatenate` in the structures that were actually obtained. This can be contrasted with the case of the MNIST handwritten number classification task, where no by-pass was observed due to restrictions on the application of mutation operations (Sect. 12.5.1). Our assumption is that the network structure reconstruc-tion algorithm for applying `add_concatenate` introduced in Sect. 12.4.2.3 has produced effective results.

The network structure of one of the two individuals that yielded a better mAP than Zou's method, which exists in the last generation shown in Fig. 12.16, is examined closely. In that structure, three by-pass lines stretch out from a layer with a $26 \times 26$ (width, height) output towards a layer having a $13 \times 13$ input located closer to the final layer. This structure is similar to the U-type structure of U-Net [14], which produced good results in medical image segmentation. As mentioned in Sect. 12.2.4, employing a by-pass from the high-resolution output of a layer with a large feature

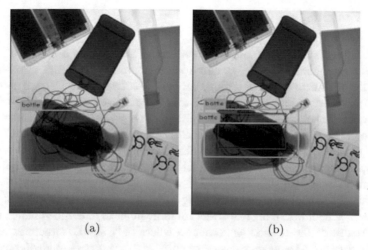

**Fig. 12.18** Example of where an image that was overlooked by the Zou's method [20] was correctly detected by the proposed method ASM+. Images of two PET bottles and a knife are overlapped in the center of the image. (**a**) Detection results in Zou's method. (**b**) Detection results by ASM+

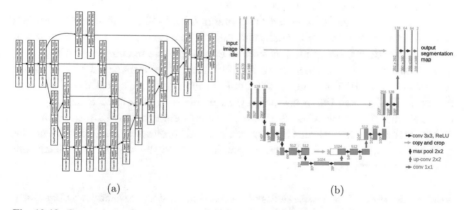

**Fig. 12.19** Comparison of evolved network structures. (**a**) Aggressive selection and mutation. (**b**) U-Net [14]

map to a layer with a small feature map acts to improve the ability to detect small objects (see Fig. 12.19).

Next, from Fig. 12.14, we observe that the fitness of the best individual of the 4th generation is lower than that of the 3rd generation. This is an indication that the high number of epochs used in the evaluation results in overfitting and deterioration in the mAP value in individuals that are cloned into the next generation as an elite. As a result, the proposed method ASM+ of increasing the number of epochs as the generations advance has the advantage of finding an optimal number of epochs that does not result in overfitting.

## 12.6   Conclusion

Optimizing the structure of convolutional neural networks is a task that requires high-level specialized knowledge and a huge amount of trial-and-error work. To enable the automatic search of optimal network structures using GA, we proposed a method ASM+ in which the number of epochs upon evaluation is increased as the number of generations advance during the evolution process in the present research.

In the MNIST handwritten number classification experiment, we showed that ASM+ exhibited a slightly lower recognition accuracy, but is relatively fast in terms of computation time. However, this might be because the algorithm was not able to search for complex structures.

In the experiment related to detecting dangerous objects in X-ray images, we introduced a method that made it easier for individuals to undergo mutation in view of the results of the MNIST experiments where only simple structures emerged from the search. Consequently, it became possible to evaluate individuals possessing several by-pass structures, resulting in an enlarged search space. Additionally, with respect to cloned individuals belonging to the elite group, a change was introduced where fitness is re-evaluated with a higher number of epochs instead of being inherited into the next generation.

Due to this change, we avoided the problem of individuals becoming overfitted with an increasing number of epochs and stuck in local solutions. Moreover, we can consequently detect the optimal number of epochs that do not cause overfitting.

A future task is the acceleration of evolution. The proposed method ASM+ requires at least 10 times more computation than the Zou's method [20], as is shown in Table 12.6. Pham et. al. [11], as in ASM+, attempted to optimize the convolutional neural network structure by using neural networks instead of GA. A considerable reduction in computation needed to evaluate individuals may be possible through a transfer learning approach where the weight parameters are shared to some extent.

**Acknowledgments** We would like to thank Shin Yokoshima and Yoji Nikaido, T&S Corporation, for providing us with X-ray data collection environment.

## References

1. Collobert, R., Weston, J., Bottou, L., Karlen, M., Kavukcuoglu, K., Kuksa, P.: Natural language processing (almost) from scratch. J. Mach. Learn. Res. **12**, 2493–2537 (2011)
2. Dahl, G.E., Yu, D., Deng, L., Acero, A.: Context-dependent pre-trained deep neural networks for large-vocabulary speech recognition. IEEE Trans. Audio Speech Lang. Process. **20**(1), 30–42 (2012)
3. Everingham, M., Van Gool, L., Williams, C.K., Winn, J., Zisserman, A.: The PASCAL Visual Object Classes Challenge 2012 (VOC2012). http://host.robots.ox.ac.uk/pascal/VOC/voc2012/index.html (2012). Accessed 07 Feb 2019

4. He, K., Zhang, X., Ren, S., Sun, J.: Deep residual learning for image recognition. In: Proceedings of the 2016 IEEE Conference on Computer Vision and Pattern Recognition (2016)
5. Krizhevsky, A., Sutskever, I., Hinton, G.: ImageNet classification with deep convolutional neural networks. In: Proceedings of the 25th International Conference on Neural Information Processing Systems (NIPS12), pp.1097–1105 (2012)
6. LeCun, Y., Bottou, L., Bengio, Y., Haffner, P.: Gradient-based learning applied to document recognition. Proc. IEEE **86**(11), 2278–2324 (1998)
7. Li, Z., Xiong, X., Ren, Z., Zhang, N., Wang, X., Yang, T.: An Aggressive Genetic Programming Approach for Searching Neural Network Structure Under Computational Constraints (2018). Preprint arXiv:1806.00851
8. Maas, A.L., Hannun, A.Y., Ng, A.Y.: Rectifier nonlinearities improve neural network acoustic models. In: Proceedings of International Conference on Machine Learning (ICML), vol. 30, no.1 (2013)
9. Nie, Y., Iba, H.: evolving convolutional neural network architectures using self-modifying Cartesian genetic programming. Master Thesis, Graduate School of Information Science and Technology, The University of Tokyo, 2019
10. Pan, S.J., Yang, Q.: A survey on transfer learning. IEEE Trans. Knowl. Data Eng. **22**(10), 1345–1359 (2010)
11. Pham, H., Guan, M.Y., Zoph, B., Le, Q.V., Dean, J.: Efficient Neural Architecture Search via Parameter Sharing (2018). Preprint arXiv: 1802.03268
12. Redmon, J., Farhadi, A.: YOLO9000: Better, Faster, Stronger (2016). Preprint arXiv: 1612.08242
13. Redmon, J., Divvala, S., Girshick, R., Farhadi, A.: You Only Look Once: Unified, Real-Time Object Detection (2015). Preprint arXiv: 1506.02640
14. Ronneberger, O., Fischer, P., Brox, T.: U-Net: Convolutional Networks for Biomedical Image Segmentation (2015). Preprint arXiv: 1505.04597
15. Russakovsky, O., Deng, J., Su, H., Krause, J., Satheesh, S., Ma, Z., Huang, Z., Karpathy, A., Khosla, A., Bernstein, M., Berg, A C , Fei-Fei, L.: ImageNet large scale visual recognition challenge. Int. J. Comput. Vis. **115**(3), 115–211 (2015)
16. Stanley, K.O., Miikkulainen, R.: Evolving neural networks through augmenting topologies. Evol. Comput. **10**(2), 99–127 (2002)
17. Suganuma, M., Shirakawa, S., Nagao, T.: A genetic programming approach to designing convolutional neural network architectures. In: Proceedings of the Genetic and Evolutionary Computation Conference 2017 (GECCO2017), pp. 497–504 (2017)
18. Szegedy, C., Liu, W., Jia, Y., Sermanet, P., Reed, S., Anguelov, D., Erhan, D., Vanhoucke, V., Rabinovich, A.: Going deeper with convolutions. In: Proceedings of the 2015 IEEE Conference on Computer Vision and Pattern Recognition (CVPR), pp.1–9 (2015)
19. Xie, L., Yuille, A.: Genetic CNN. In: Proceedings of the 2017 IEEE International Conference on Computer Vision (ICCV), pp.1388–1397 (2017)
20. Zou, L., Tanaka, Y., Iba, H.: Dangerous objects detection of X-ray images using convolution neural network. In: Yang, C.N., Peng, S.L., Jain, L. (eds.) Security with Intelligent Computing and Big-data Services (SICBS2018). Advances in Intelligent Systems and Computing, vol. 895, pp. 714–728. Springer, Cham (2019)

# Chapter 13
# Evolving the Architecture and Hyperparameters of DNNs for Malware Detection

Alejandro Martín and David Camacho

**Abstract** Deep Learning models have consistently provided excellent results in highly complex domains. Its deep architecture of layers allows to face problems where classical machine learning approaches fail, or simply are not able to provide good enough solutions. However, these deep models usually involve a complex topology and hyperparameters that have to be carefully defined, typically following a grid search, in order to reach the most profitable configuration. Neuroevolution presents a perfect instrument to perform an evolutionary search pursuing this configuration. Through an evolution of the hyperparameters (activation functions, initialisation methods and optimiser) and the topology of the network (number and type layers and the number of units) it is possible to deeply explore the space of solutions in order to find the most proper architecture. Among the multiple applications of this approach, in this chapter we focus on the Android malware detection problem. This domain, which has led to a large amount of research in the last decade, presents interesting characteristics which make the application of Neuroevolution a logical approach to determine the architecture which will better discern between malicious and benign applications. In this research, we leverage a modification of EvoDeep, a framework for the evolution of valid deep layers sequences, to implement this evolutionary search using a genetic algorithm as means. To assess the approach, we use the OmniDroid dataset, a large set of static and dynamic features extracted from 22,000 malicious and benign Android applications. The results show that the application of a Neuroevolution based strategy leads to build Deep Learning models which provide high accuracy rates, greater than those obtained with classical machine learning approaches.

A. Martín (✉) · D. Camacho
Departamento de Sistemas Informáticos, Universidad Politécnica de Madrid, Madrid, Spain
e-mail: alejandro.martin@upm.es; david.camacho@upm.es

© Springer Nature Singapore Pte Ltd. 2020      357
H. Iba, N. Noman (eds.), *Deep Neural Evolution*, Natural Computing Series,
https://doi.org/10.1007/978-981-15-3685-4_13

## 13.1   Introduction

The detection of malware constitutes a large, open and critical problem. The figures on the volume of new malware developed every year have not stopped growing. Another element worthy of consideration is the increasing complexity of these malicious programs. These two facts evidence the need for automated and sophisticated detection tools able to identify the most elaborated malware shapes. For this purpose, many approaches trust on machine learning techniques to build accurate malware detectors.

Among the different techniques proposed in the literature, Deep Neural Networks represent a widely studied and employed tool. However, unlike other supervised processes, these models require to define an architecture and a series of hyperparameters. This typically involves testing different configurations in order to find the most adequate option. Depending on the malware analysis technique followed and the representation scheme used, different layers can be involved: fully connected, convolutional, recurrent or dropout, among others.

In terms of malware analysis, there are two main approaches in order to extract relevant characteristics from both goodware and malware samples. **Static** analysis refers to the process of extracting features from the malware executable file without executing the malware. This includes API calls, permissions declared, intents or information extracted from the bytecode level. The second option is to extract **dynamic** features, those retrieved with a monitoring agent which captures all the actions performed by the suspicious sample when it is executed.

While a static analysis will mainly extract independent features (composed by binary or count values), a dynamic approach will provide temporal sequences of events (i.e. a list of disk operations with a timestamp). Once extracted the necessary set of features (either static, dynamic or both), it is necessary to find a machine learning algorithm able to construct a new model that could generalise from that information. Although some models require a small number of technical adjustments (i.e. a decision tree based algorithm), others, such as Deep Learning models, require to set a large number of hyperparameters and to define a topology where a big number of options come into play.

In addition, it is possible to identify two different tasks related to the malware detection scenario: detection and classification. The former refers to the detection of malware itself, where a suspicious sample has to be categorised as goodware or malware. A second task is related to the classification of a malicious sample to the malware family. This is an essential process where the sample is associated with a known group which shares certain common characteristics. Although the detection of malware is a vital step to detect and to prevent getting infected, the classification into families remains a major barrier to avoid the expansion of the malicious sample and to know the actions that the sample is able to take.

This chapter presents and evaluates a new method based on genetic algorithms to evolve deep neural networks architectures trained for the detection of Android malware. The main goal is to demonstrate how genetic algorithms and, in general, evolutionary algorithms can help to define the adequate architecture and hyper-parameters in order build accurate and powerful malware detection methods. For that purpose, fully connected and dropout layers are considered. In a previous work [1], focused on the classification of malware into families, an evolutionary algorithm was used to evolve both the topology and hyperparameters using a specifically designed individual encoding and genetic operators. We extend this idea by deploying EvoDeep [2].

EvoDeep leverages a genetic algorithm and a finite state machine to search for a configuration of hyperparameters and layers that is able to maximise the performance of a deep neural network in a particular problem. The genetic algorithm includes a dynamic individual encoding representing the hyperparameters and a sequence of layers of variable size. For its part, the finite state machine generates valid sequences of layers. Due to EvoDeep was originally designed for building Convolutional Neural Networks architectures, it is required to adapt the finite state machine in order to generate sequences of layers only considering fully connected and dropout layers. This tool also requires to define a broad range of values for each parameter, which will form the search space. Since each individual represents a specific network, its fitness is calculated by training and evaluating that model. In order to train the architectures and to evaluate the approach presented, the OmniDroid dataset [3] is used, a collection of feature vectors composed by static and dynamic characteristics extracted from malicious and benign Android applications. The OmniDroid dataset was built using the AndroPyTool framework.[1] In this research only static features have been considered.

The main contributions of this chapter are summarised as follows:

- An analysis of the literature focused on the use of Neuroevolution and Deep Learning architectures to build Android malware detection and classification algorithms and methods.
- A novel method based on a modification of the EvoDeep framework to automatically define Deep Learning architectures in order to classify suspicious Android applications into goodware or malware. A representation of the samples based on a wide number of static features has been followed.
- An evaluation of the approach presented using a recently published dataset containing 11,000 malware and 11,000 goodware samples. This dataset provides a plethora of already extracted features including API Calls, permissions declared, system commands, opcodes or a taint analysis performed with the FlowDroid tool.
- A thorough assessment of the architectures generated using an evolutionary search based approach.

The remaining sections of this chapter are organised as follows: Sect. 13.2 introduces the state-of-the-art literature. Then, Sect. 13.3 describes our application of Neuroevolution to the Android malware detection domain. Section 13.4 summarises

---

[1] https://github.com/alexMyG/AndroPyTool.

the experimental setup and Sect. 13.5 presents the experiments and results. Finally, a series of conclusions are provided in Sect. 13.6.

## 13.2 Deep Learning Based Android Malware Detection and Classification Approaches

In the nineties, different researches introduced the use of Evolutionary Computation to define the hyperparameter or the topology of Artificial Neural Networks. This combination was called Neuroevolution. With different representation schemes, using evolutionary strategies and focusing on different aspects of neural networks, this research line showed a great potential to generate accurate models in varied domains. However, in the case of neural networks and Deep Learning models applied to the detection of Android malware, Neuroevolution has been scarcely applied. In the following paragraphs, some of the most important approaches to address the malware detection problem using these learning models are described.

Most of the huge volume of literature presenting malware detection or classification tools has used classical machine learning algorithms [4]. However, Deep Learning based methods are increasingly present in the literature. The large number of features used to represent the behaviour of a sample, the complexity of these features or the different representation schemes (such as call graphs [5], geometric schemes [6] or Markov chains [7]) make the use of these techniques a natural alternative to build accurate classifiers.

One of the first researches evaluating the use of Deep Learning in this domain was addressed by Yuan et al [8, 9], who presented Droid-Sec. This is an approach for Android malware detection based on static and dynamic features. The learning consists of two phases. In the first one, a Deep Belief Network (DBN) is used as an unsupervised pre-training step. In the second phase, the model is fine-tuned following a supervised process. The results show that DL architectures outperform classical methods such as SVM or C4.5 with a difference of 10% accuracy. DBNs have been largely studied by many other authors. DroidDeep [10] relies on static features (permissions, API calls and components deployment) to make a search space of 30,000 characteristics. A Deep Belief Network is used to filter these features, spotting the most important ones. The detection step is then produced by a SVM classifier. DBNs are also used in DroidDelver [11], this time using API call blocks constructed by organising API calls in compliance with the smali code in which they are defined. According to the authors, these blocks represent a *complete function*, thus creating a higher level representation of the behaviour of each sample. DBNs are compared against Stacked AutoEncoders (SAEs), showing that the former leads to better results. DeepFlow [12] focuses on data flows extracted with FlowDroid [13], a static taint analysis tool which finds connections between sensitive API calls (sources and sinks). Following the approach previously used by other literature, a DBN is firstly used in an unsupervised manner to structure the features. Then, the model is trained with a supervised procedure.

Other literature has employed and compared other architectures, such as typical DNNs, CNNs or LSTMs. Regarding the first ones, Fereidooni et al. [14] tested the use of deep architectures composed of fully connected layers and static features, including intents, permissions, system commands, suspicious API calls or malicious activities. A comparison against classical methods such as AdaBoost, Random Forest or SVM is provided, showing that deep neural architectures achieve excellent results, in this case only improved by XGBoost. With a more complex features representation scheme, Deep4MalDroid [15] builds directed graphs with the system calls dynamically extracted at the kernel level. Nodes denote system kernel level calls, the size of the node indicates its frequency and connections between nodes define the probability of that sequential pair of calls. The features of the graphs are used to train a Deep Learning model with Stacked AutoEncoders. Multimodal Deep Learning has also been studied in this domain by Kim et al. [16]. By training five different deep networks with different groups of features (strings, opcodes, API calls, shared functions, and vector combining permission, components and environmental information) and finally combining these models with another DNN model.

Convolutional models have also been explored for malware detection tasks. A research by McLaughlin et al. [17] operates with opcode sequences from the disassembled code. The performance of the CNNs shows that these models are computationally efficient in comparison to n-gram features based approaches. Deep-ClassifyDroid [18] performs a three-step classification process: extraction of static features from the Android Manifest and the disassembled code, features embedding and detection. In the final step, a CNN architecture with two convolutional blocks is able to reach high results, with an improvement in comparison to kNN or linear SVM. Wang et al. [19] follow a hybrid model using static features and a deep autoencoder as a pre-training step in order to reduce the training time. A CNN is then in charge of the detection step. The use of CNNs has also been tested in comparison with Long Short Term Memory (LSTM) architectures and n-gram based methods [20]. The results show that both CNN and LSTM are able to improve the accuracy levels of n-gram based approaches, and that CNN outperforms LSTM.

Research has also focused on studying LSTM networks separately. DeepRe finer [21] concentrates on the semantic structures found in the Android bytecode and trains LSTM networks. The results show this approach is able to achieve higher results in comparison to other state-of-the-art approaches and that it is robust against obfuscation techniques. LSTM is also used as classification method by Vinayakumar et al. [22], selecting sequences of permissions as features. In this research, the authors highlight the need to test different architectures with different network parameters in order to find an appropriate model. Other authors have preferred to implement new specifically designed architectures. In MalDozer [23], samples are represented by sequences of API calls in the assembly code to perform malware detection but also to identify the malware family. In the final step of the designed deep model, a one unit output configuration is chosen in case of detection and $n$ units in case of family classification.

Although some of the previous introduced state-of-the-art literature has mentioned the difficulty of defining the structure or the hyperparameters of the deep model used, recent literature has analysed this problem in further detail. Booz et al. [24] undertake a research where only permissions are used as features. In order to define the Deep Learning architecture, the authors follow a grid search in order to set the combination of parameters. The same problem was previously addressed using a genetic algorithm to search for the optimal deep architecture, in this case focused on classifying malware into families [1]. Additionally, evolutionary approaches have also been directly applied in the Android malware detection domain to build new classifiers [25] or to improve classical machine learning algorithms [26].

In this chapter, we introduce the application of Neuroevolution to evolve Deep Learning architectures and their hyperparameters to generate accurate Android malware detection models. This combination has been scarcely studied in state-of-the-art literature. The EvoDeep framework is used to guide the genetic search, where specific crossover and mutation operators modify individuals. Simultaneously, a finite state machine defines valid sequences of layers.

## 13.3   Evolving the Architecture and Parameters of DNNs for Malware Detection

With the goal of building a Deep Neural Network model able to detect Android malware accurately, we leverage EvoDeep [2]. We have modified this framework in order to generate sequences of fully connected and dropout layers. Through an evolutionary process, this framework will create, combine and mutate different individuals, representing potential deep models where the objective is to maximise the classification accuracy. In comparison to a grid search based approach, where an excessive and unmanageable number of different topologies and hyperparameters combination would be tested, an heuristic search allows to reduce the search time and to provide near-optimal solutions. By evaluating each individual created, crossed or mutated, which implies training the whole deep model defined by the individual encoding, the evolutionary process will successively generate better solutions. In order to train these models, the OmniDroid dataset has been used, a collection of features extracted from 22,000 Android malware and goodware samples [3].

### 13.3.1   Genetic Algorithm Description

The genetic algorithm used by EvoDeep implements a $(\mu + \lambda)$ strategy, where $\mu$ individuals are selected from the previous generation, whereas new $\lambda$ individuals are created in each generation. Figure 13.1 represents the workflow of this algorithm. A

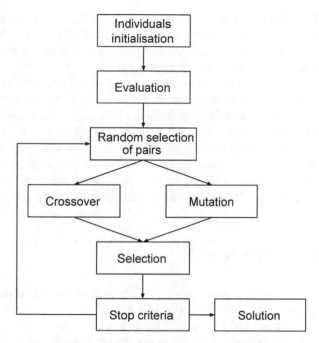

**Fig. 13.1** Scheme of the genetic algorithm used during the evolutionary search

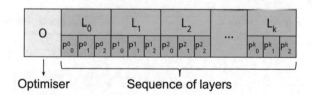

**Fig. 13.2** Representation of the individual encoding used in EvoDeep. Every individual encodes a specific Deep Neural Network configuration, including the optimiser, a structure of layers and the parameters of each layer

pool of individuals is initially randomly created. The only restriction applied in this first step is that individuals start from minimal architectures, that is to say, with the minimum number of layers predefined. This seeks to avoid unnecessarily complex and oversized architectures, which often tend towards overfitting. This decision aims to leave in hands of the genetic algorithm the selection of a higher number of layers if this leads to better solutions.

The encoding of these individuals is represented in Fig. 13.2. It is divided into two sections, one representing the optimiser $O$ and a second dynamic section representing a variable number of layers $L_i$. In addition, each layer defines a series of parameters $P_i^k$ (such as the activation or initialisation function).

For the hyperparameters of the model, the individual only encodes the optimiser, which can be one of the following algorithms: *Adam, SGD, RMSprop, Adagrad,*

*Adamax* or *Nadam*. Other hyperparameters have been fixed experimentally. Thus, the number of epochs and the batch size were fixed to 300 and 1000, respectively. In case of the number of epochs, a dynamic stop criteria ends an execution when no improvement is observed in the last 30 iterations.

In case of the layer level parameters, each type of layers involves certain variables. Fully connected or *Dense* layers define the number of units, which will be in the range between 10 and 500 with a step of 50; an activation function, where the options are *ReLU*, *softmax*, *softplus*, *softsign*, *tanh*, *sigmoid*, *hard_sigmoid* and *linear*; and also a kernel initialisation function with the following alternatives: *uniform*, *lecun_uniform*, *normal*, *zero*, *glorot_normal*, *glorot_uniform*, *he_normal*, *he_uniform*. The second type of layers engaged is dropout, where only a rate of units to drop has to be defined. This value will be in the range [0.1, 0.8] with a step of 0.1.

### 13.3.2    Genetic Operators

During the search, two specifically designed genetic operators introduce modifications in the population at each generation. Due to the two levels of parameters in the individual (optimiser and sequence of layers in the first level and the parameters of each layer in the second level), the operators are designed to operate at each

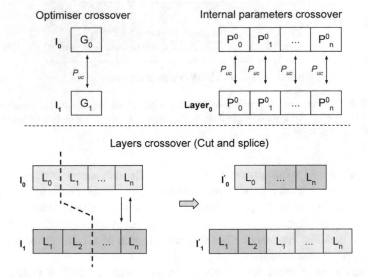

**Fig. 13.3** Representation of crossover operation. At the top, the optimiser of two different parent individuals and the parameters of each pair of coincident layers are crossed according to a $P_{uc}$ probability. At the bottom, the section of layers of two different individuals is crossed following a cut and splice scheme. A random point between layers is selected for each individual and the second section of both individuals is swapped

level individually. The **crossover** operator (see Fig. 13.3) is in charge of combining two parent individuals to create a new pair of solutions. It operates at the two levels previously mentioned. Externally, a uniform crossover is applied to the optimiser, while the set of layers of two individuals are crossed using a cut and splice approach. This last operation selects a random point in the structure of layers of each individual to create two fragments. Then, the right part of each individual is swapped, generating a new pair of structures of layers as a combination of the parents. This operation ensures the creation of new individuals of different structure of layers and sizes, always keeping intact the first and last layer and checking that the new sequence is still valid. At the internal level, the parameters of each pair of coincident layers are also uniformly crossed.

The mutation operator (see Fig. 13.4) has a similar behaviour. At the external level, each global parameter is mutated following a uniform scheme, while a new set of layers can be randomly added between any pair of the existing layers. The internal operation mutates every layer parameter according to a given probability.

**Fig. 13.4** Representation of the mutation operation. At the top, the optimiser and the internal parameters of each layers are mutated according to a $P_{um}$ probability. At the bottom, a new set of $n$ layers is introduced in a random position of the individual

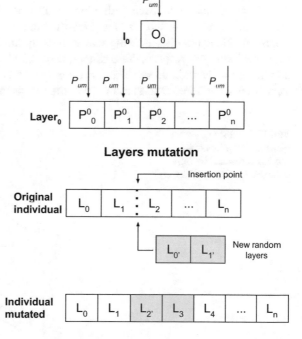

**Global and internal parameters mutation**

**Layers mutation**

### 13.3.3   Finite State Machine

EvoDeep employs a Finite State Machine (FSM) in order to generate valid layers
sequences, thus creating trainable and applicable network models. At the same
time, this FSM is also used cooperatively by the crossover and mutation operator
in order to check if new individuals encode valid layers sequences. Given the
specific characteristics of the problem faced in this chapter, we modified this FSM
as shown in Fig. 13.5, where only fully connected or dense and also dropout layers
are involved (the original implementation included the necessary layers to deploy
convolutional cycles which are not necessary in the malware detection domain).
The design of this new finite state machine limits the possible layers sequences to
a minimum size of two, where the last layer has to be unavoidably of type fully
connected. Figure 13.6 shows different examples of the layers sequences which can
be generated with the new FSM designed during the heuristic search.

### 13.3.4   Fitness Function

The fitness function of each individual defines its potential as Deep Learning
classification model in the accurate classification of Android malware. Thus, the
fitness is calculated after training and evaluating the architecture encoded by the
individual. For that purpose, the architecture defined by the genome is implemented
and trained in the training set. Then, in order to avoid overfitting, it is evaluated in a
validation set and the result is assigned to the fitness function.

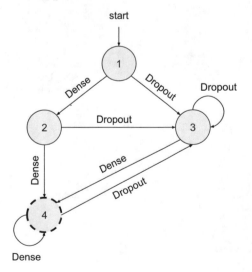

**Fig. 13.5**  FSM designed to generate Deep Learning models composed by fully connected and dropout layers

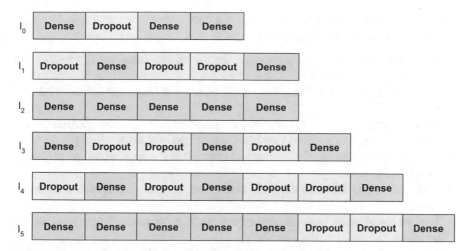

**Fig. 13.6** Examples of different layers sequences that EvoDeep is able to generate during the heuristic search with the new FSM designed for the Android malware detection problem

The next subsections present the different models generated and the results obtained.

## 13.4 Experimental Setup

This section describes a series of experiments performed to assess the performance of the evolutionary search based approach presented in the previous section. To that end, the Android malware detection domain has been chosen: given a suspicious APK (an Android Application Package), it is necessary to determine if it has malicious intentions (malware) or innocuous (goodware). Once described the dataset used in the next subsection, the results obtained from different executions, the architectures generated and a comparison against classical approaches are described.

### 13.4.1 Dataset

The OmniDroid dataset [3] has been used in this research in order to evaluate the ability of the genetic algorithm to find the most suited Deep Learning model. This dataset is a collection of static and dynamic features extracted from 22,000 Android goodware and malware samples. It contains a large number of features extracted with AndroPyTool[2] [27] which represent the behaviour of every sample. For each application, the **static features** defined are:

1. **API Calls:** Important features which determine the functions invoked of the Android operating system from the code.
2. **Opcodes:** Low level information which can reveal important behavioural patterns.
3. **Permissions:** A list of permissions declared in the Android Manifest. This information reports the functionalities that the app could use.
4. **Intents:** Data related to the actions and events that the application can trigger or being listening to.
5. **Strings:** Strings found in the code and which could reveal hidden sections of code.
6. **System commands:** Linux level commands which can be used to perform certain operations such as privilege escalation.
7. **FlowDroid:** A taint based analysis run with FlowDroid [13]. It reports connections between API calls which act as sources or sinks of data.

The OmniDroid dataset also includes **dynamic features** extracted with Droid-Box;[3] however, in this research we only considered static features.

### 13.4.2    Algorithm Parametrisation

Only a few hyperparameters need to be defined before executing the genetic algorithm. All of them have been fixed experimentally, trying not to limit the ability of the algorithm to explore the search space, and allowing, at the same time, to perform a fine-grained refining process to provide accurate solutions. The maximum number of generations was set to 200, but an early stop criteria ends the execution when no improvement was observed in the last 5 generations. $\mu$ and $\lambda$ have been experimentally fixed to 5 and 10, respectively. Although these two parameters limit the evolutionary search to a small number of individuals, it was observed that the genetic search is improved in terms of time and thoroughness. A 50% chance was chosen for the mutation and crossover probabilities and a 30% change is considered to add or remove layers in individuals.

In order to run EvoDeep it is also necessary to define limits of the search space. Table 13.1 summarises the range of values used for each hyperparameter.

---

[2]https://github.com/alexMyG/AndroPyTool.
[3]https://github.com/pjlantz/droidbox.

**Table 13.1** Range of values defined for each hyperparameter

| Hyperparameter | Range of values |
|---|---|
| Kernel initialiser | Uniform, Lecun uniform, Normal, Zero, Glorot normal, Glorot uniform, He normal, He uniform |
| Activation | Relu, Softmax, Softplus, Softsign, Tanh, Sigmoid, Hard sigmoid, Linear |
| No. units | $10 <= n <= 500$, step 50 |
| Dropout rate | $0.1 <= p <= 0.8$, step 0.1 |
| Optimiser | Relu, Softmax, Softplus, softsign, Tanh, Sigmoid, Hard Sigmoid, Linear |
| No. layers | $3 <= n <= 10$, step 1 |

### 13.4.3  Experimental Environment

All executions were run in a EVGA GeForce GTX 1060 6 GB. The code was implemented in Python, and the Keras library[4] was used to train and evaluate the deep models generated during the genetic search. For the experimental comparison against classical machine learning algorithms, the Scikit-learn library [28] was used.

## 13.5  Results

This section summarises the results obtained and the architectures generated after 40 executions of the genetic algorithm previously described. In each execution, the evolutionary search runs until the stop criteria is met or the maximum number of epochs is reached (300). Figure 13.7 shows the evolution of all the executions. As can be seen, the approach implemented, consisted in an extension of the EvoDeep algorithm, has excellent convergence characteristics, requiring only one iteration to reach close to 90% accuracy in the validation set. From this first iteration, a smooth fine-tuning process is performed. None of the executions exceeds 31 generations, which means that the stop criteria is early applied in all cases.

The approach followed allows to define the most suited architecture to solve the malware detection problem faced. Although the evolutionary search starts from minimal solutions, individuals are expected to evolve towards more complex solutions with the goal of providing accurate solutions. Figure 13.8 shows the number of layers of the best individual of each execution. As can be appreciated, there is a wide variability among solutions: while the minimum number of layers observed is 3, there are also executions where the maximum number of layers, a total of 9, is reached. This is caused by the search process, where some solutions are improved including a higher number of units or modifying some hyperparameters while other solutions are upgraded using a higher number of layers.

---

[4]https://keras.io.

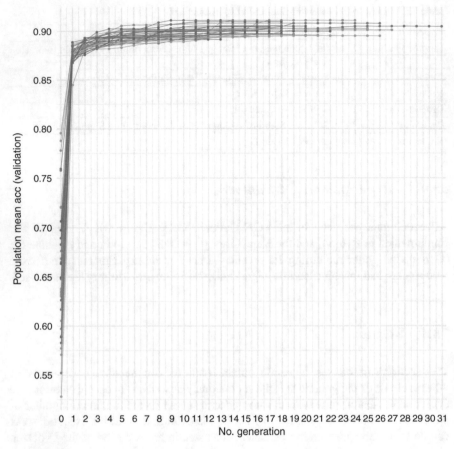

**Fig. 13.7** Evolution of the fitness function (validation set) in all the executions over generations. All executions reach high levels of accuracy after the first generation

In order to evaluate if there is a strong variation in terms of complexity, Fig. 13.9 represents the number of units of the best model generated after each execution. In contrast to Fig. 13.8, the number of units becomes more stable among executions, and remains in the range [9575, 11,165] with an average value of $10,173.65 \pm 359.12$. This evidences that although the architecture of the solutions generated can be considerably different, the final complexity is very similar.

The use of the three data splits has also been studied, aiming to assess if it is a correct procedure to evaluate individuals and to avoid overfitting. Figure 13.10 displays the accuracy reached in these sets. While the evaluation in the training set produces highly accuracy rates, with a significant difference in comparison to the other splits, the validation set allows to obtain a closer estimation of the real performance (appreciated in the test set) of each individual.

For a further assessment of the models generated using the approach based on EvoDeep, the same dataset was used to train and test classical machine learning

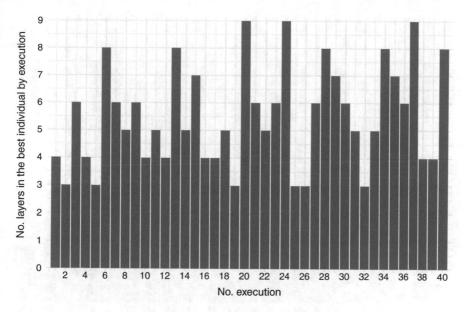

**Fig. 13.8**  Number of layers of the best individual obtained after each execution

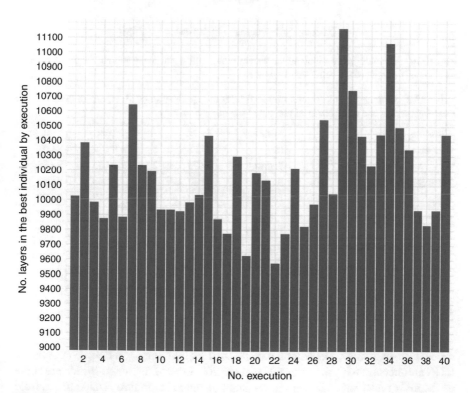

**Fig. 13.9**  Number of units present in the architecture of the best individual obtained after each execution

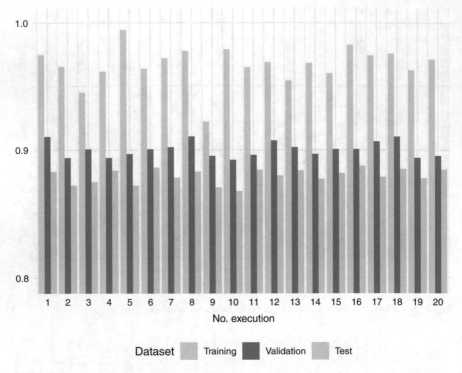

Fig. 13.10 Accuracy value obtained in the training, validation and test sets. Only 20 random executions are shows for a better visualisation

**Table 13.2** Summary of the results obtained with the application of EvoDeep to evolve Deep Learning models and other classical machine learning methods

| Classifier | Accuracy | Mean | Max. | Min. |
|---|---|---|---|---|
| Decision Tree | Test | 85.23 ± 0.32 | 85.82 | 84.61 |
| | Training | 99.79 ± 0.02 | 99.83 | 99.75 |
| Nearest Neighbors | Test | 82.32 ± 0.34 | 82.93 | 81.85 |
| | Training | 87.47 ± 0.18 | 87.77 | 87.08 |
| Random Forest | Test | 87.5  ± 0.23 | 87.93 | 87.09 |
| | Training | 99.02 ± 0.09 | 99.2 | 98.88 |
| DNN standard model | Test | 80.90 ± 10.65 | 85.38 | 50 |
| | Training | 81.43 ± 10.86 | 86.37 | 50 |
| Evolved Deep Learning model | Test | **87.92 ± 0.55** | 88.77 | 86.81 |
| | Training | 96.42 ± 1.63 | 99.43 | 92.22 |

The best value obtained in the test set is highlighted in bold

models. In particular, Decision Tree, Nearest Neighbours, Random Forest and a DNN architecture with two internal layers of 200 units were trained, which are some of the most prevalent classifiers in the state-of-the-art literature related to Android malware detection. The results are shown in Table 13.2. The average value of the best individuals delivered in the different executions of the genetic algorithm reach a 87.92% accuracy while the maximum value was 88.77%. In both cases, these are highest values obtained among classifiers. The second best alternative, Random Forest, which have shown excellent results in the literature, is not able to surpass the Deep Learning model performance.

Figure 13.11 represents the architectures and hyperparameters of the final population of one execution of the genetic search. As already stated, there is a big variability in terms of number of layers. However, all of them have similar accuracy rates. In general, it can be seen how the search introduces layers with a certain number of units which are shared among individuals. This is the case of the Dense layer of 310 units, which is present in the entire final population in different positions of the individuals. This fact reveals that those layers which show an outstanding performance will propagate to the rest of the population in the following iterations.

The use of the dropout layer, which has an important role to avoid overfitting, is also significant. In most of the solutions, this type of layer composes at least 50% of the architecture. It is noticeable that almost all individuals start with this type of layer. This can be seen as a features filtering step, in charge of reducing the number of features in subsequent layers. The only individual which does not follow this pattern (Ind. 2) has in contrast three consecutive dropout layers after the first fully connected layer. A similar pattern can be appreciated before the last layer, which in all cases is also composed by a dropout component.

Regarding the hidden layers, some individuals include up to 3 hidden components (i.e. Ind. 3 and 6). However, only one hidden layer is required to provide the best results (Ind. 8). Due to the operation of the genetic search, where a final population is provided, it is possible to select a solution applying different restrictions. In this case, it is possible to select the individual with the lower number of hidden layers.

As regards kernel initialisation functions, there are no clear trends, although the *he_uniform* function has a certain degree of significance in the first fully connected layer of all individuals. In the last layer, a broad range of functions are used, which means that this does have special relevance in achieving better results. In case of the activation function, there is no observable alignment in the first layers, while in the last layer, the prevailing functions are *sigmoid* and *softmax*.

Finally, the computational time has also been analysed. Figure 13.12 shows the mean time by generation for the 40 executions performed. As can be seen, there is an increasing evolution across generations, although the mean time by generation is always in the range between 2000 and 4500 s. The reason for these high values lies in that the evaluation of every individual implies first to train the model defined by its encoding and then to evaluate the performance in the validation set.

| | Layer 0 | Layer 1 | Layer 2 | Layer 3 | Layer 4 | Layer 5 | Layer 6 | Layer 7 |
|---|---|---|---|---|---|---|---|---|
| **Ind 0** rmsprop 88.4% | Dropout Rate: 0.4 | Dense 310 Act: sigmoid Init: zero | Dropout Rate: 0.5 | Dense 2 Act: sigmoid Init: uniform | | | | |
| **Ind 1** nadam 88.23% | Dropout Rate: 0.4 | Dense 310 Act: sigmoid Init: he_unif | Dropout Rate: 0.1 | Dense 2 Act: sigmoid Init: zero | | | | |
| **Ind 2** rmsprop 86.71% | Dense 60 Act: linear Init: he_unif | Dropout Rate: 0.2 | Dropout Rate: 0.1 | Dropout Rate: 0.7 | Dense 310 Act: sigmoid Init: he_unif | Dropout Rate: 0.4 | Dense 2 Act: sigmoid Init: normal | |
| **Ind 3** rmsprop 87.98% | Dropout Rate: 0.2 | Dense 310 Act: linear Init: he_unif | Dense 110 Act: relu Init: glorot | Dropout Rate: 0.4 | Dense 110 Act: tanh Init: glorot | Dropout Rate: 0.3 | Dropout Rate: 0.2 | Dense 2 Act: softmax Init: lecun |
| **Ind 4** nadam 87.71% | Dropout Rate: 0.4 | Dense 210 Act: sigmoid Init: normal | Dropout Rate: 0.7 | Dropout Rate: 0.4 | Dense 2 Act: softmax Init: zero | | | |
| **Ind 5** nadam 88.1% | Dropout Rate: 0.8 | Dense 310 Act: softplus Init: zero | Dropout Rate: 0.1 | Dense 2 Act: sigmoid Init: zero | | | | |
| **Ind 6** nadam 87.64% | Dropout Rate: 0.4 | Dense 310 Act: tanh Init: he_unif | Dense 110 Act: relu Init: glorot | Dropout Rate: 0.4 | Dense 110 Act: tanh Init: he_unif | Dropout Rate: 0.3 | Dropout Rate: 0.2 | Dense 2 Act: softmax Init: lecun |
| **Ind 7** rmsprop 88.33% | Dropout Rate: 0.4 | Dense 310 Act: tanh Init: he_unif | Dense 310 Act: sigmoid Init: he_unif | Dropout Rate: 0.7 | Dense 2 Act: softmax Init: normal | | | |
| **Ind 8** nadam 88.99% | Dropout Rate: 0.4 | Dense 310 Act: sigmoid Init: he_unif | Dropout Rate: 0.1 | Dense 2 Act: softmax Init: zero | | | | |
| **Ind 9** rmsprop 87.84% | Dropout Rate: 0.2 | Dropout Rate: 0.1 | Dropout Rate: 0.7 | Dense 310 Act: linear Init: he_unif | Dropout Rate: 0.3 | Dense 2 Act: sigmoid Init: normal | | |

**Fig. 13.11** Architectures and hyperparameters of the 10 individuals obtained in one execution of the genetic search. For each individual, the accuracy in the test set is provided

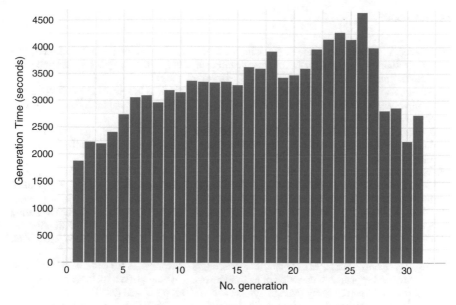

**Fig. 13.12**  Evolution of the mean time by generation for all the executions of EvoDeep

## 13.6   Conclusions

This chapter introduces the application of Neuroevolution for the design of accurate Deep Learning architectures to detect Android malware. A modification of the Finite State Machine used by the EvoDeep framework has allowed to implement a model to generate valid sequences of fully connected and dropout layers. In the experiments, it has been demonstrated that this approach is able to successfully define architectures which provide high accuracy rates and that they outperform classical machine learning approaches typically used in the state-of-the-art litera-ture. In addition, the approach followed has a strong ability to avoid overfitting, using a validation set to guide the evolutionary search. In future work, we aim to improve the genetic search by implementing a multi-objective algorithm where the accuracy of the model is maximised but where the size of the model (in terms of number of units and layers) is also minimised. This can help to create new solutions with better generalisation ability.

**Acknowledgments**  This work has been supported by several research grants: Spanish Ministry of Science and Education under TIN2014-56494-C4-4-P grant (DeepBio) and Comunidad Autónoma de Madrid under P2018/TCS-4566 grant (CYNAMON).

# References

1. Martín, A., Fuentes-Hurtado, F., Naranjo, V., Camacho, D.: Evolving deep neural networks architectures for android malware classification. In: 2017 IEEE Congress on Evolutionary Computation (CEC), pp. 1659–1666. IEEE, Piscataway (2017)
2. Martín, A., Lara-Cabrera, R., Fuentes-Hurtado, F., Naranjo, V., Camacho, D.: EvoDeep: a new evolutionary approach for automatic deep neural networks parametrisation. J. Parallel Distrib. Comput. **117**, 180–191 (2018)
3. Martín, A., Lara-Cabrera, R., Camacho, D.: Android malware detection through hybrid features fusion and ensemble classifiers: the AndroPyTool framework and the OmniDroid dataset. Inform. Fusion **52**, 128–142 (2019)
4. Martín, A., Calleja, A., Menéndez, H.D., Tapiador, J., Camacho, D.: Adroit: Android malware detection using meta-information. In: 2016 IEEE Symposium Series on Computational Intelligence (SSCI), pp. 1–8. IEEE, Piscataway (2016)
5. Zhang, M., Duan, Y., Yin, H., Zhao, Z.: Semantics-aware android malware classification using weighted contextual API dependency graphs. In: Proceedings of the 2014 ACM SIGSAC Conference on Computer and Communications Security, pp. 1105–1116. ACM, New York (2014)
6. Arp, D., Spreitzenbarth, M., Hubner, M., Gascon, H., Rieck, K., CERT Siemens: DREBIN: Effective and explainable detection of android malware in your pocket. In: 2014 Network and Distributed System Security (NDSS) Symposium, vol. 14, pp. 23–26 (2014)
7. Martín, A., Rodríguez-Fernández, V., Camacho, D.: CANDYMAN: classifying android malware families by modelling dynamic traces with Markov chains. Eng. Appl. Artif. Intell. **74**, 121–133 (2018)
8. Yuan, Z., Lu, Y., Wang, Z., Xue, Y.: Droid-Sec: deep learning in android malware detection. SIGCOMM Comput. Commun. Rev. **44**(4), 371–372 (2014)
9. Yuan, Z., Lu, Y., Xue, Y.: DroidDetector: android malware characterization and detection using deep learning. Tsinghua Sci. Technol. **21**(1), 114–123 (2016)
10. Su, X., Zhang, D., Li, W., Zhao, K.: A deep learning approach to android malware feature learning and detection. In: 2016 IEEE Trustcom/BigDataSE/ISPA, pp. 244–251. IEEE, Piscataway (2016)
11. Hou, S., Saas, A., Ye, Y., Chen, L.: DroidDelver: An android malware detection system using deep belief network based on API call blocks . In: International Conference on Web-Age Information Management, pp. 54–66. Springer, Berlin (2016)
12. Zhu, D., Jin, H., Yang, Y., Wu, D., Chen, W.: DeepFlow: Deep learning-based malware detection by mining android application for abnormal usage of sensitive data. In: 2017 IEEE Symposium on Computers and Communications (ISCC), pp. 438–443. IEEE, Piscataway (2017)
13. Arzt, S., Rasthofer, S., Fritz, C., Bodden, E., Bartel, A., Klein, J., Le Traon, Y., Octeau, D., McDaniel, P.: FlowDroid: precise context, flow, field, object-sensitive and lifecycle-aware taint analysis for android apps. Acm Sigplan Notices **49**(6), 259–269 (2014)
14. Fereidooni, H., Conti, M., Yao, D., Sperduti, A.: ANASTASIA: ANdroid mAlware detection using STatic analySIs of applications. In: 2016 8th IFIP International Conference on New Technologies, Mobility and Security (NTMS), pp. 1–5. IEEE, Piscataway (2016)
15. Hou, S., Saas, A., Chen, L., Ye, Y.: Deep4MalDroid: A deep learning framework for android malware detection based on Linux kernel system call graphs. In: 2016 IEEE/WIC/ACM International Conference on Web Intelligence Workshops (WIW), pp. 104–111. IEEE, Piscataway (2016)
16. Kim, T., Kang, B., Rho, M., Sezer, S., Gyu Im, E.: A multimodal deep learning method for android malware detection using various features. IEEE Trans. Inform. Foren. Sec. **14**(3), 773–788 (2018)

17. McLaughlin, N., Martinez del Rincon, J., Kang, B., Yerima, S., Miller, P., Sezer, S., Safaei, Y., Trickel, E., Zhao, Z., Doupé, A., et al.: Deep android malware detection. In: Proceedings of the Seventh ACM on Conference on Data and Application Security and Privacy, pp. 301–308. ACM, New York (2017)
18. Zhang, Y., Yang, Y., Wang, X.: A novel android malware detection approach based on convolutional neural network. In: Proceedings of the 2nd International Conference on Cryptography, Security and Privacy, pp. 144–149. ACM, New York (2018)
19. Wang, W., Zhao, M., Wang, J.: Effective android malware detection with a hybrid model based on deep autoencoder and convolutional neural network. J. Amb. Intel. Hum. Comp. **10**(8), 3035–3043 (2019)
20. Nix, R., Zhang, J.: Classification of android apps and malware using deep neural networks. In: 2017 International Joint Conference on Neural Networks (IJCNN), pp. 1871–1878. IEEE, Piscataway (2017)
21. Xu, K., Li, Y., Deng, R.H., Chen, K.: DeepRefiner: Multi-layer android malware detection system applying deep neural networks. In: 2018 IEEE European Symposium on Security and Privacy (EuroS&P), pp. 473–487. IEEE, Piscataway (2018)
22. Vinayakumar, R., Soman, K.P., Poornachandran, P.: Deep android malware detection and classification. In: 2017 International Conference on Advances in Computing, Communications and Informatics (ICACCI), pp. 1677–1683. IEEE, Piscataway (2017)
23. Karbab, E.B., Debbabi, M., Derhab, A., Mouheb, D.: MalDozer: automatic framework for android malware detection using deep learning. Digit. Invest. **24**, S48–S59 (2018)
24. Booz, J., McGiff, J., Hatcher, W.G., Yu, W., Nguyen, J., Lu, C.: Tuning deep learning performance for android malware detection. In: 2018 19th IEEE/ACIS International Conference on Software Engineering, Artificial Intelligence, Networking and Parallel/Distributed Computing (SNPD), pp. 140–145. IEEE, Piscataway (2018)
25. Martín, A., Menéndez, H.D., Camacho, D.: MOCDroid: multi-objective evolutionary classifier for Android malware detection. Soft Comput. **21**(24), 7405–7415 (2017)
26. Martin, A., Menéndez, H.D., Camacho, D.: Genetic boosting classification for malware detection. In: 2016 IEEE Congress on Evolutionary Computation (CEC), pp. 1030–1037. IEEE, Piscataway (2016)
27. Martín, A., Lara-Cabrera, R., Camacho, D.: A new tool for static and dynamic android malware analysis. In: Data Science and Knowledge Engineering for Sensing Decision Support, pp. 509–516 (2018)
28. Pedregosa, F., Varoquaux, G., Gramfort, A., Michel, V., Thirion, B., Grisel, O., Blondel, M., Prettenhofer, P., Weiss, R., Dubourg, V., et al.: Scikit-learn: machine learning in python. J. Machine Learn. Resea. **12**(Oct), 2825–2830 (2011)

# Chapter 14
# Data Dieting in GAN Training

Jamal Toutouh, Erik Hemberg, and Una-May O'Reilly

**Abstract** We investigate training Generative Adversarial Networks, GANs, with less data. Subsets of the training dataset can express empirical sample diversity while reducing training resource requirements, e.g., time and memory. We ask how much data reduction impacts generator performance and gauge the additive value of generator ensembles. In addition to considering stand-alone GAN training and ensembles of generator models, we also consider reduced data training on an evolutionary GAN training framework named Redux-Lipizzaner. Redux-Lipizzaner makes GAN training more robust and accurate by exploiting overlapping neighborhood-based training on a spatial 2D grid. We conduct empirical experiments on Redux-Lipizzaner using the MNIST and CelebA data sets.

## 14.1  Introduction

In Generative Adversarial Network(GAN) training pathologies such as mode and discriminator collapse can be overcome by using an evolutionary approach [25, 27]. In particular, an evolutionary GAN training method called Lipizzaner has been used for creating robust and accurate generative models [25]. We work with Redux-Lipizzaner, a descendant of Lipizzaner. Per Lipizzaner, Redux-Lipizzaner operates on spatially distributed populations of generators and discriminators. It executes an asynchronous competitive coevolutionary algorithm on an abstract 2D spatial grid of cells organized into overlapping Moore neighborhoods. On each cell there is a subpopulation of generators and the other of discriminators, aggregated from the cell and its adjacent neighbors. The neural network models' parameters are updated with stochastic gradient descent following conventional machine learning. Between training epochs, the subpopulations are

J. Toutouh (✉) · E. Hemberg · U.-M. O'Reilly
MIT CSAIL, Cambridge, MA, USA
e-mail: toutouh@mit.edu; jamal@lcc.uma.es; hembergerik@csail.mit.edu;
unamay@csail.mit.edu

© Springer Nature Singapore Pte Ltd. 2020          379
H. Iba, N. Noman (eds.), *Deep Neural Evolution*, Natural Computing Series,
https://doi.org/10.1007/978-981-15-3685-4_14

reinitialized by requesting copies of best neural network models from the cell's neighborhood. This implicit asynchronous information exchange relies upon overlapping neighborhoods. In contrast to `Lipizzaner`, only after the final epoch, in `Redux-Lipizzaner`, the probability weights for each generator ensemble, consisting of a cell and its neighbors, are optimized using an evolutionary strategy. One model in the ensemble is selected probabilistically, on the basis of the weights, to generate the sample.

While it has been shown that mixtures of GANs perform well [6], one drawback of relying upon multiple generators is that it can be resource intensive to train them. A simple approach to reduce resource use during training is to use less data. For example, different GANs can be trained on different subsamples of the training data set. The use of less training data reduces the storage requirements, both disk and RAM while depending on the ensemble of generators to limit possible loss in performance from the reduction of training data. In the case of `Redux-Lipizzaner` there is also the potential benefit of the implicit communication that comes from the training on overlapping neighborhoods and updating the cell with the best generator after a training epoch. This leads to the following research questions:

1. How does the accuracy of generators change in spatially distributed grids when the dataset size is decreased?
2. How do ensembles support training with less data in cases where models are trained independently or on a grid with implicit communication?

The contributions of this chapter are:

- `Redux-Lipizzaner`, a resource efficient method for evolutionary GAN training,
- a method for optimizing GAN generator ensemble mixture weights via evolutionary strategies
- analysis of the impact of data size on GAN training on the MNIST and CelebA data sets
- analysis of the value of ensembling after GAN training on subsets of the data.

We proceed as follows. Notation for this chapter is in Sect. 14.2. In Sect. 14.3 we describe related work. The `Redux-Lipizzaner` is described in Sect. 14.4. Empirical experiments are reported in Sect. 14.5. Finally conclusions and future work are in Sect. 14.6.

## 14.2  General GAN Training

In this study, we adopt the notation similar to [5, 17]. Let $\mathcal{G} = \{G_g, g \in \mathcal{U}\}$ and $\mathcal{D} = \{D_d, d \in \mathcal{V}\}$ denote the class of generators and discriminators, where $G_g$ and $D_d$ are functions parameterized by $g$ and $d$. $\mathcal{U}, \mathcal{V} \subseteq \mathbb{R}^p$ represent the respective parameters space of the generators and discriminators. Finally, let $G_*$ be the target unknown distribution to which we would like to fit our generative model.

Formally, the goal of GAN training is to find parameters $g$ and $d$ in order to optimize the objective function

$$\min_{g \in \mathcal{U}} \max_{d \in \mathcal{V}} \mathcal{L}(g, d) \ , \ \text{where}$$

$$\mathcal{L}(g, d) = e_{x \sim G_*}[\phi(D_d(x))] + e_{x \sim G_g}[\phi(1 - D_d(x))] \ , \tag{14.1}$$

and $\phi : [0, 1] \rightarrow \mathbb{R}$, is a concave *measuring function*. In practice, we have access to a finite number of training samples $x_1, \ldots, x_m \sim G_*$. Therefore, an empirical version $\frac{1}{m} \sum_{i=1}^{m} \phi(D_d(x_i))$ is used to estimate $e_{x \sim G_*}[\phi(D_d(x))]$. The same also holds for $G_g$.

## 14.3    Related Work

**Evolutionary Computing and GANs** Competitive coevolutionary algorithms have adversarial populations (usually two) that simultaneously evolve [12] population solutions against each other. Unlike classic evolutionary algorithms, they employ fitness functions that rate solutions relative to their *opponent* population. Formally, these algorithms can be described with a minimax formulation [2, 9] which makes them similar to GANs.

**Spatial Coevolutionary Algorithms** Spatial (toroidal) coevolution is an effective means of controlling the mixing of adversarial populations in coevolutionary algorithms. Five cells per neighborhood (one center and four adjacent cells) are common [15]. With this notion of distributed evolution, each neighborhood can evolve in a different direction and more diverse points in the search space are explored. Additional investigation into the value of spatial coevolution has been conducted by [20, 29].

**Scaling Evolutionary Computing for Machine Learning** A team from OpenAI [22] applied a simplified version of Natural Evolution Strategies (NES) [28] with a novel communication strategy to a collection of reinforcement learning (RL) benchmark problems. Due to better parallelization over thousand cores, they achieved much faster training times (wall-clock time) than popular RL techniques. Likewise, a team from Uber AI [23] showed that deep convolutional networks with over four million parameters trained with genetic algorithms can also reach results competitive to those trained with OpenAI's NES and other RL algorithms. OpenAI ran their experiments on a computing cluster of 80 machines and 1440 CPU cores [22], whereas Uber AI employed a range of hundreds to thousands of CPU cores (depending on availability). EC-Star [14] is another example of a large scale evolutionary computation system. By evaluating population individuals only on a small number of training examples per generation, Morse et al. [21] showed that a simple evolutionary algorithm can optimize neural networks of over 1000 dimensions as effectively as gradient descent algorithms. FCUBE, see https://flexgp.github.io/FCUBE/, is a cloud-based modeling system that uses genetic programming [4].

**Ensembles—Evolutionary Computation and GANs** Evolutionary model ensembling has been explored with the aforementioned FCUBE system. FCUBE factors different data splits to cloud instances that model with symbolic regression. These instances draw subsets of variables and fitness functions and learn weakly. After learning the best models are filtered to eliminate the weakest ones and ensemble fusion is used to unify the prediction.

Bagging applies a weighted average to the outputs of a model set for prediction and assumes that all models use the same input variables. Random forests combine bagging with decision trees that use randomized subsets of the input variables. The ensemble technique of `Redux-Lipizzaner` has weights that bias probabilistic selection of one model in the ensemble to generate a sample in contrast to these techniques which consider all model outputs and average them. There are alternative methods of combining GANs into ensembles. For example, "self-ensembles" of GANs were introduced by Wang et al. [26] and are constructed with models based on the same network initialization while training for different numbers of iterations. The same authors introduced also cascade GANs where the part of the training data which is badly modeled by one GAN is redirected to a follow-up GAN. Other examples include boosting such as [7] and [24] who present AdaGAN, which adds a new component into a mixture model at each step by running a GAN algorithm on a reweighted sample. MD-GAN [8] distributes GANs so that they can be trained over datasets that are spread on multiple workers. It proposes a novel learning procedure to fit this distributed setup, whereas `Lipizzaner` uses conventional gradient-based training and a probabilistic mixture model. In K-GANS [3] an ensemble of GANs is trained using semi-discrete optimal transport theory. Quoting the authors, "each generative network models the transportation map between a point mass (Dirac measure) and the restriction of the data distribution on a tile of a Voronoi tessellation that is defined by the location of the point masses. We iteratively train the generative networks and the point masses until convergence." MGAN [13] trains with multiple generators given the specific goal of overcoming mode collapse. They add a classifier to the architecture and use it to specify which generator a sample comes from. Essentially, internal samples are created from multiple generators and then one of them is randomly drawn to provide the sample. With the specific aim to provide complete guaranteed mode coverage, [30] constructing the generator mixture with a connection to the multiplicative weights update rule.

The next section presents `Redux-Lipizzaner`: a scalable, distributed framework for coevolutionary GAN training with reduced training data use.

## 14.4    Data Reduction in Evolutionary GAN Training

This section describes Redux-Lipizzaner which is a spatially distributed coevolutionary GANs training method in which GANs at each cell are trained by using subsets of the whole training data set. The key output of Redux-Lipizzaner is the best performing ensemble (mixture) of generators. First we describe the spatial topology used to evolutionary train GANs in Sect. 14.4.1. Next we present how we subsample the training data in Sect. 14.4.2. Then in Sect. 14.4.3 we describe how the final generator mixture weights are determined. Finally, we formalize the Redux-Lipizzaner algorithm in Sect. 14.4.4.

### 14.4.1    Overview of Redux-Lipizzaner

Redux-Lipizzaner is an extension of Lipizzaner and addresses the robust training of GANs by employing adversarial arms races between two populations, one of generators, and one of discriminators. Going forward, we use the term *adversarial populations* to denote these two populations. Thus, we define a population of generators $\mathbf{g} = \{g_1, \ldots, g_Z\}$ and a population of discriminators $\mathbf{d} = \{d_1, \ldots, d_Z\}$, where $Z$ is the size of the population. These two populations are trained one against the other. The use of populations is one source of diversity that has shown to be adequate to deal with some of the GAN's training pathologies [1].

Redux-Lipizzaner defines a toroidal grid. In each cell, it places a GAN (a pair generator-discriminator), which is named *center*. Each cell has a neighborhood that forms a subpopulations of models: $\mathbf{g}$ (generators) and $\mathbf{d}$ (discriminators). The size of these subpopulations is denoted by $s$. In this study, Redux-Lipizzaner uses five-cell Moore neighborhood ($s = 5$), i.e., the neighborhoods include the cell itself (center) and the cells in the *west*, *north*, *east*, and *south*.

For the $k$-th neighborhood in the grid, we refer to the generator in its center cell by $\mathbf{g}^{k,1} \subset \mathbf{g}$ and the set of generators in the rest of the neighborhood cells by Redux-Lipizzaner $\mathbf{g}^{k,2}, \ldots, \mathbf{g}^{k,s}$, respectively. Furthermore, we denote the union of these sets by $\mathbf{g}^k = \cup_{i=1}^s \mathbf{g}^{k,i} \subseteq \mathbf{g}$, which represents the $k$th generator neighborhood. Note that given a grid size $m \times m$, there are $m^2$ neighborhoods. Figure 14.1 illustrates some examples of the overlapping neighborhoods on a $4 \times 4$ toroidal grid and how the subpopulations of each cell are built ($G_{1,1}$ and $D_{1,1}$). The use of this grid for training the models addresses the quadratic computational complexity of the basic adversarial competitions based algorithms. Without loss of generality, we consider square grids of $m \times m$ size in this study.

The overlapping neighborhoods define the possible exchange of information among the different cells during the training process due to the *selection* and *replacement* operators applied in coevolutionary algorithms.

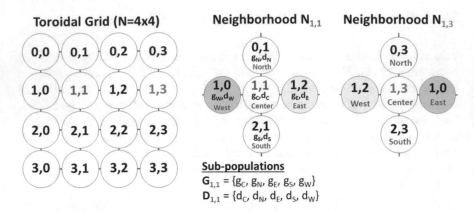

**Fig. 14.1** Illustration of overlapping neighborhoods on a toroidal grid. Note how a cell update at $N_{1,2}$ can be communicated to $N_{1,1}$ and $N_{1,3}$ when they gather their neighbors. If $N_{1,1}$ is then updated with the updated value from $N_{1,2}$, the value has propagated. If the propagation continues one more cell to the left to $N_{1,0}$, the value will come into the range of both $N_{0,0}$ and $N_{2,0}$. Propagation runs laterally and vertically. We also show an example of a cell's generator and discriminator subpopulations (based on its neighborhood) for $N_{1,1}$

Redux-Lipizzaner is built on the following basis: *selection and replacement, fitness evaluation,* and *reproduction based on GAN training*.

**Selection and Replacement** Selection promotes high performing solutions when updating a subpopulation. Redux-Lipizzaner applies *tournament selection* of size $\tau$ to update the center of the cell. First, the subpopulations with the updated copies of the neighbors evaluate all GAN generator-discriminator pairs, then $\tau$ generators and $\tau$ discriminators are randomly picked, and the center of the cell is set as the fittest generator and discriminator from the $\tau$ selected ones (lines from 1 to 6 of Algorithm 2). After all GAN training is completed all models are evaluated again, and the tournament selection is applied to replace the least fit generator and discriminator in the subpopulations with the fittest ones and sets them as the center of the cell (lines from 20 to 27 of Algorithm 2).

**Fitness Evaluation** The search and optimization in evolutionary algorithms are guided by the evaluation of the *fitness*, a measure that evaluates how good a solution is at solving the problem. In Redux-Lipizzaner, an adversarial method, the performance of the model depends on the adversary. The performance of a given generator (discriminator) is evaluated in terms of some loss function $M$. Redux-Lipizzaner uses *Binary cross entropy (BCE) loss* (see Eq. 14.2), where the model's objective is to minimize the Jensen–Shannon divergence (JSD) between the real ($p$) and fake ($q$) data distributions, i.e., $JSD(p \parallel q)$. In Redux-Lipizzaner, fitness $\mathcal{L}$ of a model ($g_i \in \mathbf{g}$ or $d_j \in \mathbf{d}$) is its average performance against all its adversaries.

$$M^{BCE} = \frac{1}{2}e_{x \sim G_g}[\log(1 - D_d(x))].\qquad(14.2)$$

**Variation—GAN Training** Model variation is done via GAN training, which is applied in order to update the parameters of the models. Stochastic Gradient Descent training performs gradient-based updates on the parameters (network weights) the models. Moreover, Gaussian-based updates create new learning rate values $n_\delta$.

The center generator (discriminator) is trained against a randomly chosen adversary from the subpopulation of discriminators (generators) (lines 9 and 14 of Algorithm 2, respectively).

### 14.4.2   Dataset Sampling in Redux-Lipizzaner

Instead of training each subpopulation with the whole training dataset, per Lipizzaner, Redux-Lipizzaner applies random sampling with replacement over the training data to define $m^2$ different subsets (partitions) of data that will be used as training dataset for each cell (see Fig. 14.2). Thus, each cell has its own *training subset* of data.

### 14.4.3   Evolving Generator Mixture Weights

Redux-Lipizzaner searches for and returns a mixture of generators composed from a neighborhood. The mixture of generators is the fusion of the different generators in the neighborhood trained by subsets of the training data set. The selection of the *best* weights that define mixture ensemble is difficult. Redux-Lipizzaner applies an ES-(1+1) algorithm [19, Algorithm 2.1] to evolve a mixture weight vector **w** for each neighborhood in order to optimize the performance of the fused generative model, see Algorithm 3.

**Fig. 14.2** Illustration of how the training dataset is sampled to generate training data subsets to train the different neighborhoods on the grid ($N_{1,1}$ and $N_{1,3}$)

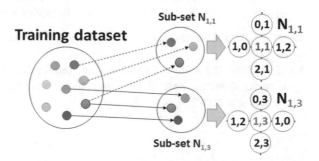

When using `Redux-Lipizzaner` for images Fréchet Inception Distance (FID) score [10] is used to assess the accuracy of the generative models. Note, nothing prevents the use of different metrics, e.g., inception score.

The $s$-dimensional mixture weight vector $\mathbf{w}$ is defined as follows:

$$\mathbf{g}^*, \mathbf{w}^* = \underset{\mathbf{g}^k, \mathbf{w}^k : 1 \le k \le m^2}{\operatorname{argmin}} \sum_{g_i \in \mathbf{g}^k w_i \in \mathbf{w}^k} w_i FID_{g_i}, \tag{14.3}$$

where $w_i$ represents the probability that a data point comes from the $i$th generator in the neighborhood, with $\sum_{w_i \in \mathbf{w}^k} w_i = 1$.

### 14.4.4 Algorithms of Redux-Lipizzaner

`Algorithm 1` formalizes the main steps of `Redux-Lipizzaner`. First, it starts the parallel execution of the training on each cell by initializing their own learning hyper-parameters (i.e., the *learning rate* and the mixture weights) and by assigning them their own training subsets (Lines 2 and 3). Then, the training process, see `Algorithm 2`, consists of a loop with two main steps: first, gather the GANs (neighbors) to build the subpopulations (neighborhood) and, second, update the center by applying the coevolutionary GANs training method for all mini-batches in the training subset. These steps are repeated $T$ (generations or training epochs) times. After that, each cell optimizes its mixture weights by applying an Evolutionary Strategy in order to optimize the performance of the ensemble defined by the neighborhood, see `Algorithm 3`. Finally, the best performing ensemble is selected across the entire grid and returned, including its probabilistic weights, as the final solution.

Section 14.5 next presents results regarding the question of how `Redux-Lipizzaner` performs given data reduction with the support of ensembles.

## 14.5  Experimental Analysis

In this section we proceed experimentally. We use Sect. 14.5.1 to present our experimental setup. We then investigate the following research questions:

**RQ1:**  How robust are spatially distributed grids when training with less of the dataset?

**RQ2:**  Given the use of ensembles, if we reduce the data quantity at each cell, at what point will the ensemble fail to fuse the resulting models towards achieving sufficient accuracy?

---

**Algorithm 1** `Redux-Lipizzaner`: In parallel, for each cell, initialize settings then iterate over each generation. Each generation, retrieve neighbor cells to build generator and discriminator subpopulations, evolve generators and discriminators trained with SGD, replace worst with best and update self with best, and finally evolve weights for a neighborhood mixture model.

**Input:** T : Total generations, $E$ : Grid cells, $k$ : Neighborhood size, $\theta_D$ : Training dataset, $\theta_p$ : Sampling size in terms of dataset portion, $\theta_{COEV}$ : Parameters for `CoevolveAndTrainModels`, $\theta_{EA}$: Parameters for `MixtureEA`

**Return:** $n$ : neighborhood, $\omega$ : mixture weights

---

1:  **parfor** $c \in E$ **do**                          ▷ Asynchronous parallel execution of all cells in grid
2:      $ds \leftarrow$ getDataSubset($\theta_D, \theta_p$)                          ▷ Creates a sub-set of the dataset
3:      $n, \omega \leftarrow$ initializeCells($c, k, ds$)                          ▷ Initialization of cells
4:      **for** generation **do** $\in [0, \ldots, T]$                          ▷ Iterate over generations
5:          $n \leftarrow$ copyNeighbours($c, k$)                  ▷ Collect neighbor cells for the subpopulations
6:          $n \leftarrow$ LipizzanerTraining $(n, \theta_{COEV})$  ▷ Coevolve GANs using `Algorithm 2`
7:      **end for**
8:      $\omega \leftarrow$ MixtureEA($\omega, n, \theta_{EA}$)                          ▷ Evolve mixture weights, `Algorithm 3`
9:  **end parfor**
10: **return** $(n, \omega)^*$                          ▷ Cell with best generator mixture

---

## 14.5.1    Experimental Setup

We use two common image datasets from the GAN literature: MNIST [16] and CelebA [18]. MNIST has been widely used and it consist of low dimensional handwritten digits images. The larger CelebA dataset contains more than 200,000 images of faces. To obtain an absolute measure of model accuracy, we draw fake image samples from the generative models computed and score them with Frechet inception distance (FID) [11]. FID score is a black box, discriminator-independent, metric and expresses image similarity to the samples used in training.

The process of sampling the data is independent for each cell of the grid and it consist on randomly selecting different mini-batches of the training dataset. In the context of a grid, given grid size, there is an expectation that every sample will be drawn at least once. This can be considered 100% coverage, *over the grid*, though not at any cell. When the subset size is lower and/or the grid is smaller, this expected coverage of the complete dataset is nonetheless higher than that of a subset drawn for a single GAN trained independently of others.

For a fixed budget of training samples, when a GAN is trained with a larger dataset and the batch size of a smaller dataset is maintained, the gradient is estimated more often because there are more mini-batches per generation. (Given the standard terminology that an epoch is one forward pass and one backward pass of all the training examples, one epoch is one generation.) In contrast, in the same circumstances, if the number of mini-batches is held constant, and the mini-batch size increased, we incur a cost increase in RAM to store the mini-batch and the

---

**Algorithm 2** `LipizzanerTraining`: Select a new neighborhood from the current one. Each mini-batch train discriminators against a randomly drawn generator and generators against a randomly drawn discriminator, using SGD. Evaluate all against each other, using minimum loss as value to choose best to replace worst and update center. Return this new neighborhood.

**Input:** $\tau$ : Tournament size, $X$ : Input training dataset, $\beta$ : Mutation probability, $n$ : Cell neighborhood subpopulation, $ds$ : Sub-set of the training dataset

**Return:** $n$ : Cell neighborhood subpopulations

---

1: $\mathbf{B} \leftarrow$ getMiniBatches($ds$) ▷ Load mini-batches
2: $B \leftarrow$ getRandomMiniBatch($\mathbf{B}$) ▷ Get a random mini-batch to evaluate GAN pairs
3: **for** $g, d \in \mathbf{g} \times \mathbf{d}$ **do** ▷ Evaluate all GAN pairs
4:     $\mathcal{L}_{g,d} \leftarrow$ evaluate($g, d, B$) ▷ Evaluate GAN
5: **end for**
6: $\mathbf{g}, \mathbf{d} \leftarrow$ select($n, \tau$) ▷ Tournament selection with minimum loss($\mathcal{L}$) as fitness
7: **for** $B \in \mathbf{B}$ **do** ▷ Loop over batches
8:     $n_\delta \leftarrow$ mutateLearningRate($n_\delta, \beta$) ▷ Update neighborhood learning rate
9:     $d \leftarrow$ getRandomOpponent($\mathbf{d}$) ▷ Get uniform random discriminator
10:     **for** $g \in \mathbf{g}$ **do** ▷ Evaluate generators and train with SGD
11:         $\nabla_g \leftarrow$ computeGradient($g, d$) ▷ Compute gradient for neighborhood center
12:         $g \leftarrow$ updateNN($g, \nabla_g, B$) ▷ Update with gradient
13:     **end for**
14:     $g \leftarrow$ getRandomOpponent($\mathbf{g}$) ▷ Get uniform random generator
15:     **for** $d \in \mathbf{d}$ **do** ▷ Evaluate discriminator and train with SGD
16:         $\nabla_d \leftarrow$ computeGradient($d, g$) ▷ Compute gradient for neighborhood center
17:         $d \leftarrow$ updateNN($d, \nabla_d, B$) ▷ Update with gradient
18:     **end for**
19: **end for**
20: **for** $g, d \in \mathbf{g} \times \mathbf{d}$ **do** ▷ Evaluate all updated GAN pairs
21:     $\mathcal{L}_{g,d} \leftarrow$ evaluate($g, d, B$) ▷ Evaluate GAN
22: **end for**
23: $\mathcal{L}_g \leftarrow \min(\mathcal{L}_{\cdot,d})$ ▷ Fitness for generator is the average loss value ($\mathcal{L}$)
24: $\mathcal{L}_d \leftarrow \min(\mathcal{L}_{g,\cdot})$ ▷ Fitness for discriminator is the average loss value ($\mathcal{L}$)
25: $n \leftarrow$ replace($n, \mathbf{g}$) ▷ Replace the generator with worst loss
26: $n \leftarrow$ replace($n, \mathbf{d}$) ▷ Replace the discriminator worst loss
27: $n \leftarrow$ setCenterIndividuals($n$) ▷ Best generator and discriminator are placed in the center
28: **return** $n$

---

gradient is estimated on better information but less frequently. To date, there is no clear well-founded procedure or even a heuristic for setting mini-batch size.

We place all experiments on equal footing by training them with the same budget of mini-batches while keeping mini-batch size, i.e., the number of examples per mini-batch, constant. We experimentally vary the training set size per cell or GAN and adjust the number of generations to arrive at the mini-batch budget. See Eq. 14.4.

$$\text{batches\_to\_train} = \frac{\text{training\_dataset\_size}}{\text{mini-batch\_size}} \times \text{data\_portion} \times \text{generations}. \quad (14.4)$$

**Algorithm 3** MixtureEA: Evolve mixture weights $\omega$ with a ES-(1+1).

**Input:** $GT$ : Total generations to evolve the weights, $\mu$ : Mutation rate, $n$ : Cell neighborhood subpopulation, $\omega$ : Mixture weights

**Return:** $\omega$ : mixture weights

| | | |
|---|---|---|
| 1: | **for** generation **do** $\in [0, \ldots, GT]$ | ▷ Loop over generations |
| 2: | $\omega' \leftarrow$ mutate$(\omega, \mu)$ | ▷ Gaussian mutation of mixture weights |
| 3: | $\omega'_f \leftarrow$ evaluateMixture$(\omega', n)$ | ▷ Evaluate generator mixture score, e.g. FID for images |
| 4: | **if** $\omega'_f < \omega_f$ **then** | ▷ Replace if new mixture weights are better |
| 5: | $\omega \leftarrow \omega'$ | ▷ Update mixture weights |
| 6: | **end if** | |
| 7: | **end for** | |
| 8: | **return** $\omega$ | |

**Table 14.1** Batches and generations used in experimental comparisons under equalization to the same computational budget (expressed as batches)

| Portion of data | 100% | 75% | 50% | 25% |
|---|---|---|---|---|
| *MNIST (Computation budget = $1.20 \times 10^5$)* | | | | |
| Number of mini-batches | 600 | 450 | 300 | 150 |
| Number of generations | *200* | *267* | *400* | *800* |
| *CelebA (Computation budget = $31.66 \times 10^3$)* | | | | |
| Number of mini-batches | 1583 | 1187 | 792 | 396 |
| Number of generations | *20* | *27* | *40* | *80* |

For example, given a budget of $1.2 \times 10^5$ mini-batches and a mini-batch size of 100, when the training set size per cell is 60,000, there will be 600 mini-batches per generation. We therefore train for 200 generations to reach the $1.2 \times 10^5$ mini-batches budget. When we reduce the training set size to 30,000 (50%), there will be only 300 mini-batches per generation so we train for 400 generations to reach the training budget of $1.2 \times 10^5$ mini-batches.

Considering the dataset sizes in terms of images (60,000 in MNIST and 202,599 in CelebA), a constant batch size of 100 (Table 14.2), and the relative training data subset size, we provide the number of generations executed in Table 14.1. The total number of batches used to train MNIST is $1.20 \times 10^5$ and CelebA is $31.66 \times 10^3$ when training with the 100% of the data.

In this analysis, we compare Redux-Lipizzaner by using different grid sizes ($4 \times 4$ and $5 \times 5$ for MNIST and $3 \times 3$ for CelebA) with a *Single GAN* training method. These different grid sizes allow us to explore the performance of Redux-Lipizzaner according to different degrees of cell overlap. The datasets selected, MNIST and CelebA, represent different challenges for GANs training due to: first, the size of each sample of MNIST (vector of 784 real numbers) is smaller than the same of CelebA (vector of 12,288 real numbers); second, MNIST dataset has fewer number of samples than CelebA, and third, the size of the models (generator-discriminator) is much larger for CelebA generation than for MNIST. This makes the computational resources required to address CelebA higher than for MNIST. Thus, we have defined our experimental analysis taking into account

**Table 14.2** Setup for experiments conducted with `Redux-Lipizzaner` on MNIST and CelebA

| Parameter | MNIST | CelebA |
|---|---|---|
| *Coevolutionary settings* | | |
| Generations | *See Table 14.1* | |
| Population size per cell | 1 | 1 |
| Tournament size | 2 | 2 |
| Grid size | $1 \times 1$, $4 \times 4$, and $5 \times 5$ | $3 \times 3$ |
| *Mixture evolution* | | |
| Mixture mutation scale | 0.01 | 0.01 |
| Generations | 5000 | 5000 |
| *Hyper-parameter mutation* | | |
| Optimizer | Adam | Adam |
| Initial learning rate | 0.0002 | 0.00005 |
| Mutation rate | 0.0001 | 0.0001 |
| Mutation probability | 0.5 | 0.5 |
| *Network topology* | | |
| Network type | MLP | DCGAN |
| Input neurons | 64 | 100 |
| Number of hidden layers | 2 | 4 |
| Neurons per hidden layer | 256 | 16,384–131,072 |
| Output neurons | 784 | $64 \times 64 \times 3$ |
| Activation function | tanh | tanh |
| *Training settings* | | |
| Mini-batch size | 100 | 128 |

different overlapping patterns and datasets, but also the computational resources available.

All these methods are configured according to the parameterization shown in Table 14.2. In order to extend our analysis, we apply a *bootstrapping* procedure to compare the *Single GANs* with `Redux-Lipizzaner`. Therefore, we randomly generated 30 populations (grids) of 16 and 25 generators from the 30 generators computed by using *Single GAN* method. Then, we compute their FIDs and create mixtures to compare these results against `Redux-Lipizzaner` $4 \times 4$ and `Redux-Lipizzaner` $5 \times 5$, respectively. We name these variants *Bootstrap* $4 \times 4$ and *Bootstrap* $5 \times 5$.

All methods have been implemented in `Python3` and `pytorch`.[1] The experiments are performed on a cloud that provides 16 Intel Cascade Lake cores up to 3.8 GHz with 64 GB RAM and a GPU which are either NVIDIA Tesla P4 GPU with 8 GB RAM or NVIDIA Tesla P100 GPU with 16 GB RAM. All implementations

---

[1] Pytorch Website—https://pytorch.es/.

use the same `Python` libraries and versions to minimize computational differences
that could arise from using the cloud.

## 14.5.2    Research Question 1: How Does the Accuracy of Generators Change in Spatially Distributed Grids When the Dataset Size Is Decreased?

We first establish a non-grid baseline by training a single GAN with decreasing
amounts of data and examining the resulting FID scores, see Table 14.3, and
Fig. 14.3 for pairwise statistical significance with a Wilcoxon Rank Sum test with
$\alpha = 0.01$. What we see is obvious, FID score increases (performance worsens)
as the GAN is trained on less data. In 30 runs of single GAN training on 25% of
the data, the mean FID score is very high: 574.6 while the standard deviation of FID
score is 51.3% including the best FID score of 35.1. When the data subset is doubled
to 50%, the mean FID score drops to 71.2 but the observed standard deviation is
higher (104.6%). The best FID score falls to 30.1. Mean FID score improves with
75% of the data significantly (from 71.2 to 39.8) but minimally in terms of the best
FID score (30.1 vs. 30.2), see Table 14.4. A marked decrease in standard deviation
(104.6–12.4%) occurs. In all cases, the smaller training subsets do not match the
performance when training with 100% of the data where the best FID score is 27.4
and the mean FID score is 38.8, see Tables 14.3 and 14.4. These results are straight
forwardly explained by smaller quantities of data failing to sufficiently cover the
latent distribution.

We can now consider competitive coevolutionary grid-trained GANs where there
is one GAN per cell, the best of that cell's training, at the end of execution (see
Table 14.3). This data allows us to isolate the value of the evolutionary training's
communication in contrast to (1) the independently trained GANs we previously
evaluated and (below) (2) the performance impact of ensemble. Recall that the
overlapping neighborhoods facilitate signal propagation. GANs which perform well
in one neighborhood migrate to their adjacent and overlapping neighborhoods in a
form of communication. The grid-trained GANs achieve better FID scores, given the
same training budget, than independently trained GANs [25]. In the case of a 4 ×
4 grid, the experimental mean FID score is 37.3 with a standard deviation of 15.1%
and the best generator has a FID score of 26.4. The improvement over independently
trained GANs is present with the 5 × 5 grid, where the experimental mean FID score
is 34.3 with a standard deviation of 19.9% and the best generator has a FID score
of 20.8. One possible explanation is that the communication indirectly leads to a
mixing of the data subsamples (that are drawn independently and with replacement)
that effectively improves the coverage of the data.

Figure 14.3 illustrates the statistical analysis of different methods evaluated here
when using the same amount of data. When using the smallest training datasets
(MNIST 25%), the use of `Redux-Lipizzaner` with larger grids and allow signif-

**Table 14.3** Mean(±std) of the *best FID in the grid* for 30 independent runs

| Dataset | Variant | 25% | 50% | 75% | 100% |
|---|---|---|---|---|---|
| MNIST | Single GAN | 574.6±51.3% | 71.2±104.6% | 39.8±12.4% | 38.8±17.0% |
| MNIST | Single GAN Ensemble | 44.2 ± 9.5% | 35.4 ± 8.0% | 34.4±10.0% | 38.6±12.3% |
| MNIST | Bootstrap 4 × 4 | 578.0±12.3% | 73.5 ± 25.3% | 39.7 ± 3.0% | 38.9 ± 4.0% |
| MNIST | Bootstrap 4 × 4 Ensemble | 44.2 ± 9.0% | 35.4 ± 5.1% | 34.5 ± 6.6% | 35.9 ± 7.4% |
| MNIST | Redux-Lipizzaner 4 × 4 | 47.0±19.2% | 42.0 ± 16.5% | 36.5±19.2% | 37.3±15.1% |
| MNIST | Redux-Lipizzaner 4 × 4 Ensemble | 44.1±21.9% | 40.5 ± 15.4% | 33.6±16.7% | 30.7±17.3% |
| MNIST | Bootstrap 5 × 5 | 573.2 ± 9.5% | 74.8 ± 22.6% | 39.7 ± 2.3% | 38.7 ± 3.3% |
| MNIST | Bootstrap 5 × 5 Ensemble | 43.3 ± 8.3% | 33.3 ± 6.3% | 33.0 ± 5.4% | 34.6 ± 6.7% |
| MNIST | Redux-Lipizzaner 5 × 5 | 39.9±15.6% | 34.4 ± 9.1% | 32.9±14.2% | 34.3±19.9% |
| MNIST | Redux-Lipizzaner 5 × 5 Ensemble | 36.2±16.9% | 31.8 ± 12.6% | 30.1±15.9% | 26.3±16.7% |
| CelebA | Redux-Lipizzaner 3 × 3 | 51.9±29.8% | 50.3 ± 25.7% | 51.3±26.6% | 46.5 ± 7.3% |
| CelebA | Redux-Lipizzaner 3 × 3 Ensemble | 49.1 ± 1.4% | 44.7 ± 6.8% | 40.3 ± 3.9% | 43.2 ± 0.9% |

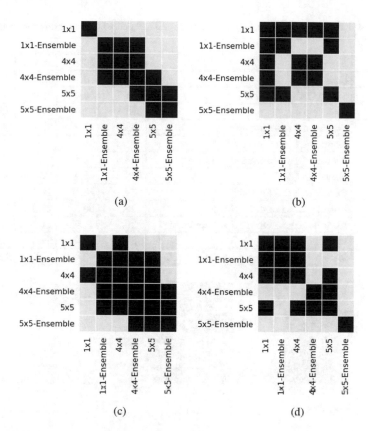

**Fig. 14.3** Statistical analysis comparing the same amount of data applying different methods. Black indicates no significance with $\alpha = 0.01$. (**a**) MNIST 25%. (**b**) MNIST 50%. (**c**) MNIST 75%. (**d**) MNIST 100%

icant improvements of the results. The results provided by Redux-Lipizzaner are also improved when the optimization of the mixture weights is applied. With larger datasets (MNIST 75%), Redux-Lipizzaner $4 \times 4$ provides results as competitive as the same with $5 \times 5$. Something similar is observed when the whole data is used to rain the GANs.

Figure 14.4 shows the statistical analysis of the impact on the performance of the methods analyzed here when using different size of training data. Redux-Lipizzaner $4 \times 4$ provides similar results when reducing the training data in 25% (i.e., for MNIST 100% and MNIST 75%). When the grid size increases, i.e., Redux-Lipizzaner $5 \times 5$, the use of the training datasets with the half of the data or larger does not show statistical differences in the results. However, the application of the mixtures drives the results with the 100% of the data to be the most competitive ones.

Scrutinizing Table 14.3, Figs. 14.3, and 14.4 indicate that Redux-Lipizzaner on a large grid, $5 \times 5$, performs among the best for all training data sizes.

**Table 14.4** Min(best) of the *best FID in the grid* for 30 independent runs for different data diets

| Dataset | Variant | Data diet 25% | 50% | 75% | 100% |
|---------|---------|------|------|------|------|
| MNIST | Single GAN | 35.1 | 30.1 | 30.2 | 27.4 |
| MNIST | Single GAN Ensemble | 33.7 | 30.0 | 26.9 | 27.1 |
| MNIST | Bootstrap 4 × 4 | 395.5 | 47.0 | 37.0 | 36.1 |
| MNIST | Bootstrap 4 × 4 Ensemble | 33.6 | 32.7 | 30.7 | 28.0 |
| MNIST | Redux-Lipizzaner 4 × 4 | 31.8 | 28.8 | 27.1 | 26.4 |
| MNIST | Redux-Lipizzaner 4 × 4 Ensemble | 26.5 | 28.1 | 24.6 | 21.1 |
| MNIST | Bootstrap 5 × 5 | 440.8 | 48.0 | 37.6 | 34.9 |
| MNIST | Bootstrap 5 × 5 Ensemble | 34.8 | 28.1 | 28.0 | 29.8 |
| MNIST | Redux-Lipizzaner 5 × 5 | 30.5 | 26.8 | 27.3 | 26.3 |
| MNIST | Redux-Lipizzaner 5 × 5 Ensemble | 26.3 | 21.9 | 21.2 | 20.8 |
| CelebA | Redux-Lipizzaner 3 × 3 | 39.3 | 39.0 | 39.4 | 42.0 |
| CelebA | Redux-Lipizzaner 3 × 3 Ensemble | 48.3 | 42.4 | 38.9 | 42.7 |

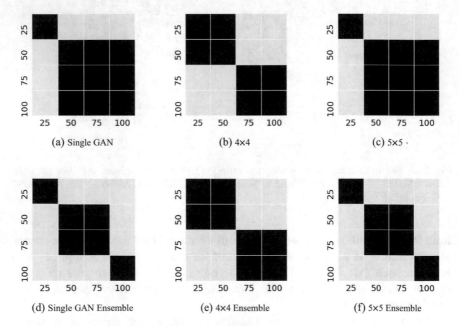

**Fig. 14.4** Statistical analysis comparing the same method with different size of data for MNIST experiments. Black indicates no significance with $\alpha = 0.01$. Figures (b), (c), (e), and (f) illustrate the results of Redux-Lipizzaner and their mixtures. Figures (a) and (d) shows the results of bootstrapping and its mixture

Furthermore, the Redux-Lipizzaner 4 × 4 can perform well with reduced training data size. The CelebA results indicate that Redux-Lipizzaner has similar response to data dieting on another data set. Future work on applying Redux-Lipizzaner to address CelebA with larger grid sizes will allow us to confirm this statement. These results lend support to the hypothesis that communication during training can accelerate the impact of small training datasets. We can answer affirmatively, for the datasets we examined, for `Redux-Lipizzaner`.

### 14.5.3 Research Question 2: Given We Use Ensembles, If We Reduce the Data Quantity at Each Cell, at What Point Will the Ensemble Fail to Unify the Resulting Models Towards Achieving High Accuracy?

We can now consider two directions of inquiry, their order not being important, in the context of the evolutionary GAN training that occurs on a grid. As well, each cell's neighborhood on the grid defines a subpopulation of generators and a subpopulation of discriminators which are trained against each other. This naturally suggests the neighborhood to be the set of GANs in each ensemble of the grid, that is, the fusing of each center cell's generator with its North, South, West, and East cell neighbors. We, therefore, can compare the impact of a neighborhood-based ensemble to each cell's FID score.

We thus isolate communication from grid-based ensembles and we also isolate co trained ensembles from ensembles arising from independently trained GANs. For tabular results, see Table 14.3. To measure the impact of ensembles in this context, i.e., independently trained GANs, we sample sets of 5 GANS from the 30 different training runs and train an mixture. We formulate experiments with different portions of the training dataset (i.e., 25%, 50%, 75%, and 100%. of the samples). We train GANs individually (non-population based) with the same training algorithm as when we train the GANs within `Lipizzaner`.

We first measure the improvement of a given method $m$ ($\Delta(m)$) attributable to using mixtures ($m_{ensemble}$). This metric is evaluated as a percentage in terms of the difference between the average FID of $m$, $\overline{FID(m)}$ and the average FID of the same method when applying the mixtures $\overline{FID(m_{ensemble})}$, see Eq. 14.5.

$$\Delta(m) = \frac{\overline{FID(m)} - \overline{FID(m_{ensemble})}}{\overline{FID(m)}} \%. \tag{14.5}$$

We expect these results to be consistent with [6] who observed an advantage with mixtures of generators. We start with the sets of generators obtained from the 30 runs of independent training for each data subset. For each data subset's set, we optimize the weights of 5 generators randomly drawn from it via ES-(1+1) for 5000 generations, and report the mean, min, and std of FID scores for 30 independent draws. We see statistically significant improvements for some subsets of data (25 and 75%), see Table 14.5. The mean ensemble FID score with 25% subsets is 44.2 versus the single generator's mean FID score of 574.6, an improvement of 92.3%.

**Table 14.5** Mean FID improvement $\Delta(m)$ by weighted ensembles for 30 independent runs (Eq. 14.5) for different data diets

| Data set | Variant | Data diet | | | |
|---|---|---|---|---|---|
| | | 25% | 50% | 75% | 100% |
| MNIST | Single GAN | 92.3% | 50.2% | 13.7% | 0.7% |
| MNIST | Bootstrap 4 × 4 | 92.4% | 51.8% | 13.1% | 7.6% |
| MNIST | Redux-Lipizzaner 4 × 4 | 6.2% | 3.6% | 8.0% | 17.5% |
| MNIST | Bootstrap 5 × 5 | 92.4% | 55.5% | 16.9% | 10.6% |
| MNIST | Redux-Lipizzaner 5 × 5 | 9.3% | 7.4% | 8.5% | 23.5% |
| CelebA | Redux-Lipizzaner 3 × 3 | 5.3% | 11.1% | 21.5% | 7.1% |

**Table 14.6** Min(best) of the *mean FID in the grid* for 30 independent runs for different data diets

| Dataset | Variant | Data diet | | | |
|---|---|---|---|---|---|
| | | 25% | 50% | 75% | 100% |
| MNIST | Bootstrap 4 × 4 | 395.5 | 47.0 | 37.0 | 36.1 |
| MNIST | Redux-Lipizzaner 4 × 4 | 38.3 | 33.9 | 32.9 | 28.6 |
| MNIST | Bootstrap 5 × 5 | 440.8 | 48.0 | 37.6 | 34.9 |
| MNIST | Redux-Lipizzaner 5 × 5 | 34.0 | 29.3 | 30.3 | 27.2 |
| CelebA | Redux-Lipizzaner 3 × 3 | 58.2 | 58.6 | 59.4 | 49.0 |

The improvement is diminishes at 50–50.2% and again at 75–13.7% and finally only 0.7% for 100%. When all the data is used for training, the least improvement but still an improvement is observed.

These results can be anticipated because different subsets were used in training and the fusion of the generators. An interesting note is that Redux-Lipizzaner almost has an inverse progression of mixture effect, see Table 14.5, with the mixture improving the performance of Redux-Lipizzaner more the more training data is available.

Moreover, we study the capacity of the generative models created by using the fusion method presented in Sect. 14.4.3 (see Algorithm 3) from GANs individually trained. In Tables 14.6 and 14.7 the best and the mean FIDs of each generator are shown. The FIDs improve as the training data size increases. This again highlights the improvement of the accuracy on the generated samples for the separate generator with more training data.

Finally, the results in Tables 14.6 and 14.7 are less competitive (higher FIDs) than the ones presented in Tables 14.4 and 14.3, respectively. This delves on the idea of the improvements on the results when mixtures are used. Figure 14.5 illustrates the FID scores distribution at the end of an independent run of MNIST-4 × 4. We cannot compare the results among the different data sizes because we have selected a random independent run, and therefore, these results do not follow the general observations discussed above. The impact of the mixture optimization is shown in this figure. Here, we can observe how the ES(1+1) optimizes the mixture FID values

**Table 14.7** Mean(±std) of the *mean FID in the grid* for 30 independent runs for different data diets

| Dataset | Variant | Data diet | | | |
| --- | --- | --- | --- | --- | --- |
| | | 25% | 50% | 75% | 100% |
| MNIST | Bootstrap4 × 4 | 578.0 ± 12.3% | 73.5 ± 25.3% | 39.7 ± 3.0% | 38.9 ± 4.0% |
| MNIST | Redux-Lipizzaner 4 × 4 | 55.3 ± 21.4% | 49.7 ± 16.9% | 43.4 ± 14.8% | 40.4 ± 20.0% |
| MNIST | Bootstrap 5 × 5 | 573.2 ± 9.5% | 74.8 ± 22.6% | 39.7 ± 2.3% | 38.7 ± 3.3% |
| MNIST | Redux-Lipizzaner 5 × 5 | 46.3 ± 16.7% | 37.0 ± 15.2% | 38.7 ± 12.0% | 34.2 ± 14.1% |
| CelebA | Redux-Lipizzaner 3 × 3 | 64.4 ± 9.8% | 60.9 ± 3.8% | 64.5 ± 7.3% | 51.6 ± 4.6% |

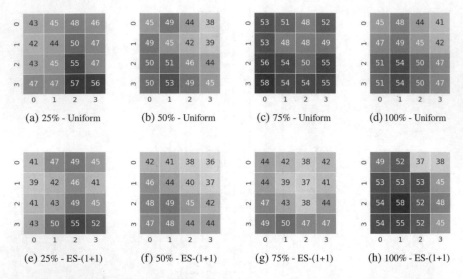

**Fig. 14.5** FID score distribution through a given grid at the end of an independent run for MNIST-4 × 4. Lighter blues represent lower (better) FID scores. The first row (a, b, c, and d) illustrates the FIDs of each ensemble by using uniform mixture weights. The third row (e, f, g, and h) shows the FIDs of each ensemble by using the weights computed by the ES-(1+1)

for each cell and manages to improve most of them (the best cell of the grid is always improved).

## 14.6  Conclusions and Future Work

The use of less training data reduces the storage requirements, both disk and RAM while depending on the ensemble of generators to limit possible loss in performance from the reduction of training data. In the case of Redux-Lipizzaner there is also the potential benefit of the implicit communication that comes from the training on overlapping neighborhoods and updating the cell with the best generator after a training epoch. Our method, Redux-Lipizzaner, for spatially distributed evolutionary GAN training makes use of information exchange between neighborhoods to generate high performing generator mixtures. The spatially distributed grids allow training with less of the dataset because of signal propagation leading to exchange of information and improved performance when training data is reduced compared to ordinary parallel GAN training. In addition, the ensembles lose performance when the training data is reduced, but they are surprisingly robust with 75% of the data.

Future work will investigate the impact of distributing different modes (e.g., classes) of the data to different cells. In addition, more data sets will be evaluated, as well as more fine grained reductions in amount of training data.

# References

1. Al-Dujaili, A., Schmiedlechner, T., Hemberg, E., O'Reilly, U.M.: Towards distributed coevolutionary GANs. In: AAAI 2018 Fall Symposium (2018)
2. Al-Dujaili, A., Srikant, S., Hemberg, E., O'Reilly, U.M.: On the application of Danskin's theorem to derivative-free minimax optimization. In: International Workshop on Global Optimization (2018)
3. Ambrogioni, L., GüÃğlü, U., van Gerven, M.: k-GANs: Ensemble of generative models with semi-discrete optimal transport (2019). arXiv:abs/1907.04050
4. Arnaldo, I., Veeramachaneni, K., Song, A., O'Reilly, U.M.: Bring your own learner: a cloud-based, data-parallel commons for machine learning. IEEE Comput. Intel. Mag. **10**(1), 20–32 (2015)
5. Arora, S., Ge, R., Liang, Y., Ma, T., Zhang, Y.: Generalization and equilibrium in generative adversarial nets (GANs) (2017). Preprint arXiv:1703.00573
6. Arora, S., Risteski, A., Zhang, Y.: Do GANs learn the distribution? some theory and empirics. In: International Conference on Learning Representations (2018). https://openreview.net/forum?id=BJehNfW0-
7. Grover, A., Ermon, S.: Boosted generative models (2017). arXiv:abs/1702.08484
8. Hardy, C., Merrer, E.L., Sericola, B.: MD-GAN: Multi-discriminator generative adversarial networks for distributed datasets. In: 2019 IEEE International Parallel and Distributed Processing Symposium (IPDPS), pp. 866–877 (2018)
9. Herrmann, J.W.: A genetic algorithm for minimax optimization problems. In: CEC, vol. 2, pp. 1099–1103. IEEE, Piscataway (1999)
10. Heusel, M., Ramsauer, H., Unterthiner, T., Nessler, B.: GANs trained by a two time-scale update rule converge to a local Nash equilibrium (2017). Preprint arXiv:1706.08500
11. Heusel, M., Ramsauer, H., Unterthiner, T., Nessler, B., Klambauer, G., Hochreiter, S.: GANs trained by a two time-scale update rule converge to a local nash equilibrium. In: Advances in Neural Information Processing Systems 30 (NIPS 2017), pp. 6626–6637 (2017)
12. Hillis, W.D.: Co-evolving parasites improve simulated evolution as an optimization procedure. Physica D Nonlinear Phenomena **42**(1), 228–234 (1990). https://doi.org/10.1016/0167-2789(90)90076-2
13. Hoang, Q., Nguyen, T.D., Le, T., Phung, D.Q.: MGAN: Training generative adversarial nets with multiple generators. In: 6th International Conference on Learning Representations, ICLR 2018, (2018)
14. Hodjat, B., Hemberg, E., Shahrzad, H., O'Reilly, U.M.: Maintenance of a long running distributed genetic programming system for solving problems requiring big data. In: Genetic Programming Theory and Practice XI, pp. 65–83. Springer, Berlin (2014)
15. Husbands, P.: Distributed coevolutionary genetic algorithms for multi-criteria and multi-constraint optimisation. In: AISB Workshop on Evolutionary Computing, pp. 150–165. Springer, Berlin (1994)
16. LeCun, Y.: The MNIST database of handwritten digits (1998). http://yann.lecun.com/exdb/mnist/
17. Li, J., Madry, A., Peebles, J., Schmidt, L.: Towards understanding the dynamics of generative adversarial networks (2017). Preprint arXiv:1706.09884
18. Liu, Z., Luo, P., Wang, X., Tang, X.: Deep learning face attributes in the wild. In: Proceedings of International Conference on Computer Vision (ICCV) (2015)
19. Loshchilov, I.: Surrogate-assisted evolutionary algorithms. Ph.D. Thesis, University Paris South Paris XI; National Institute for Research in Computer Science and Automatic-INRIA (2013)
20. Mitchell, M.: Coevolutionary learning with spatially distributed populations. In: Computational Intelligence: Principles and Practice (2006)

21. Morse, G., et al.: Simple evolutionary optimization can rival stochastic gradient descent in neural networks. In: Proceedings of the Genetic and Evolutionary Computation Conference 2016, pp. 477–484. ACM, New York (2016)
22. Salimans, T., Ho, J., Chen, X., Sutskever, I.: Evolution strategies as a scalable alternative to reinforcement learning (2017). arXiv:1703.03864
23. Stanley, K.O., Clune, J.: Welcoming the era of deep neuroevolution - uber engineering blog (2017). https://eng.uber.com/deep-neuroevolution/
24. Tolstikhin, I.O., Gelly, S., Bousquet, O., Simon-Gabriel, C.J., Schölkopf, B.: ADAGAN: Boosting generative models. In: Advances in Neural Information Processing Systems (2017)
25. Toutouh, J., Hemberg, E., O'Reilly, U.M.: Spatial evolutionary generative adversarial networks. In: Proceedings of the Genetic and Evolutionary Computation Conference, GECCO '19, pp. 472–480. ACM, New York (2019)
26. Wang, Y., Zhang, L., van de Weijer, J.: Ensembles of generative adversarial networks (2016). arXiv:abs/1612.00991
27. Wang, C., Xu, C., Yao, X., Tao, D.: Evolutionary generative adversarial networks (2018). Preprint arXiv:1803.00657
28. Wierstra, D., Schaul, T., Peters, J., Schmidhuber, J.: Natural evolution strategies. In: 2008 IEEE Congress on Evolutionary Computation (IEEE World Congress on Computational Intelligence), pp. 3381–3387. IEEE, Piscataway (2008)
29. Williams, N., Mitchell, M.: Investigating the success of spatial coevolution. In: Proceedings of the 7th Annual Conference on Genetic and Evolutionary Computation, pp. 523–530. ACM, New York (2005)
30. Zhong, P., Mo, Y., Xiao, C., Chen, P., Zheng, C.: Rethinking generative coverage: A pointwise guaranteed approach (2019). arXiv:abs/1902.04697

# Chapter 15
# One-Pixel Attack: Understanding and Improving Deep Neural Networks with Evolutionary Computation

Danilo Vasconcellos Vargas

**Abstract** Recently, the one-pixel attack showed that deep neural networks (DNNs) can misclassify by changing only one pixel. Beyond a vulnerability, by demonstrating how easy it is to cause a change in classes, it revealed that DNNs are not learning the expected high-level features but rather less robust ones. In this chapter, recent findings further confirming the affirmations above will be presented together with an overview of current attacks and defenses. Moreover, it will be shown the promises of evolutionary computation as both a way to investigate the robustness of DNNs as well as a way to improve their robustness through hybrid systems, evolution of architectures, among others.

## 15.1 Introduction

In the moment of writing this chapter, the world is dominated by deep neural networks (DNNs) in what some researchers consider the third wave of intelligent systems. Results in deep learning showed that DNNs can get extremely accurate when provided with huge datasets. They are the first and currently the only systems capable of dealing with high-dimensional problems such as image and speech recognition. Moreover, the image recognition accuracy of DNNs surpasses human beings in some tasks and there are many challenging tasks in which DNNs outperform us (e.g., Go, Atari games, etc.).

In 2017, however, a paper appeared on arXiv showing that deep neural networks, albeit extremely accurate, can be fooled by changing a single pixel [1]. It is surprising that a high accuracy is possible to achieve even with such a low robustness. Making a parallel for human beings, it would be similar to say that a person can correctly answer most questions in mathematics but may fail if some typos are inserted in the text.

D. V. Vargas (✉)
Faculty of Information Science and Electrical Engineering, Kyushu University, Fukuoka, Japan
e-mail: vargas@inf.kyushu-u.ac.jp

© Springer Nature Singapore Pte Ltd. 2020
H. Iba, N. Noman (eds.), *Deep Neural Evolution*, Natural Computing Series,
https://doi.org/10.1007/978-981-15-3685-4_15

The one-pixel attack, which is the name given to the attack that can fool DNNs by changing only one pixel, is part of different types of attacks. Before diving deep into the one-pixel attack and other different types of attacks, I will describe briefly some concepts used in the area that study these attacks called adversarial machine learning.

## 15.2  Adversarial Machine Learning: A Brief Introduction

Adversarial machine learning is the area which studies attacks to all types of machine learning. An attack is basically a search for closer inputs that can fool a DNN. In other words, attacks can be seen as a constrained optimization problem, i.e., a maximization of error constrained by a region around the original sample.

Let $f(x) \in \mathbb{R}^k$ be the output of a machine learning algorithm denoted by function $f$ in which $x \in \mathbb{R}^{m \times n \times 3}$ is the input of the algorithm for input and output of respective sizes $m \times n \times 3$ (images with three channels are considered) and $k$. For untargeted attacks, adversarial machine learning can be defined as the following optimization problem:

$$\underset{\delta}{\text{minimize}} \quad f(x + \delta)_c$$
$$\text{subject to} \quad \|\delta\| \leq \varepsilon \tag{15.1}$$

in which $f()_c$ denotes the soft label for the correct class $c$, $\varepsilon$ is a threshold value, and $\delta \in \Delta$ is a small perturbation added to the input. The definition above regards black-box untargeted attacks, for targeted attacks there are a couple of possible objective functions. One could, for example, maximize the targeted class soft label $tc$:

$$\underset{\delta}{\text{maximize}} \quad f(x + \delta)_{tc}$$
$$\text{subject to} \quad \|\delta\| \leq \varepsilon \tag{15.2}$$

To verify if a given attack is successful, one must check if the soft label for the correct class decreased enough, causing the classification to change. To this modified sample that caused the change in class is given the name of adversarial sample. Adversarial samples $x'$ are explicitly defined as follows:

$$x' = x + \delta$$
$$\{x' \in \mathbb{R}^{m \times n \times 3} \mid \underset{j}{\text{argmax}}(f(x')) \neq \underset{i}{\text{argmax}}(f(x))\}, \tag{15.3}$$

in which $i$ and $j$ are the indices with the maximum value inside the output, corresponding to the chosen class. There are also white-box attacks. They are based on using information from gradients of DNNs to find adversarial samples.

However, these types of attacks are less realistic in the sense that in real-world scenarios one would not have access to such information. Moreover, gradients can be masked to defend against such types of attacks. Lastly, evolutionary computation is more suitable to multi-modal black-box problems which need a well-balanced exploration-exploitation trade-off rather than a direct search strategy.

### 15.2.1  The Constraint

The constraint ($\|\delta\| \leq \varepsilon$) in the optimization problem has a very important meaning. It has the objective of disallowing perturbations which could make $x$ unrecognizable or change its correct class. Therefore, the constraint is itself a mathematical definition of what constitutes an imperceptible perturbation.

Moreover, the norm in Eq. 15.1 can be of any type and each of them has completely different meaning. Many different norms are used in the literature (e.g., $L_0$, $L_1$, $L_2$, and $L_\infty$). This results in different definition of what is an imperceptible perturbation, influencing the search space and allowing for different types of attacks. Intuitively, the norms allow for different types of attacks. $L_0$ allows attacks to perturb a few pixels strongly, $L_\infty$ allows all pixels to change slightly while both $L_1$ and $L_2$ allow for a mix of both strategies.

## 15.3  One-Pixel Attack

One-pixel attack aims to fool DNNs by changing only one pixel (Fig. 15.1).

Therefore, the one-pixel attack can be defined by slightly changing the initial adversarial machine learning formulation. By modifying the norm to a $L_0$ one and limiting it to a maximum of 1, i.e., $\varepsilon = 1$, it is possible to define the one-pixel attack

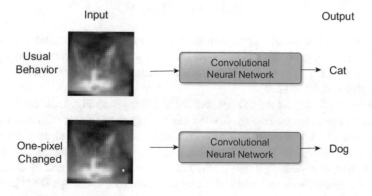

**Fig. 15.1** Illustration of the one-pixel attack

**Fig. 15.2** Examples of
one-pixel attacks created with
one-pixel attack that
successfully fooled three
types of DNNs trained on
CIFAR-10 dataset. The
original class labels are in
black color, while the target
class labels and the
corresponding confidence are
given below [1]

mathematically:

$$\underset{\delta}{\text{minimize}} \quad f(x + \delta)_c$$

$$\text{subject to} \quad \|\delta\|_0 \le 1 \tag{15.4}$$

To solve the optimization problem above, the one-pixel attack used a differential evolution (DE) algorithm. DE is a type of evolutionary algorithm which optimize functions by using a population of candidate solutions (called individuals). For more details please refer to [2, 3].

Examples of attacks on CIFAR dataset are shown on Fig. 15.2. Each of these adversarial samples successfully fooled various types of DNNs with similar accuracy (Table 15.1). Generally speaking, for all networks tested, circa a third of the samples were found to have at least one pixel in which if perturbed correctly can cause a change in class. Moreover, the results are not only limited to small input. Figure 15.3 shows some adversarial examples found by attacking a DNN learned on ImageNet (Table 15.2).

**Table 15.1** Accuracy of the one-pixel attack on original CIFAR-10 test set for both targeted and non-targeted attacks

|              | AllConv | NIN    | VGG16  |
|--------------|---------|--------|--------|
| Targeted     | 3.41%   | 4.78%  | 5.63%  |
| Non-targeted | 22.60%  | 35.20% | 31.40% |
| Confidence   | 56.57%  | 60.08% | 53.58% |

**Cup(16.48%)**
Soup Bowl(16.74%)

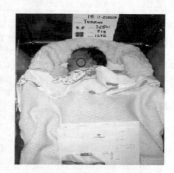

**Bassinet(16.59%)**
Paper Towel(16.21%)

**Teapot(24.99%)**
Joystick(37.39%)

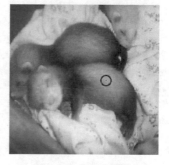

**Hamster(35.79%)**
Nipple(42.36%)

**Fig. 15.3** One-pixel attacks on ImageNet dataset [1]. To facilitate the visualization of the modified pixels, they are highlighted with red circles. The original class labels are in black color, while the target class labels and their corresponding confidence are given below

**Table 15.2** Comparison of non-targeted attack effectiveness between one-pixel attack and other works

| Method     | Success rate | Confidence | Number of pixels | Network |
|------------|--------------|------------|------------------|---------|
| Our method | 35.20%       | 60.08%     | 1 (0.098%)       | NIN     |
| Our method | 31.40%       | 53.58%     | 1 (0.098%)       | VGG     |
| LSA[4]     | 97.89%       | 72%        | 33 (3.24%)       | NIN     |
| LSA[4]     | 97.98%       | 77%        | 30 (2.99%)       | VGG     |
| FGSM[5]    | 93.67%       | 93%        | 1024 (100%)      | NIN     |
| FGSM[5]    | 90.93%       | 90%        | 1024 (100%)      | VGG     |

**Table 15.3** Success rate of one-pixel attack on both nearby pixels and a single randomly chosen pixel

|                                         | LENET  | RESNET |
|-----------------------------------------|--------|--------|
| ORIGINAL ONE-PIXEL ATTACK               | 59%    | 33%    |
| ONE-PIXEL ATTACK ON **Random Pixels**   | 4.9%   | 3.1%   |
| ONE-PIXEL ATTACK ON **Nearby Pixels**   | 33.1%  | 31.3%  |

This experiment is conducted in the following manner. First the one-pixel attack is executed. Afterwards, the same perturbation is used to modify one random or nearby pixel of the original image and evaluate success of the method. To obtain a statistically relevant result, both random and nearby pixel attack are repeated once per image for each successful attack in 5000 samples of the CIFAR-10 dataset (in which there are 1638 successful attacks for ResNet and 2934 successful attacks for LeNet)

One-pixel attack became widely known by its capability of fooling DNNs with an extremely small perturbation. However, to what extent is the position of the perturbed pixel important? In Table 15.3, it is shown that nearby pixels when perturbed randomly have roughly a 30% chance to fool the network. In other words, the exact position of the pixel itself is not the main factor. To explain this experiment we must look into the inner workings of the initial layers. Recall that DNNs process the initial input using convolution, therefore the influence of nearby pixels on the convolution, which is a linear function, is similar. Therefore, there is a high chance of finding nearby pixels which would cause the same vulnerability. Thus, the vulnerable part of a neural network which is highlighted by the one-pixel attack is the convolution. This is further evidenced by the similar accuracy rate on different architectures [6].

### 15.3.1 How Is It Possible?

Notice that all networks attacked have state-of-the-art performance accuracy on the tests. The obvious question is, how is it possible for highly accurate models to be vulnerable to such a small perturbation? There are many other questions that arise from recent results.

- **Are DNNs chaotic systems?**—The definition of a chaotic system is to have high sensitivity to input. Therefore, are we dealing with some sort of chaotic learning systems?
- **Do DNNs learn high-level features?**—Recall that DNN's success originates from the ability to automatically develop features which are better suited than manually coded ones. It was believed and to some extent showed that high-level features develop inside DNNs; however, recent results are showing counter-examples. Thus, what features are really being learned?

In the following sections, we will dive deeper in to the questions above. To understand as well as try to envision the future before us.

## 15.3.2 Are Deep Neural Networks Chaotic Systems?

"Does the flap of a butterfly's wings in Brazil set off a tornado in Texas?" In chaos theory, this is a question that somewhat delineate a counter-intuitive puzzle. The puzzle that small perturbations can lead to great differences in final state. That is why weather predictions are not very accurate for long periods of time. In chaos theory, this is what could be said for dynamical systems that have one positive Lyapunov exponent (Lyapunov exponent basically calculates if predictions have an error that, with time, differs exponentially from the target system).

As showed previously, small perturbations to the input are enough to make algorithms change completely their output. Thus, are machines chaotic systems?

If we consider chaotic systems as defined originally in dynamical systems, the answer is no. First, machine learning systems are not, in their majority, dynamical systems. This is true even when dealing with dynamic problems such as games. Therefore, most of the definition and other related concepts cannot be applied here.

However, what can be said of non-dynamical systems that are somewhat sensitive to perturbations? The point here is that the origin of chaos can be tracked down to many complex iterations. This many iterations were connected to unlimited time and some iterations in a limited space (small number of nodes interacting with each other). What if, instead of unlimited time, we consider unlimited space and limited time. The results could be a different type of chaos. A lot has to be rethought since, for example, trajectories are less meaningful when time is limited. Moreover, as time is limited, defining chaos as trajectories that exponentially separate is also, naturally, non-applicable.

Thus, they are not chaotic machines by following strictly the current definition. However, they may fit a different kind of chaotic system defined on unlimited space rather than time. Further discussion on this goes beyond the topic of this chapter.

## 15.4 Do Deep Neural Networks Learn High-Level Features?

The main reason for the use of DNNs is their capability of learning features which are not feasible to be coded manually. DNNs learn these features automatically and based on them achieve high accuracy on many tasks. Many of these features were believed to be high-level ones, such as, for example, the recognition of body parts to classify a portion of an image as a human being or the recognition of eyes and nose to classify an image patch as a face. However, high-level features should not change by only a pixel. In other words, if a single pixel can change the final classification, high-level features are probably not being recognized by the DNN.

The following subsections evaluate as well as try to understand DNNs to answer the question of "Do DNNs learn high-level features?"

### 15.4.1 Propagation Maps

To evaluate the effects of one pixel change throughout the layers of a given DNN, let us define a map of the differences between layers of an adversarial perturbed input and for the original input. To the map of the difference between feature maps is given the name of propagation map [6]. Consider an element-wise maximum of a three-dimensional array $O$ for indices $a$, $b$, and $k$ to be described as:

$$M_{a,b} = \max_k(O_{a,b,0}, O_{a,b,1}, \ldots, O_{a,b,k}), \tag{15.5}$$

where $M$ is the resulting two-dimensional array. Specifically, propagation map $PM_i$ is defined for a layer $i$ as:

$$PM_i = \max_k(FM_{i,k} - FM_{i,k}^{adv}), \tag{15.6}$$

where $FM_{i,k}$ and $FM_{i,k}^{adv}$ are, respectively, the feature maps for layer $i$ and kernel $k$ of the natural (original) and adversarial samples.

Figure 15.4 shows an example of a propagated map for a given adversarial sample. The difference between feature maps shows that the change of one pixel not only stays strong but also spread throughout the layers. This behavior reveals that DNNs pay a lot of attention to small changes in the input. Moreover, such changes are important for the processing even on deeper layers which were supposed to process higher-level features.

### 15.4.2 Texture-Based Features

Results on ImageNet showed that no shape information is required to achieve an accuracy that rivals the state-of-the-art [7]. The authors showed that a simple bag-of-local-features is enough to have high accuracy on ImageNet. In fact, this result suggests that current DNNs do not need more than the recognition of texture patterns and may be limited to it. In [8], the authors show that current DNNs do not learn well when only the contour or the silhouette of an object is given. Figure 15.5 shows how accuracy changes from usual images to images in which only the shape is preserved. Interestingly, the results observed of current DNNs do not change much when only the texture of objects are shown. Thus, both results suggest that current DNNs only process textures. The reliance on textures might be the culprit of the lack of robustness, since some textures may change with a few pixels or some weak noise but the shape would not.

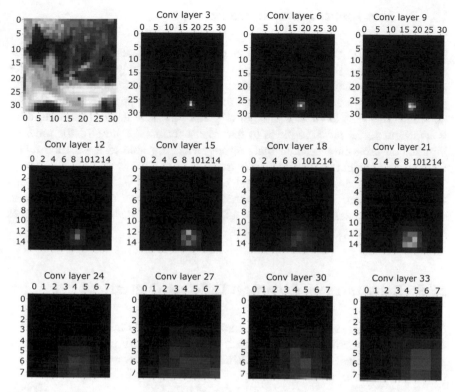

**Fig. 15.4** Propagation Map (PM) for ResNet using a sample from CIFAR. The sample above is incorrectly classified as automobile after one pixel is changed in the image. Values are scaled with the maximum value for each layer of the feature maps being the maximum value achievable in the color map. Therefore, bright values show that the difference in activation is close to the maximum value in the feature map, demonstrating the strength of the former [6]

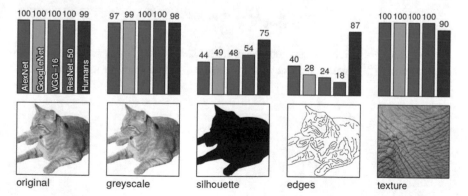

**Fig. 15.5** Change of accuracy from original images to images in which only the shape is preserved or images in which only the texture is preserved [8]

### 15.4.3  Non-robust Features Are Enough

In [9] the authors further demonstrate that DNNs use non-robust features. Moreover, in an experiment in which the training set is composed of only adversarial samples and their corresponding incorrect classes, it is showed that training in this dataset still results in high accuracy. Based on these results, it is possible to say that the non-robust features learned is enough to predict the true label. Therefore, there is no incentive for the algorithms to find robust features. Naturally, the previous information takes into account that current DNNs are able to learn such robust features which might not be necessarily true.

Before going deeper into the understanding of DNNs' vulnerabilities and limitations, it is important to dive deeper into the meaning of robustness. The next section explicitly define robustness as a different objective which may sometimes be at odds with accuracy.

## 15.5  Robustness vs. Accuracy: Different Objectives

For human beings, robustness against adversarial samples and accuracy are one and the same objective. However, such is not true for algorithms. In the canonical classification setting, the goal of a classifier is to achieve low expected loss:

$$\mathop{\mathbb{E}}_{(x,y)\sim D} [\mathcal{L}(x, y; \theta)]. \tag{15.7}$$

Robustness against adversarial attacks is a slightly different setting. To achieve a high robustness in this setting, a classifier should have a lower adversarial loss in noise $\delta \in \Delta$ [1]:

$$\mathop{\mathbb{E}}_{(x,y)\sim D} [\mathcal{L}(x + \delta, y; \theta)]. \tag{15.8}$$

These two objectives are sometimes at odds. This is why many solutions end up with a trade-off between accuracy and robustness rather than an improvement of both.

---

[1]Here we use the error in noise to be the adversarial loss instead of the worst-case error, for a discussion of the relationship between error in noise and adversarial samples please refer to [10].

## 15.6  Attacks on Deep Neural Networks

There are many attacks which were shown to be effective, here some of them will be reviewed briefly. All attacks can be divided into two types of attacks:

- Evasion Attacks—In this scenario, the attacker has only access to the machine learning algorithm after it has already learned. Thus, attacks involve using small modifications in the input to cause changes in the behavior of the algorithm. Targeted type of attacks have aims to change the class to a specifically defined class while untargeted attacks do not have a target, aiming only to fool the algorithm. This is the most common type of attack and the one this chapter will cover in detail.
- Poison Attacks—Most machine learning algorithms need to be trained in a given dataset beforehand. The dataset used is sometimes open and may be modified by attackers. Moreover, there are algorithm which use information throughout the application to improve themselves. In both cases, poison attacks refer to when attackers introduce vulnerabilities by modifying the dataset. This type of attack will not be covered in this chapter.

The first paper on the subject dates back to 2013, when DNNs were shown to behave strangely for nearly the same images [11]. Figure 15.6 shows examples of the vulnerabilities found in the first experiments, in which a small amount of noise added to all pixels of the image.

Afterwards, a series of vulnerabilities were found In [12], the authors demonstrated that DNNs output a high confidence when presented with textures and random noise (Fig. 15.7). This suggests that confidence in current DNNs is less meaningful than previously thought. A solution to this problem lies in changing the problem formulation slightly to include a new class for unknown classes. This solution will be discussed later in the section about defenses.

Single adversarial perturbations which can be added to most of the samples to fool a DNN were shown to be possible [13]. To this type of adversarial perturbation was given the name of universal adversarial sample. Figure 15.8 shows examples of universal adversarial sample. Notice that the universal perturbations found differ from network to network. They are only constant inside a single DNN.

Until now, perturbation took the form of a small noise added throughout the image or in a single pixel. In 2017, researchers found that the addition of small patches could fool the classifier [14] (Fig. 15.9). It might be less surprising to think about a patch after even one pixel was able to fool the classifier. However, the patches developed are targeted attacks and can therefore be a real threat to DNNs. In other words, somebody could use an adversarial patch to disguise as an object or somebody else.

**Fig. 15.6** First experiments on adversarial attacks. Original sample (left), added mask (middle), and resulting image (right). The resulting image on the right was misclassified by the algorithm [11]

Until now, examples were shown with relation to machine learning algorithms which require images as input. However, this is in no way a requirement. For example, text input can also be manipulated by changing letters slightly. Figure 15.10 shows an example of an adversarial sample for text input. In text input, there is an equivalent to the one-pixel attack which is the swap of two letters (Fig. 15.11).

In fact, attacks do not need to be constrained to the development of single perturbations. Automatic and smarter models could be created to change inputs and create adversarial samples with high confidence. Figure 15.12, for example, shows an example of how a rule can be developed to automatically create adversarial examples for text input. The rule is composed of many pattern matching segments, each segment composed of a search and change procedure. For more information please refer to [16], the technique used to evolve such a rule is a type of co-evolutionary algorithm.

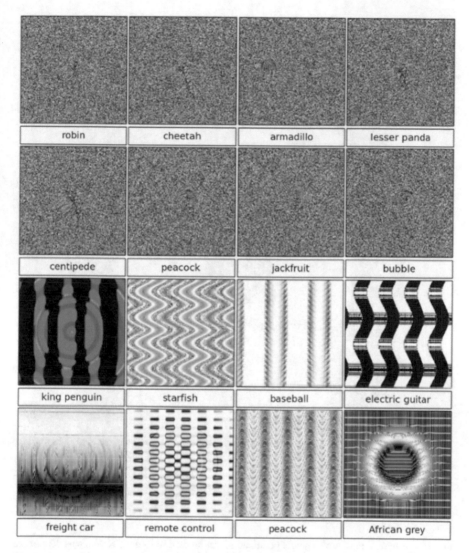

**Fig. 15.7** Random noise which is recognized with high confidence by DNNs [12]. The names below the image represent the class in which they were classified by the DNN

## 15.6.1 Are These Attacks Feasible in the Real World?

All attacks showed here were executed in a lab. Some of these attacks are white-box attacks in which attackers have access to even more information such as the gradient of the DNN. Thus, the natural question is "can these attacks be used in the real world?" Or even, to what extent are the lack of robustness of DNNs critical and dangerous?

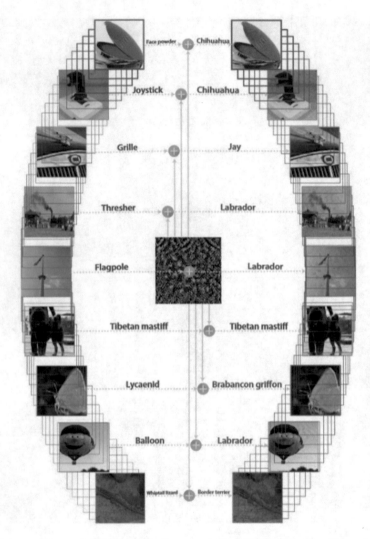

**Fig. 15.8** Illustration of how a single perturbation can be used to fool a DNN independent of the sample [13]

In [17], the authors showed that current attacks can be translated to real-world threats by merely printing them out. They demonstrated that printed out adversarial samples still work because many adversarial samples are robust against different light conditions. Moreover, carefully crafted glasses can also be made into attacks [18] or even general 3d adversarial objects were shown possible [19]. 3d adversarial objects can be built with 3d printing machines and revealed that such objects exist and can be explored by malicious users.

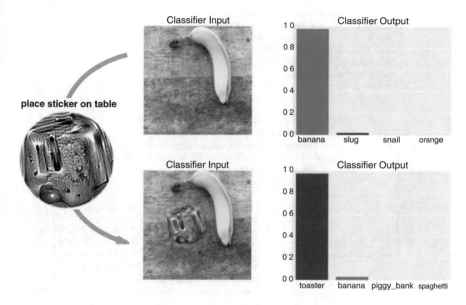

**Fig. 15.9** Example of an adversarial patch in which the classifier changed the classification from banana to toaster [14]

> The Old Harbor Reservation Parkways are three ~~historic~~ roads in the Old Harbor area of Boston. *Some exhibitions of Navy aircrafts were often held here.* They are part of the Boston parkway system designed by Frederick Law Olmsted. They include all of William J. Day Boulevard running from *Castle* Island to Kosciuszko Circle along Pleasure Bay and the Old Harbor shore. The part of Columbia Road from its northeastern end at Farragut Road west to Pacuska Circle (formerly called Preble Circle). Old Harbor Reservation

**Fig. 15.10** An adversarial text sample generated with the perturbations proposed in [15]. By inserting an irrelevant fact: Some exhibitions of Navy aircrafts were often held here. (red), removing an HTP: historic (blue), and modifying an HSP: Castle (red). The output classification is successfully changed from Building to Means of Transportation. Liang et al. [15]

> The rcok is destined to be the 21st century's new 'conan' and that he's going to make a splash even greater than Arnold Schwarzenegger, jean-claud van damme or steven segal.

**Fig. 15.11** An adversarial text sample generated by swapping the two letters (two elements of perturbation, shown in red)

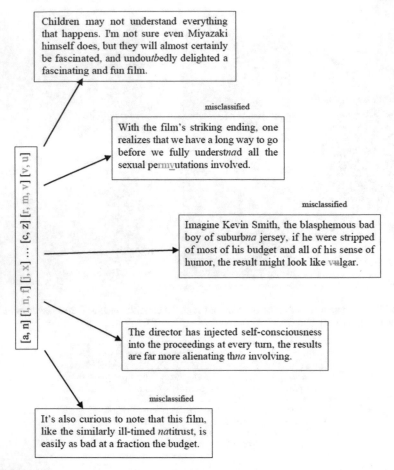

**Fig. 15.12** Example of a crafted universal rule (sequence of prototype-matching based perturbation procedures) for fooling text classification. Using the technique proposed in this paper it is possible to craft a universal rule which automatically create adversarial samples, i.e., once the universal rule is crafted no search is needed anymore. In fact, the universal rule will only do a few perturbations which is imperceptible to typos. In the figure, adversarial samples are generated by one universal adversarial rule (changes to the original sample are shown with a different color. Swapping is shown in red, deletion shown in green, insertion shown in blue.) [16]

## 15.7 Defense Systems

Many defensive systems and detection systems were proposed to mitigate the attacks described previously. However, there are still no current solutions or promising ones. Regarding defensive systems, defensive distillation in which a smaller neural network squeezes the content learned by the original one was proposed as a defense [20]; however, it was shown not to be robust enough in [21]. Adversarial training was also proposed in which adversarial samples are used to augment the training dataset. The main idea of adversarial training is to include adversarial samples in the dataset to allow DNN to train and then be able to correctly classify them. Indeed, this type of defense was showed to increase the robustness [22–24]. The resulting neural network, however, is still vulnerable to attacks [25]. Moreover, it was shown in [26] that adversarial samples insert a bias to the type of adversarial sample used. For example, adversarial samples related to $L_\infty$ attacks will increase the robustness against $L_\infty$. However, robustness against $L_0$ types of attacks would not be increased and may even decrease.

There are many other variations of defenses [27, 28] which were carefully analyzed and many of their shortcomings explained in [29, 30]. In brief, many defenses are based on hiding the gradients from DNNs, called gradient masking. These modifications result in DNNs that are more difficult to attack for white-box attacks which rely on this information. However, it was showed that gradient masking gives a false sense of security, since they do not remove the vulnerability but they only hide it. Methods that do not rely on gradients can bypass gradient masking techniques resulting in no improvement of robustness.

Another type of defense systems are not based on variations of previous DNNs but structural or even revolutionary modifications. For example, the addition of the unknown class to correctly classify examples that are just pure noise. OpenMax layer was proposed as a layer that is able to learn with this additional class [31]. Actually, there are many new types of machine learning algorithms which may be robust but were never tested. An example of this is the CapsNet [32] which was shown to be relatively more robust recently. Thus, partial solutions may lie on less usual places or even already exist but still be hard to find.

## 15.7.1   Detection Systems

Detection systems are a subfield of defenses. They aim to identify the presence of a malicious sample rather than removing the vulnerability. Regarding detection systems, a study from [33] revealed that adversarial samples in general have different statistical properties which could be exploited for detection. In [34], the authors proposed a system which compares the prediction of a classifier with the prediction of the same input but "squeezed" (either color or spatial smoothing). This technique allowed classifiers to detect adversarial samples with small perturbations. Having said that, many detection systems might fail when adversarial samples differ from test conditions [35, 36]. Thus, the clear benefits of detection systems remain inconclusive.

## 15.8  Overview of the Current State of Robustness: Evaluating Algorithms and Defenses

Previous sections showed many attacks and defenses but failed to give an overview of the current state in terms of their robustness. To realize this, two attacks based on different objectives will be defined based on [26]. The difference between these attacks will serve the purpose of analyzing the state-of-the-art DNNs and defenses from two different perspectives. We will see that robustness changes depending on the type of attack used. Notice that the objective here is not to achieve 100% attack accuracy (which can be achieved provided enough noise is inserted) but rather to evaluate DNNs in very restricted $L_0$ and $L_\infty$ attacks. These restricted $L_0$ and $L_\infty$ attacks have several advantages such as (a) evaluating all DNNs and defenses with the same fixed amount of perturbation and (b) possessing no false adversarial samples. Figure 15.13 shows an example of a false adversarial sample with very few amount of perturbation which would be allowed in most attacks based on $L_1$ and $L_2$.

### 15.8.1  Threshold Attack ($L_\infty$ Black-Box Attack)

The threshold attack optimizes the constrained optimization problem defined in Eq. 15.1 with the constraint $\|\delta\|_\infty \leq \varepsilon$, i.e., it uses the $L_\infty$ norm. $\varepsilon$ is a threshold which is set here to be one of the following values $\{1, 3, 5, 10\}$.

Threshold attack encode solutions as a matrix of pixel variations. Therefore, the search space is the same as the input space because the variables can be any variation

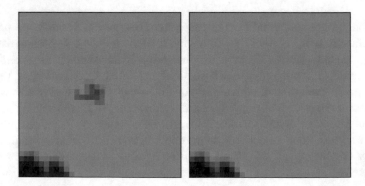

**Fig. 15.13** Example of a false adversarial sample (right) and its respective original sample (left) [26]. The false adversarial sample is built with few total perturbations (i.e., low $L_1$ and $L_2$) but with unrecognizable final image (false adversarial sample). This is a result from non-constrained spatial distribution of perturbations which is prevented if low $L_0$ or $L_\infty$ is used. In fact, this attack has a $L_2$ of merely 356, well below the maximum $L_2$ for the one-pixel attack

of the input as long as the threshold is respected. Therefore, the algorithm search happens in $\mathbb{R}^{m \times n \times 3}$ space.

The threshold attack uses the state-of-the-art black-box optimization algorithm called Covariance Matrix Adaptation Evolution Strategy (CMA-ES) [37]. To satisfy the constraint, a clipping function is used to keep the values inside the feasible region, i.e., a simple repair method is employed in which pixels that surpass the minimum/maximum are brought back to the minimum/maximum value.

### 15.8.2 Few-Pixel Attack ($L_0$ Black-Box Attack)

The few-pixel attack is a variation of the one-pixel attack [1]. It optimizes the constrained optimization problem defined in Eq. 15.1. However, the constraint used is $\|\delta\|_0 \leq \varepsilon$, i.e., it uses the $L_0$ norm. $\varepsilon$ is a threshold which may also admit one of the following values $\{1, 3, 5, 10\}$.

Few pixel attack uses a search variable which is a combination of pixel values (3 values) and position (2 values) for all of the pixels ($\varepsilon$ pixels). Therefore, the search space is smaller than the threshold attack with dimensions of $\mathbb{R}^{5 \times \varepsilon}$. To conduct the optimization, we use the CMA-ES.[2] The constraint is always satisfied because the number of parameters is itself modeled after the constraint. In other words, when searching for one pixel perturbation, the number of variables are fixed to pixel values (three values) plus position values (two values). Therefore it will always modify only one pixel, respecting the constraint. Since the optimization is done in real values, to force the values to be within range, a simple clipping function is used for pixel values. For position values a modulo operation is performed.

### 15.8.3 Analysis and Discussion

Table 15.4 shows results on various state-of-the-art DNNs: WideResNet [38], DenseNet [39], ResNet [40], Network in Network (NIN)[41], All Convolutional Network (AllConv) [42], CapsNet[32], and LeNet [43]. It reveals that they are vulnerable to all types of attacks in all levels. This demonstrates that although robustness may differ between current DNNs, none of them is able to completely overcome even the lowest level of perturbation possible.

Results within a Hamming distance of five from the lowest are considered to be equally good. These results are written in bold. CapsNet and AllConv can be

---

[2]Notice that we do not use DE optimization algorithm which was used in the original version of the one-pixel attack. The reason for this is that DE performed worse than CMA-ES in the threshold attack and both threshold and few-pixel attack should ideally have the same optimization algorithm.

**Table 15.4** Attack accuracy results for few-pixel attack ($L_0$ black-box attack) and threshold attack ($L_\infty$ black-box attack) [26]

| Model (Accuracy) | $L_0$ Attack's $\varepsilon$ | | | |
|---|---|---|---|---|
| | 1 | 3 | 5 | 10 |
| WideResNet (95.12) | **11%** | 55% | 75% | 94% |
| DenseNet (94.54) | **9%** | 43% | 66% | 78% |
| ResNet (92.67) | **12%** | 52% | 73% | 85% |
| NIN (90.87) | 18% | 62% | 81% | 90% |
| AllConv (88.46) | **11%** | **31%** | 57% | 77% |
| CapsNet (79.03) | 21% | 37% | **49%** | **57%** |
| LeNet (73.57) | 58% | 86% | 94% | 99% |
| Model (Accuracy) | $L_\infty$ Attack's $\varepsilon$ | | | |
| | 1 | 3 | 5 | 10 |
| WideResNet (95.12) | 15% | 97% | 98% | 100% |
| DenseNet (94.54) | 23% | 68% | **72%** | **74%** |
| ResNet (92.67) | 33% | 71% | **76%** | 83% |
| NIN (90.87) | **11%** | 86% | 88% | 92% |
| AllConv (88.46) | **9%** | 70% | **73%** | **75%** |
| CapsNet (79.03) | **13%** | **34%** | **72%** | 97% |
| LeNet (73.57) | 44% | 96% | 100% | 100% |

Left column shows the model attacked with the classification accuracy between brackets. The attack is performed over 100 random samples from the CIFAR dataset using the CMA-ES optimization algorithm. Results in bold are the lowest attack accuracy and other results which are within a distance of five from the lowest one

considered the most robust with five bold results. The third place in robustness achieves only three bold results and therefore is far away from the top performers.

The behavior of $L_0$ and $L_\infty$ differ specially in the most robust DNNs, showing that the robustness is achieved with some trade-offs. Moreover, this further justifies the importance of using both metrics to evaluate DNNs.

Defenses can also be attacked with similar accuracy. Table 15.5 shows the few-pixel and threshold attacks' accuracy on adversarial training (AT) [24], total variance minimization (TVM) defenses [28], and feature squeezing (FS) [34]. Regarding the adversarial training, it is easier to attack with the few-pixel attack than with threshold attack. This result should derive from the fact that the adversarial samples used in the training contained mostly images from $L_\infty$ type of attacks. This happens because Projected Gradient Descent (PGD), which was used to create the adversarial samples for the adversarial training, is a $L_\infty$ attack. Therefore, it suggests that *given an attack bias that differs from the invariance bias used to train the networks, the attack can easily succeed*. Regarding TVM, the attacks were less successful but the original accuracy of the model trained with TVM is also not great. Therefore, even with a small attack percentage of 24% the resulting model accuracy is 35%. Attacks on feature squeezing had a relatively high accuracy. This is true

**Table 15.5** Accuracy results for few-pixel ($L_0$) and threshold attack ($L_\infty$ black-box attack) on adversarial training (AT) [24], total variance minimization (TVM) [28], and feature squeezing (FS) [34] defenses

| | $L_0$ Attack's $\varepsilon$ | | | |
|---|---|---|---|---|
| | 1 | 3 | 5 | 10 |
| AT (87%) | 22% (67%) | 52% (41%) | 66% (29%) | 86% (12%) |
| TVM (47%) | 16% (39%) | 12% (41%) | 20% (37%) | 24% (35%) |
| FS (92%) | 17% (72%) | 49% (44%) | 69% (26%) | 78% (19%) |
| | $L_\infty$ Attack's $\varepsilon$ | | | |
| | 1 | 3 | 5 | 10 |
| AT (87%) | 3% (84%) | 12% (76%) | 25% (65%) | 57% (37%) |
| TVM (47%) | 4% (45%) | 4% (45%) | 6% (44%) | 14% (40%) |
| FS (92%) | 26% (64%) | 63% (32%) | 66% (29%) | 74% (22%) |

In the leftmost column, the original attack is between brackets. The numbers between brackets in the other columns are the resulting accuracy of the defenses when under attack. For the modified accuracy, the value is calculated by multiplying the original accuracy by one minus attack accuracy. The attack is performed over 100 correctly classified samples from the CIFAR dataset [26]

for both $L_0$ and $L_\infty$ attacks. Moreover, both types of attacks had similar accuracy, revealing a lack of bias in the defense system.

In summary, defense systems proposed until now do not give a relevant improvement to robustness when all types of attacks are taken into consideration. In this section, the current results after 7 years of research were presented, i.e., since the first paper on adversarial machine learning was presented. This open problem is still a recent one but it is part of a very active research area with huge socioeconomic impact (autonomous cars and other applications need the solution to this to be reliable). Therefore, there is a chance that the solution for DNNs, if it exists, will not be a trivial one.

## 15.9   Down the Rabbit Hole: The Representation Problem

Researchers do not agree upon what could be the cause for DNN's lack of robustness. A recent result, however, suggests that the problem may lie in the faulty representation learned by DNNs [44]. The results reveal that the quality of representation is well aligned with the attack accuracy. Before going into the details of the results, the representation metric and its relationship to the error of the learning algorithm will be introduced.

### 15.9.1  Representation Metric

Take Eq. 15.8 and consider the loss to be the mean squared error (MSE), we have

$$\underset{(x,y)\sim D}{\mathbb{E}}[\mathcal{L}(x+\delta, y; \theta)] = \underset{(x,y)\sim D}{\mathbb{E}}[(f(x+\delta)-h(x+\delta))^2]+ \underset{(x,y)\sim D}{\mathbb{E}}[(h(x+\delta)-\hat{y}(x+\delta))^2],$$

where $h(x) = E[h_D(x)]$ is the expected behavior of the prediction when averaged over many datasets, $f(x)$ is the ground-truth, and $\hat{y}(x) = h_D(x)$ is the output after learned on a given dataset $D$. For robustness to increase, adversarial training requires that datasets should have many noisy samples, i.e., $x + \delta \in D$. However, the more noise is added to images the more $D$ becomes close to all possible images of size $M \times N$, i.e., $\mathbb{R}^{M*N}$:

$$\lim_{\Delta\to\infty} D = \mathbb{R}^{M*N}. \tag{15.9}$$

However, $f(x)$ is undefined[3] for $D \in \mathbb{R}^{M*N}$ in which $y \notin C$ for the set of known classes $C$. Even a small amount of noise may be enough to cause $y \notin C$ and thus $f(x)$ undefined. Therefore, the following question arises. Would it be possible to evaluate the robustness and/or the quality of a model without a well defined $y$?

To answer this question we take into account an ideal representation $z$ and the representation learned by the model $\hat{z}$:

$$\mathbb{E}[(f(x + \delta; z) - h(x + \delta; \hat{z}))^2] + \mathbb{E}[(h(x + \delta; \hat{z}) - \hat{y}(x + \delta; \hat{z}))^2]. \tag{15.10}$$

Interestingly, although $y$ is undefined, $z$ represents the features learned and is well defined for any input. Moreover, by considering learned classes to be clustered in $z$ space, unsupervised learning evaluation can be used to evaluate $z$ even without a well defined $y$. We use here a famous clustering analysis index to evaluate clusters in $z$ by their intracluster distance. In $z$, it is also possible to evaluate the representation of known and unknown classes which should have common features. Moreover, we hypothesize here that unknown classes should evaluate $z$ with less bias because a direct map from input to output is nonexistent. Any projection of the input in any of the feature maps or the output layer could be used as $z$. To take the entire projection into account, we use here $z$ as the final projection of the input to the classes, i.e., $z$ is the soft label array **e**.

Thus, to evaluate representation quality we define the Davies–Bouldin metric based on the clustering analysis of the soft label array for unknown classes. It is specifically defined as follows.

---

[3]Alternatively, $f(x)$ could be defined for any noise if an additional unknown class is defined, such as with an OpenMax layer [31].

#### 15.9.1.1 Davies–Bouldin Metric: Clustering Hypothesis

Soft labels of a classifier compose a space in which a given image would be classified as a weighted vector concerning the previous classes learned. Considering that a cluster in this space would constitute a class, we can use clustering validation techniques to evaluate the representation. Here we choose for simplicity one of the most used metric in internal cluster validation, Davies–Bouldin Index (DBI). DBI is defined as follows:

$$
DBI = \left( \frac{1}{n_e} \sum_{j=1}^{n_e} |\mathbf{e}_j - \mathbf{cn}|^2 \right)^{1/2}, \tag{15.11}
$$

in which $\mathbf{cn}$ is the centroid of the cluster, $\mathbf{e}$ is one soft label, and $n_e$ is the number of samples.

Interestingly, this metric resembles the variance from Eq. 15.10 which is the only part that is defined (the bias error is nonexistent for undefined $y$). If the centroid $\mathbf{cn}$ does not vary with the dataset, both equations are one and the same. Therefore, the DBI metric can be seen as an approximation of the variance. Here, however, only unknown classes will be used to avoid any bias present from learning directly the input-output map.

### 15.9.2 Analysis

To evaluate the DBI metric on unknown classes, a classifier is trained in $c-1$ classes and tested on the remaining one. This process is repeated for all $c$ classes. Table 15.6 shows the Pearson correlation between attacks and the DBI metric in the CIFAR dataset. The majority of the results reveal a strong correlation. Moreover, they show that both are inversely correlated which is expected since a higher DBI means sparse clusters, consequently worse representation, and higher L2 means higher robustness.

## 15.10 The Role of Evolutionary Computation

Evolutionary computation was shown to be able to attack DNNs. Evolutionary algorithms are well suited for black-box attacks in which very few, if any, information about the neural network is known. In fact, those are the types of attacks that are more realistic because attackers have little knowledge about the structure and what is inside DNNs.

Moreover, attacks can also be used to study DNNs. Thus, evolutionary computation can be used, as already shown by previous examples, to attack and understand DNNs.

**Table 15.6** Pearson correlation between DBI and L2 score of attacks

| Model | Newton Fool | PGD | Virtual | FGM | Deep fool | BIM | Carlini | JSMA | All |
|---|---|---|---|---|---|---|---|---|---|
| Pearson correlation between DBI and L2 score ($p$-value) | | | | | | | | | |
| WideResNet | −0.58 (0.08) | −0.61 (0.06) | −0.62 (0.06) | −0.64 (0.05) | −0.62 (0.06) | −0.68 (0.03) | −0.69 (0.03) | −0.68 (0.03) | −0.11 (0.31) |
| DenseNet | −0.51 (0.13) | −0.68 (0.03) | −0.58 (0.08) | −0.71 (0.02) | −0.61 (0.06) | −0.76 (0.01) | −0.74 (0.02) | −0.60 (0.07) | −0.10 (0.35) |
| ResNet | −0.55 (0.10) | −0.60 (0.06) | −0.56 (0.09) | −0.55 (0.10) | −0.55 (0.10) | −0.65 (0.04) | −0.66 (0.04) | −0.70 (0.02) | −0.10 (0.37) |
| NetInNet | −0.64 (0.05) | −0.71 (0.02) | −0.59 (0.07) | −0.66 (0.04) | −0.61 (0.06) | −0.74 (0.01) | −0.76 (0.01) | −0.73 (0.02) | −0.13 (0.25) |
| AllConv | −0.70 (0.02) | −0.65 (0.04) | −0.67 (0.03) | −0.74 (0.01) | −0.71 (0.02) | −0.75 (0.01) | −0.75 (0.01) | −0.59 (0.07) | −0.12 (0.28) |
| CapsNet | −0.80 (0.01) | −0.78 (0.01) | −0.81 (0.00) | −0.81 (0.00) | −0.76 (0.01) | −0.74 (0.01) | −0.73 (0.02) | −0.46 (0.18) | −0.18 (0.11) |
| LeNet | −0.57 (0.08) | −0.59 (0.07) | −0.57 (0.09) | −0.60 (0.07) | −0.56 (0.09) | −0.56 (0.09) | −0.60 (0.07) | −0.02 (0.95) | −0.10 (0.38) |

In this section, the objective is to give an overview of the possibilities allowed by evolutionary computation to solve adversarial machine learning, going beyond the understanding and attacking of DNNs. In other words, here, many possibilities of learning systems will be discussed. These learning system might shape the path to a partial or even complete solution to inherently robust learning system.

### 15.10.1   Evolving Robust Architectures

There are many possible DNN architectures. It is, however, difficult to search manually and test each one of them. To solve this, in [45] the authors proposed to search for architectures using evolutionary computation. The search space includes many types of layers, multiple branches, and even any combinations of convolution and full connected layers, i.e., multiple bottlenecks are possible.

The architecture found had its vulnerability decreased to around half the value of other architecture search methods (Table 15.7). Moreover, these results are only based on modifying the architecture, i.e., without adversarial training. Thus, when other defenses are included, it is possible to achieve even higher robustness. This is a promising research line which is still merely beginning.

### 15.10.2   Neuroevolution

Previously, it was shown that a faulty representation might be the main culprit for the lack of robustness. Neuroevolution, which is the area concerned with the evolution of neural networks both in weights as well as in topology [48]. In other words, neuroevolution methods are not only limited to search for better weights, some of them also search for better topologies. And a couple of these methods also search for neuron models and connection types. One of the few of these methods is the Spectrum-diverse Unified Neuroevolution Architecture (SUNA) [49].

**Table 15.7** Error Rate (ER) on both the testing dataset and adversarial samples for the best architecture find when the *evaluation function has both accuracy on the testing data and accuracy on the adversarial samples*

| Architecture search | Testing ER | ER on adversarial samples |
|---|---|---|
| DeepArchitect[a] [46] | 25% | 75% |
| Smash[a] [47] | 23% | 82% |
| Ours | 18% | 42% |

The test was done over the CIFAR dataset and the adversarial examples were also created using some samples from the test dataset of the CIFAR dataset
[a]Both DeepArchitect and Smash had their evaluation function modified to be the sum of accuracy on the testing and adversarial samples

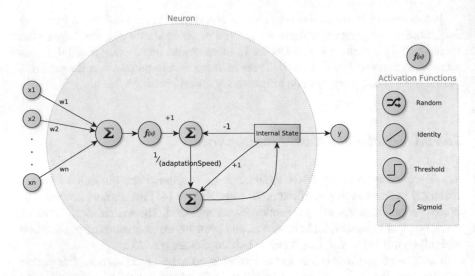

**Fig. 15.14** Neuron activation process and possible activation functions for SUNA

Figure 15.14 shows the unified neuron model proposed and used by SUNA. Notice that this unified neuron model has an internal state, allowing for slower or faster neurons to be defined. Moreover, the architecture of SUNA also enable connections and neurons to be neuromodulated. The explanation of all these functions and more details about the architecture and results go beyond the scope of this chapter. To learn more about SUNA and how it tackled five complete different hard problems without any parameter change, please refer to [49].

### 15.10.3  Self-Organizing Classifiers

Learning complex representations may help as mentioned in Sect. 15.10.2. However, these representations are not easy to adapt to newer problems. They are also not easy to learn and increment. This is a consequence of its monolithic complexity. Self-Organizing Classifiers (SOC) introduce a map that divides input into different groups (subpopulations) [50]. Each of these groups can change whenever a input is presented that is different enough from the past experience. Moreover, in each group there is population of neural networks that evolve to adapt to the type of inputs it receives. Since the type of inputs of each group differs, neural networks can be specialized and, consequently, less complex. Figure 15.15 illustrates SOC's architecture.

This type of divide and conquer approach with neural networks allows for a collection of relatively simple models to tackle complex problems. Interestingly, since each group is specialized, more knowledge can be added in a modular manner,

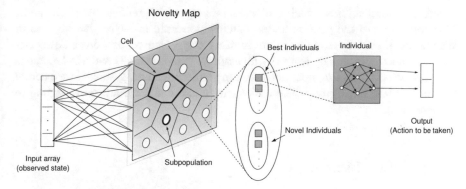

**Fig. 15.15** SOC's structure is illustrated. Novelty map can be exchanged with Self-Organizing Map; however, novelty map is known to give better results [51]

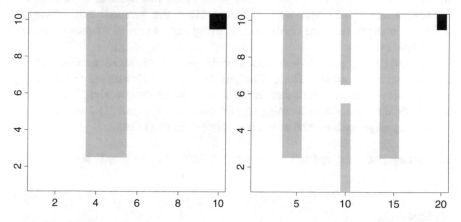

**Fig. 15.16** Dynamic maze problem in which the dimensions also change, growing from $10 \times 10$ to $10 \times 20$ state and vice versa. The black square is the objective while the gray ones are the wall. Further details of the experiment and the performance of the algorithm can be found in [53]

by merely adding new groups. Many variations of SOC were released improving further on its first versions [51, 52]. For example, including more robust group updates among other upgrades. These improvements allowed SOC to tackle very challenging problems such as dynamic mazes (mazes that change their configuration with time, similar to mazes used to test the learning abilities of mice). Figure 15.16 shows an example of maze solved [53].

### 15.10.4 Hybrids

All methods described in previous subsections are purely using evolutionary algorithms for learning (in the case of SUNA and SOC) or using evolution to search for architectures which are trained solely with gradient based methods. However, most of the gradient based optimization used in DNNs can be used in the middle of evolutionary algorithms in many ways. Neuroevolution can use parts of neural networks already learned with gradient based methods and SOC methods can even benefit from a competition between gradient and non-gradient based neural networks inside each group. The options are endless and we are only starting to explore them.

## 15.11   What Lies Beyond

In this chapter, adversarial machine learning was reviewed exposing its current state-of-the-art attacks and defenses. We saw that, albeit much effort of researchers, robustness has barely improved. However, since the research area started in 2013, we understand more deeply the meaning and causes of robustness against adversarial samples.

Moreover, many paths to solve many of the problems from adversarial machine learning were pointed out. These solutions involve evolutionary machine learning and hybrids which are not limited in the complexity of their models. Examples of such evolutionary machine learning algorithms are SUNA and SOC which possess more complex models as well as more adaptive (modular) ones.

**Acknowledgments**   This work was supported by JST, ACT-I Grant Number JP-50166, Japan.

## References

1. Su, J., Vargas, D.V., Kouichi, S.: One pixel attack for fooling deep neural networks. IEEE Trans. Evol. Comput. **23**, 828–841 (2019)
2. Storn, R., Price, K.: Differential evolution–a simple and efficient heuristic for global optimization over continuous spaces. J. Glob. Optim. **11**(4), 341–359 (1997)
3. Das, S., Suganthan, P.N.: Differential evolution: a survey of the state-of-the-art. IEEE Trans. Evol. Comput. **15**(1), 4–31 (2011)
4. Narodytska, N., Kasiviswanathan, S.: Simple black-box adversarial attacks on deep neural networks. In: 2017 IEEE Conference on Computer Vision and Pattern Recognition Workshops (CVPRW), pp. 1310–1318. IEEE, Piscataway (2017)
5. Goodfellow, I.J., Shlens, J., Szegedy, C.: Explaining and harnessing adversarial examples (2014). Preprint arXiv:1412.6572
6. Vargas, D.V., Su, J.: Understanding the one-pixel attack: Propagation maps and locality analysis (2019) Preprint arXiv:1902.02947
7. Brendel, W., Bethge, M.: Approximating CNNs with bag-of-local-features models works surprisingly well on ImageNet (2019). Preprint arXiv:1904.00760
8. Geirhos, R., Rubisch, P., Michaelis, C., Bethge, M., Wichmann, F.A., Brendel, W.: ImageNet-trained CNNs are biased towards texture; increasing shape bias improves accuracy and robustness. In: International Conference on Learning Representations 2019 (2019)

9. Ilyas, A., Santurkar, S., Tsipras, D., Engstrom, L., Tran, B., Madry, A.: Adversarial examples are not bugs, they are features (2019). Preprint arXiv:1905.02175

10. Gilmer, J., Ford, N., Carlini, N., Cubuk, E.: Adversarial examples are a natural consequence of test error in noise. In: International Conference on Machine Learning, pp. 2280–2289 (2019)

11. Szegedy, C., et al.: Intriguing properties of neural networks. In: International Conference on Learning Representations. Citeseer (2014)

12. Nguyen, A., Yosinski, J., Clune, J.: Deep neural networks are easily fooled: High confidence predictions for unrecognizable images. In: Proceedings of the IEEE Conference on Computer Vision and Pattern Recognition, pp. 427–436 (2015)

13. Moosavi-Dezfooli, S.-M., Fawzi, A., Fawzi, O., Frossard, P.: Universal adversarial perturbations. In: 2017 IEEE Conference on Computer Vision and Pattern Recognition (CVPR), pp. 86–94. IEEE, Piscataway (2017)

14. Brown, T.B., Mané, D., Roy, A., Abadi, M., Gilmer, J.: Adversarial patch (2017). Preprint arXiv:1712.09665

15. Liang, B., Su, M., You, W., Shi, W., Yang, G.: Cracking classifiers for evasion: a case study on the Google's phishing pages filter. In: Proceedings of the 25th International Conference on World Wide Web. International World Wide Web Conferences Steering Committee, pp. 345–356 (2016)

16. Li, D., Vargas, D.V., Kouichi, S.: Universal rules for fooling deep neural networks based text classification. In: Proceeding of the IEEE CEC 2019 (2019)

17. Kurakin, A., Goodfellow, I., Bengio, S.: Adversarial examples in the physical world. Preprint arXiv:1607.02533 (2016)

18. Sharif, M., Bhagavatula, S., Bauer, L., Reiter, M.K.: Accessorize to a crime: Real and stealthy attacks on state-of-the-art face recognition. In: Proceedings of the 2016 ACM SIGSAC Conference on Computer and Communications Security, pp. 1528–1540. ACM, New York (2016)

19. Athalye, A., Sutskever, I.: Synthesizing robust adversarial examples. In: International Conference on Machine Learning (2018)

20. Papernot, N., McDaniel, P., Wu, X., Jha, S., Swami, A.: Distillation as a defense to adversarial perturbations against deep neural networks. In: 2016 IEEE Symposium on Security and Privacy (SP), pp. 582–597. IEEE, Piscataway (2016)

21. Carlini, N., Wagner, D.: Towards evaluating the robustness of neural networks. In: 2017 IEEE Symposium on Security and Privacy (SP), pp. 39–57. IEEE, Piscataway (2017)

22. Goodfellow, I.J., Shlens, J., Szegedy, C.: Explaining and harnessing adversarial examples (2014). Preprint arXiv:1412.6572.

23. Huang, R., Xu, B., Schuurmans, D., Szepesvári, C.: Learning with a strong adversary (2015). Preprint arXiv:1511.03034

24. Madry, A., Makelov, A., Schmidt, L., Tsipras, D., Vladu, A.: Towards deep learning models resistant to adversarial attacks. In: International Conference on Machine Learning (2018)

25. Tramèr, F., Kurakin, A., Papernot, N., Goodfellow, I., Boneh, D., McDaniel, P.: Ensemble adversarial training: Attacks and defenses. In: International Conference on Learning Representations (2018)

26. Vargas, D.V., Kotyan, S.: Robustness assessment for adversarial machine learning: Problems, solutions and a survey of current neural networks and defenses (2019). Preprint arXiv:1906.06026

27. Ma, X., Li, B., Wang, Y., Erfani, S.M., Wijewickrema, S., Schoenebeck, G., Song, D., Houle, M.E., Bailey, J.: Characterizing adversarial subspaces using local intrinsic dimensionality (2018). Preprint arXiv:1801.02613

28. Guo, C., Rana, M., Cisse, M., van der Maaten, L.: Countering adversarial images using input transformations. In: International Conference on Learning Representations (2018)

29. Athalye, A., Carlini, N., Wagner, D.: Obfuscated gradients give a false sense of security: Circumventing defenses to adversarial examples. In: International Conference on Machine Learning (2018)

30. Uesato, J., O'Donoghue, B., Kohli, P., Oord, A.: Adversarial risk and the dangers of evaluating against weak attacks. In: International Conference on Machine Learning, pp. 5032–5041 (2018)
31. Bendale, A., Boult, T.E.: Towards open set deep networks. In: Proceedings of the IEEE Conference on Computer Vision and Pattern Recognition, pp. 1563–1572 (2016)
32. Sabour, S., Frosst, N., Hinton, G.E.: Dynamic routing between capsules. In: Advances in Neural Information Processing Systems, pp. 3856–3866 (2017)
33. Grosse, K., Manoharan, P., Papernot, N., Backes, M., McDaniel, P.: On the (statistical) detection of adversarial examples (2017). Preprint arXiv:1702.06280
34. Xu, W., Evans, D., Qi, Y.: Feature squeezing: Detecting adversarial examples in deep neural networks. In: Network and Distributed Systems Security Symposium (NDSS) (2018)
35. Carlini, N., Wagner, D.: Adversarial examples are not easily detected: Bypassing ten detection methods. In: Proceedings of the 10th ACM Workshop on Artificial Intelligence and Security, pp. 3–14. ACM, New York (2017)
36. Carlini, N., Wagner, D: Magnet and "efficient defenses against adversarial attacks" are not robust to adversarial examples (2017). Preprint arXiv:1711.08478
37. Hansen, N., Müller, S.D., Koumoutsakos, P.: Reducing the time complexity of the derandomized evolution strategy with covariance matrix adaptation (CMA-ES). Evol. Comput. **11**(1), 1–18 (2003)
38. Zagoruyko, S., Komodakis, N.: Wide residual networks (2016). Preprint arXiv:1605.07146
39. Iandola, F., Moskewicz, M., Karayev, S., Girshick, R., Darrell, T., Keutzer, K.: DenseNet: Implementing efficient ConvNet descriptor pyramids (2014). Preprint arXiv:1404.1869
40. He, K., Zhang, X., Ren, S., Sun, J.: Deep residual learning for image recognition. In: Proceedings of the IEEE Conference on Computer Vision and Pattern Recognition, pp. 770–778 (2016)
41. Lin, M., Chen, Q., Yan, S.: Network in network (2013). Preprint arXiv:1312.4400
42. Springenberg, J.T., Dosovitskiy, A., Brox, T., Riedmiller, M.: Striving for simplicity: The all convolutional net (2014). Preprint arXiv:1412.6806
43. LeCun, Y., Bottou, L., Bengio, Y., Haffner, P., et al.: Gradient-based learning applied to document recognition. Proc. IEEE **86**(11), 2278–2324 (1998)
44. Vargas, D.V., Kotyan, S., Matsuki, M.: Representation quality explains adversarial attacks. Preprint arXiv:1906.06627 (2019)
45. Vargas, D.V., Kotyan, S.: Evolving robust neural architectures to defend from adversarial attacks (2019). Preprint arXiv:1906.11667
46. Negrinho, R., Gordon, G.: DeepArchitect: Automatically designing and training deep architectures (2017). Preprint arXiv:1704.08792
47. Brock, A., Lim, T., Ritchie, J.M., Weston, N.: Smash: One-shot model architecture search through hypernetworks (2017). Preprint arXiv:1708.05344
48. Floreano, D., Dürr, P., Mattiussi, C.: Neuroevolution: from architectures to learning. Evol. Intell. **1**(1), 47–62 (2008)
49. Vargas, D.V., Murata, J.: Spectrum-diverse neuroevolution with unified neural models. IEEE Trans. Neural Netw. Learn. Syst. **28**(8), 1759–1773 (2017)
50. Vargas, D.V., Takano, H., Murata, J.: Self organizing classifiers and niched fitness. In: Proceedings of the Fifteenth Annual Conference on Genetic and Evolutionary Computation Conference, pp. 1109–1116. ACM, New York (2013)
51. Vargas, D.V., Takano, H., Murata, J.: Novelty-organizing team of classifiers-a team-individual multi-objective approach to reinforcement learning. In: 2014 Proceedings of the SICE Annual Conference (SICE), pp. 1785–1792. IEEE, Piscataway (2014)
52. Vargas, D.V., Takano, H., Murata, J.: Novelty-organizing team of classifiers in noisy and dynamic environments. In: 2015 IEEE Congress on Evolutionary Computation (CEC), pp. 2937–2944. IEEE, Piscataway (2015)
53. Vargas, D.V., Takano, H., Murata, J.: Self organizing classifiers: first steps in structured evolutionary machine learning. Evol. Intel. **6**(2), 57–72 (2013)

# Index

## A

Abstract representation, 41
Acceleration coefficient, 165, 167, 168
Acoustic feature vector, 98
Acoustic model, 99, 100, 103, 113
Activation function, 37, 41, 43, 58, 216, 220,
    227, 256, 259, 300, 301, 304, 308,
    310, 312, 335, 337, 342, 363, 364,
    373, 390, 426
Adam, 247, 363
Adam optimizer, 192, 200, 306, 390
Adversarial machine learning, 402–403, 421,
    425, 428
Aggressive selection and mutation, 328–331,
    333, 336, 348, 353
AlexNet, 45, 156, 254
Android, 358–362, 366–368, 372, 375
Android application package (APK), 367
Ant colony optimization (ACO), 16–18, 29,
    263
API call, 358–361, 368
$\varepsilon$-Archive, 160, 161
Artificial bee colony optimization (ABC),
    21–24
Artificial intelligence (AI), v, 4, 36, 37, 41, 60,
    132
Artificial neural network (ANN), 37, 38, 42,
    43, 49, 132, 135, 145, 148, 149, 210,
    253, 254, 260–262, 287, 360, vi
ASM+, 333, 335, 346, 347, 352–354
Asynchronous competitive coevolutionary, 379
Asynchronous island-based evolution strategy,
    264
Autoencoder (AE), 40, 49–51, 57, 59, 60, 187,
    197, 201, 212, 360, 361

Automatic music transcription (AMT), 241
AutoML, 182
Autoregressive integrated moving average
    (ARIMA), 148, 149
Average-pooling, 44, 158, 160, 189–191, 193,
    215, 222
Aviation flight recorder data, 272–273

## B

Backpropagation, vii, 38–40, 44, 47, 50, 54,
    55, 58, 71, 116, 119, 134, 136,
    186, 188, 212, 237, 261, 275–276,
    296
Backpropagation through structure (BPTS), 58
Backpropagation through time (BPTT), 47,
    261, 262, 275, 276
Backtracking search optimization algorithm
    (BSA), 73, 76, 78–83, 86–92
Bagging, 382
Bat algorithm (BA), 72, 76, 78–81, 85–92
Batch normalization (BN), 44, 56, 57, 160,
    189, 190, 215, 217, 222, 301, 326,
    336–338, 342, 348, 352
Batch updating, 40
Bayesian modeling, 243
Bayesian optimization, 108–110, 112
Bernoulli distribution, 69
Binary cross entropy (BCE), 384
Black-box optimization, 105, 113, 116, 419
BLEU, 104
Boosting, 45, 276, 382
Bootstrapping, 390, 394
Butterfly effect, 141

Printed in the United States
by Baker & Taylor Publisher Services